John P. Beaumont

Runner and Gating Design Handbook

John P. Beaumont

Runner and Gating Design Handbook

Tools for Successful Injection Molding

HANSER

Hanser Publishers, Munich • Hanser Gardner Publications, Cincinnati

The Author:
John P. Beaumont, Penn State Erie, The Behrend College, Erie,
Pennsylvania, 16563-1702

Distributed in the USA and in Canada by
Hanser Gardner Publications, Inc.
6915 Valley Avenue, Cincinnati, Ohio 45244-3029, USA
Fax: (513) 527-8801
Phone: (513) 527-8977 or 1-800-950-8977
Internet: http://www.hansergardner.com

Distributed in all other countries by
Carl Hanser Verlag
Postfach 86 04 20, 81631 München, Germany
Fax: +49 (89) 98 12 64
Internet: http://www.hanser.de

Library of Congress Cataloging-in-Publication Data
Beaumont, John P., 1952-
Runner and gating design handbook : tools for successful injection molding /
John P. Beaumont.-- 1st ed.
p. cm.
ISBN 1-56990-347-6 (hardcover)
1. Injection molding of plastics-Handbooks, manuals, etc. 2. Molding
(Chemical technology)--Handbooks, manuals, etc. I. Title.
TP1150.B389 2004
668.4'12--dc22
2004014993

Bibliografische Information Der Deutschen Bibliothek
Die Deutsche Bibliothek verzeichnet diese Publikation in der Deutschen Nationalbibliografie;
detaillierte bibliographische Daten sind im Internet über <http://dnb.ddb.de> abrufbar.
ISBN 3-446-22672-9

Production Management: Oswald Immel
Coverconcept: Marc Müller-Bremer, Rebranding, München, Germany
Coverdesign: MCP • Susanne Kraus GbR, Holzkirchen, Germany
Coverillustration with kind support by Husky Injection Molding Systems, Ltd.
Typeset by Kösel, Germany
Printed by Appl, Germany

Preface

Quality management methods, such as *Design for Six Sigma,* stress the critical review of fundamentals in order to identify and eliminate potential problems before they take their toll on the manufacturing process. In developing a mold design to produce an injection molded plastic part, one of the most fundamental and influential components is its melt delivery system. It also turns out that the melt delivery, or runner, system is probably the most misunderstood component of the injection mold. This makes it a prime candidate for critical review, particularly for the conscientious molder striving to improve his bottom line.

The melt delivery system begins with the injection molding machine nozzle and continues into the mold, progressing through the sprue, runner, and gate. Though the melt may only experience these flow channels for a fraction of a second, their effects are dramatic and result in the most extreme conditions experienced by the plastic melt in any phase of nearly any plastics processing method. Shear rates in gates commonly exceed 100,000 s^{-1} and localized melt temperature in high shear laminates can spike at as much as 200 °C, at rates of 2000 °C/s. Due to the extremity of these conditions, the actual effect of these conditions on the melt is not well understood. Most material characterization methods do not even come close to measuring melt conditions under these extremes. Viscosity vs. shear rate data is generally developed at a maximum of 10,000 s^{-1}, DSC data at less than 32 °C/min, and PVT data at 3 °C/min. As a result of limitation of material characterization as well as solution modeling and meshing issues, today's injection molding and fluid flow simulation programs are unable to accurately predict the extreme asymmetric melt conditions developed in a branching runner. The challenge of dealing with these conditions has generally been underestimated and seems to exceed the capabilities of today's technology.

The influences of these extreme melt conditions developed in the runner are just beginning to be understood. One of the most significant is the realization that the combination of laminar flow and high perimeter shear in a runner results in extreme non-homogenous melt conditions across a runner. Not only can a 200 °C variation in melt temperature exist but, as a result of the non-Newtonian characteristics of the melt, the viscosity may easily vary 100 fold from the zero shear conditions in the center of a flow channel to the extreme shear conditions around the perimeter. This creates significant asymmetric melt conditions when the melt branches in a runner or part-forming cavity. The conditions developed in the runner continue into the part corrupting the expected filling pattern and influencing how the part is packed, its mechanical properties, shrinkage, and warpage. These are all factors that are hardly known by most in the molding industry and their dramatic effects are rarely fully appreciated. The influence can be particularly acute in two-stage injection processes such as gas assist, structural foam, MuCell, and co-injection.

As stated earlier, the melt delivery system consists of the molding machine's nozzle, sprue, runner, and gate. Each of these components, or regions, can have a significant influence on both the process and the molded part. Process effects include the ability to fill and pack the part, the injection fill rate, the clamp tonnage, and the cycle time. Effects on the part include size, weight, mechanical properties, and variations in these characteristics between parts formed in different cavities within a multi-cavity mold.

Despite the significant influence that the melt delivery system has on the molding process, its various components are generally poorly designed relative to the time, effort, and cost put into the other components/regions of a mold and molding machine. This book bridges the critical gap left by other publications dealing with injection molding, which generally touch only briefly on the design of the melt delivery system and its relationship to successful injection molding. In particular, the lack of information on cold runners needed to be addressed. Though a fair amount of published data on hot runners is available, these data are generally heavily influenced by the bias of companies that sell these systems. There are over 50 companies offering hot runner systems and components commercially, while there is no company at all offering cold runner systems. As a result, one can imagine the lackluster image of cold runners, as there is no company commercially promoting them.

Evidence of the lack of understanding of runners includes the fact that the significant effects of shear-induced flow imbalances in runners was not documented, or clearly understood, until 1997 when I published the first journal article on this phenomenon. For the first time, it became obvious that the industry standard "naturally balanced" runners were creating significant imbalances. Melt filling imbalance, developed from shear-induced melt variations, were found to be the norm in most of the industry standard geometrically balanced runner designs being used. This phenomenon was being overlooked by nearly the entire molding industry for both cold and hot runner molds. In addition, the industry's leading state-of-the-art mold filling simulation programs had been developed without the realization of the shear-induced imbalance. As a result, these programs did not predict the imbalance and left the analyst with a false impression that these runners provided uniform melt, filling, and packing conditions. The problem still exists today and should be considered when using analysis programs.

This book takes an independent view of both hot and cold runners, trying not to make a judgment as to which is best for a given application. Rather, it addresses some of the critical design issues unique and common to both. The early chapters lay a foundation for designing runners by establishing an understanding of the rheological characteristics of plastic melt and how the influence of runner design and gating positions can affect the molded part. Chapter 4 provides important strategies for runner designs and gating position, which are critical to the successful molding of a plastic part. Chapter 5 provides an overview of the melt delivery system followed by Chapter 6 and 7, which teach the development and solutions to shear-inducted imbalances. These three chapters (5, 6 and 7) address issues, which are common to both cold and hot runners, blending basic geometrical channel issues with melt rheology. Chapter 8 focuses on cold runner designs including specific guidelines for runner and a wide variety of gate designs. Chapters 9 through 13 provide a close look at the design of hot runner systems and their unique capabilities and challenges. Chapter 14 provides a summary on the process of designing and selecting a runner system. Finally, the book concludes with an extensive trouble-shooting chapter with contributions from John Bozzelli and Brad Johnson.

This book is intended to provide the reader a better understanding of the critical role the runner plays in successful injection molding. It is hoped that this understanding should go a long way toward reducing mold commissioning times, improving product realization, increasing productivity, improving customer satisfaction, and achieving quality goals such as Six Sigma.

Acknowledgements

I would like to thank and acknowledge John Bozzelli, John Kleese, and Brad Johnson who either directly or indirectly contributed toward material contained in this book. Each of them regularly consults for the plastics industry and conducts industry training seminars. First, I would like to thank John Bozzelli (of Injection Molding Solutions) and Brad Johnson (from Penn State University, Erie) for their significant contributions to Chapter 15, *Troubleshooting*. John Bozzelli's contribution of an extensive trouble-shooting guide (Section 15.2 – "*Injection Molding Troubleshooting Guidelines for Scientific Injection Molding*") contains a wealth of information based on his extensive industry experience. Johnson's "*Two Stage Molding Set-Up*", Section 15.3, provides a practical guide to setting up an optimum injection process. Additionally, some of the material contained in the hot runner portion of this book is based on material presented by John Klees of Klees Interprise and is part of regular training seminars that he provides to the industry. John's willingness to share this information is deeply appreciated.

I would also like to thank the various students that assisted with research, editing and illustration development for this book. In particular, I would like to thank Scott Cleveland and Amanda Neely as well as my son Alex Beaumont. Further, I would like to thank both INCOE and Husky who provided both technical information and a number of the figures found in the hot runner sections of this book.

Contents

1 Overview of Runners, Gates, and Gate Positioning

In many cases, the mold design dictates the gating position, although ideally, the optimum gate position should be determined based on part requirements and afterwards the mold design should be selected to provide for the gate desired position. Available gating positions, and gate designs, are significantly influenced by whether the runner travels along the primary parting plane of the mold (the parting plane where the part forming cavity is defined) or whether it *does not* travel along this plane.

This chapter provides only a brief introduction and orientation of basic runner types and their influence on gate design and gating location. More detail on each of these subjects is presented later in the book.

1.1 Primary Parting Plane Runners

In the dominant runner type used in the industry the runner and part forming cavities are located along the same primary parting plane. Primary parting planes, often referred to as the parting lines, are where the mold opens and closes to allow ejection of the molded part and/or of the runner. The primary parting plane is the one where the molded part is formed and ejected. The *primary parting plane runner* is used in *two plate cold runner molds*. A cold runner mold is defined as a mold in which the plastic material in the runner is cooled and ejected from the mold during each mold cycle. Molten plastic material is injected through the runner, the gate, and then into the part-forming cavity. This molten plastic is then cooled by the mold, and when sufficiently solidified, the mold opens and the runner, gate, and part are ejected along the same primary parting plane. Figure 1.1 illustrates the position of the runner within the mold and its ejection from the primary parting plane. Notice that the part and runner are formed and ejected along the same parting plane.

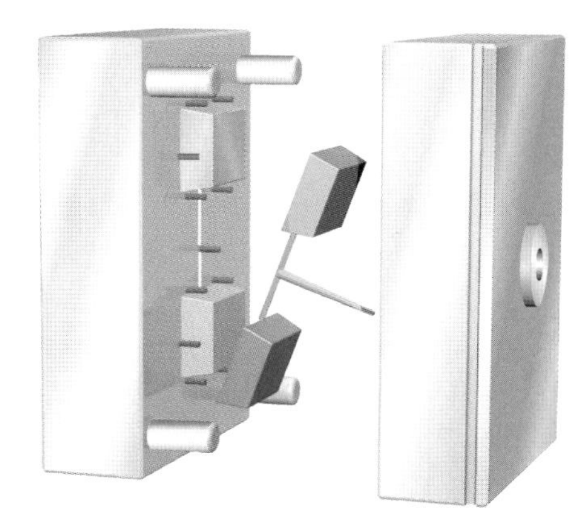

Figure 1.1 2-plate mold open and ejecting parts and runner

After the molded part and runner are ejected, the mold again closes, creating a flow channel (runner path) between the injection molding machine nozzle to the part forming cavity. As the primary parting plane runner is located along the same parting plane as the part forming cavity, gating into the part is limited to its perimeter, or very near its perimeter. Sub gates, such as the tunnel, cashew and jump gates, allow gating to be positioned within a short distance from the actual perimeter of the part (for gate designs see Section 8.4).

1.2 Sub Runners

A second runner type does not travel along the primary parting plane of the mold. This *sub-runner* generally travels parallel to the primary parting plane, but not along it. The sub-runner can be used in either a cold runner or a hot runner mold.

1.2.1 Cold Sub Runners

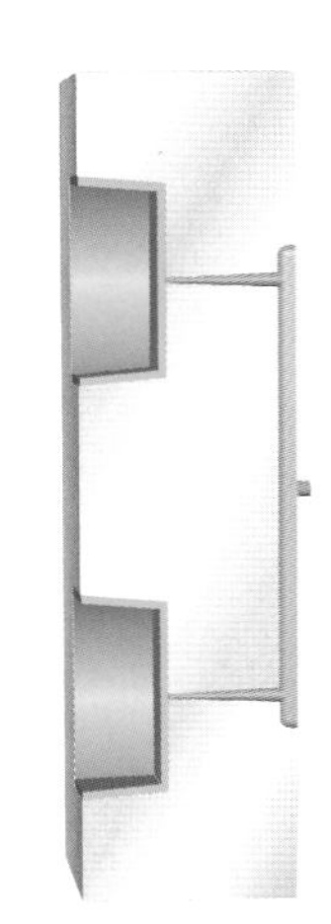

Figure 1.2 Cold runner with secondary sprue feeding the part forming cavities in a 3-plate cold runner mold

In a cold runner mold, the sub-runner travels along a second parting plane other than the primary parting plane where the part is formed. The two parting planes are normally parallel to each other and are separated, and partially defined, by at least one mold plate. The sub-runner and part forming cavities are connected by an extension of the sub-runner referred to as a *secondary sprue*. The bridging secondary sprue passes though the at least one separating mold plate and connects to the part-forming cavity through a small gate opening. The secondary sprues are normally parallel to the opening direction of the mold and perpendicular to the sub-runner (see Fig. 1.2).

During molding, after the plastic melt in the runner and part forming cavity solidify, the mold will open along the two parting planes. The part is ejected from the opened primary parting plane and the runner (which includes the secondary sprue and gate) is ejected from the opened second parting plane as seen in Figure 1.3.

This type of mold is commonly referred to as a *three-plate cold runner mold*. The terms two-plate and three-plate cold runner molds refer to the minimum number of mold plates required to form and to allow removal of both the part and the solidified runner. With the two-plate cold runner mold, the part and runner are formed and removed between at least a first and second mold plate. With the three-plate cold runner mold, the part is formed and removed between at least a first and second plate and the runner and gate are formed and removed between at least a third plate and often the same second plate used to help form the part.

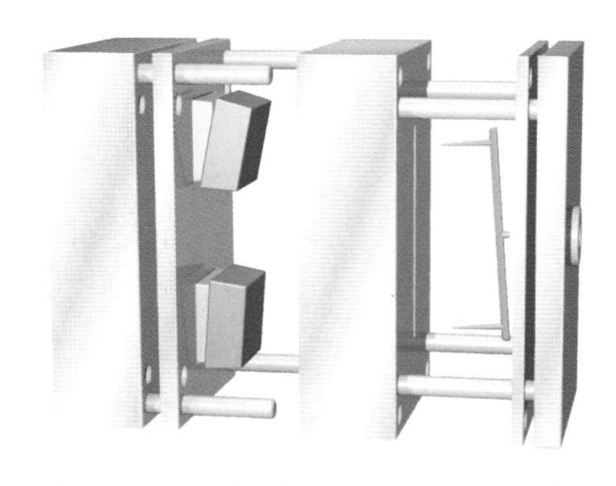

Figure 1.3 Typical 3 plate cold runner mold open and ejecting parts off the primary parting plane and ejecting the cold runner along the secondary parting plane

This type of mold is used when it is desirable to gate the part in a location other than the perimeter. It is commonly used for molding gears where it is desirable to gate in the center hub of the gear.

1.2.2 Hot Sub Runners

A second variation of the sub-runner mold is the *hot runner mold*. This type of runner provides the same gating flexibility as the three-plate cold runner mold. However, unlike a cold runner mold, the melt that travels through the runner remains molten, and is not ejected between molding cycles. The design of hot runner systems is more complex than that of cold runners. Their design and contrast to cold runners are discussed later in Chapter 9.

Two variations of a hot runner are illustrated in Fig. 1.4. Here the melt travels in a *hot manifold* along a path, which is normally parallel with the platens of the molding machine. A hot drop, or nozzle, is then used to deliver the melt from the manifold to the part-forming cavity. Special attention is required to isolate the heat from the hot manifold and drop from the part-forming cavity, which requires good cooling. This figure illustrates both a valve gated nozzle (top) and a more conventional open gated nozzle (bottom).

Unlike the cold runner molds, the runner in the hot runner mold remains molten during processing and is not ejected each cycle. Like the three plate cold runner mold, this type of mold provides more gating options than the two-plate cold runner mold.

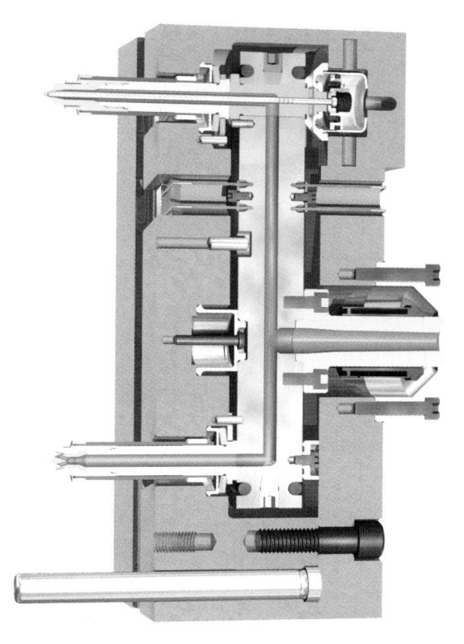

Figure 1.4 Externally heated hot runner illustrating manifold and drops. The figure illustrates two types of nozzles: the top nozzle is valve gated and the bottom nozzle has a more conventional open gate design (Courtesy: Husky)

1.3 Hybrid Sub-Runner and Parting Line Runner

It is common for a mold to contain both a sub runner and a parting plane runner. This is most common when a hot runner is used. Here, the hot runner would deliver the melt to a cold runner, or gate, along the primary parting line. An example is a two-cavity mold, used is to produce the flat donut shaped part shown in Fig. 1.5. The hot drop delivers the melt to a cold runner located within the center region of the part. The cold runner then radiates out and gates the part along its inner edge.

1.4 Gate Designs

Parting line runners are the most restrictive on gating position but provide the greatest flexibility in gating design. The three-plate cold runner mold is limited to restrictive pin point gates, which must allow the gate to be separated, or torn, from the part as the mold opens. Parting line runners can use similar restrictive gates to provide for automatic degating during mold opening, but they are *not required* to use these types of gates. Parting line runners provide for gates to be formed along the parting plane of the mold. This provides significant flexibility in their design to achieve a desired effect. Some of these gates include fan, film, tab, edge, and diaphragm gates. The effects that can be achieved with these gates could include keeping the runner and gate attached to the part to facilitate post-mold handling, using broader gates to limit shear rates and gate region shear stress during mold filling, using broader gates to improve flow patterns across a cavity, and using thicker gates to improve packing.

The hybrid sub-runner and parting plane runner, presented above, increases gating opportunities. An example where this hybrid design might be used is when molding a cylindrical part, where a diaphragm gate is desirable (see

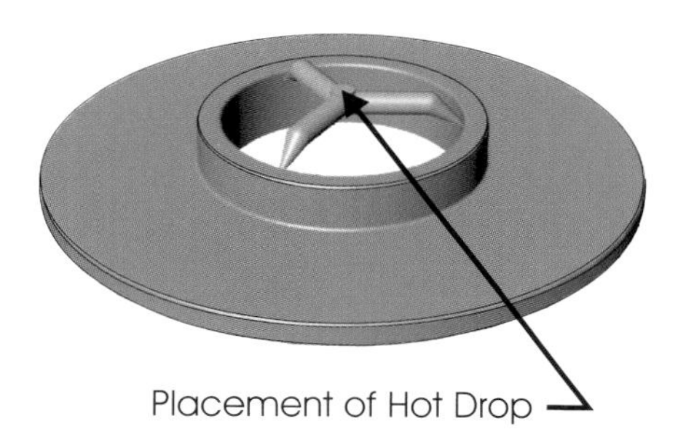

Figure 1.5 Round part internally gated by a sub-runner which is fed by a hot drop

Fig. 8.30 in Chapter 8). Here, a hot drop feeds directly into a diaphragm gate, which in turn feeds the cavity. The above-mentioned gates, as well as additional gating options, such as valve gates used in hot runners and gates providing automatic degating in cold runners, will be presented in Chapters 8 and 11.

Though a vast majority of hot runner molds have restrictive gates, there are additional options available. These options include valve gates and edge gates.

2 Rheology of Plastics

2.1 Laminar vs. Turbulent Flow

Owing to the relatively high viscosity of commercial polymers, it is generally expected that flow is laminar during injection molding, even in the case when melt is passing through small restrictive pinpoint gates. As a result of this laminar flow behavior, processes such as co-injection are made possible.

Although the term turbulence is often used to describe the cause of flaws in injection molded parts, actual turbulence is very rare. Turbulence can be calculated using the Reynolds number. The Reynolds number is a dimensionless term that considers the velocity and viscosity of a fluid flowing through a channel. The Reynolds number is given by:

$$\text{Re\#} = \frac{\text{velocity} \times \text{diameter}}{\text{kinematic viscosity}} \tag{2.1}$$

Where:

$$\text{kinematic viscosity} = \frac{\text{dynamic viscosity}}{\text{melt density}} \tag{2.2}$$

$$\text{velocity} = \frac{\text{flow rate}}{\text{area}} \tag{2.3}$$

Turbulence usually develops at a Reynolds number of 2300 and above. The Reynolds number calculated for most polymers during injection molding is less than 10, even at high flow rates through small gates. This same calculation can be repeated to determine the Reynolds number for plastic melt flowing through a runner channel. The Reynolds number of the plastic melt is dependent on the shape of the flow channel. An example of how to determine the Reynolds number for an ABS resin flowing through a round gate is provided in the following:

Example:

Given is a four-cavity hot runner mold with 1 mm diameter gates. Each cavity has a volume of 8 cm³. The molding machine injects material at a rate of 32 cm³/s. This results in a fill time of 1 second/cavity and a flow rate through each gate of 8 cm³/s.

Shear Rate through the gates is:

$$\dot{\gamma} = \frac{32Q}{\pi d^3} = \frac{32 \times 8\ \text{cm}^3/\text{s}}{\pi \times 0.1^3} = 81{,}528\ \text{s}^{-1} \tag{2.4}$$

This is a rather high shear rate for most materials.

- Flow rate/gate = 8.0 cm^3/s ($8.0 \times 10^{-6}\ m^3/s$)
- Diameter of each gate = 1.0 mm (0.0010 m)
- Cross-sectional area of each gate = 0.783 mm^2 ($7.83 \times 10^{-7}\ m^2$)
- Dynamic viscosity = 8 Pa-s = 8 kg/m-s

$$\text{Gate velocity} = \frac{8.0 \times 10^{-6}\ m^3/s}{7.83 \times 10\ m^2} = 10.22\ m/s$$

$$\text{Kinematic viscosity} = \frac{8\ kg/m \times s}{890\ kg/m^3} = 0.008989\ m^2/s$$

$$Re\# = \frac{10.22 m/s \times 0.001\ m}{0.008989\ m^2/s} = 1.14$$

This value for the Reynolds Number is a small fraction of the 2300 required for turbulence to occur.

Figure 2.1 In injection molding, plastics exhibit both laminar and fountain flow during processing. In a cold runner mold, thermoplastic materials flow within a perimeter of frozen material

2.2 Fountain Flow

During injection molding, independent of the kind of runner system used, plastics exhibit a specific behavior known as "fountain flow" (see Fig. 2.1). The first material into a mold is deposited on the cavity walls and the following material flows through this material, advancing to the flow front. Fountain flow results from drag of the flowing plastics laminates on the channels walls, causing the center laminates to flow faster. This results in a somewhat parabolic velocity, or fountains, profile at the flow front. The material at the center of the flow front is continually pushed to the perimeter of the flow channel as new material moves forward. The material that is pushed to the outer wall experiences very little shearing. The shear rate of the material at the plastics-metal interface is expected to be near zero. In cold runner systems, a frozen layer is developed where the plastic melt is deposited on the cavity walls. In most hot runner systems, a frozen layer is not developed, yet fountain flow still exists.

Fountain flow provides for processes such as co-injection molding. In co-injection molding, the primary material that is to form the part's exterior surface (material "A" in Fig. 2.2) is injected first. When the cavity is partially filled, a second material is injected into the mold. This second material travels between the outer frozen layers of material "A", pushing additional primary material forward where it continues to be deposited on the walls at the flow front. It is important to meter in a sufficient amount of the primary material to make sure that the secondary material does not pass along the flow channel between the outer frozen layers and emerge at the flow front and appear on the walls.

Beginning of injecting material "B":

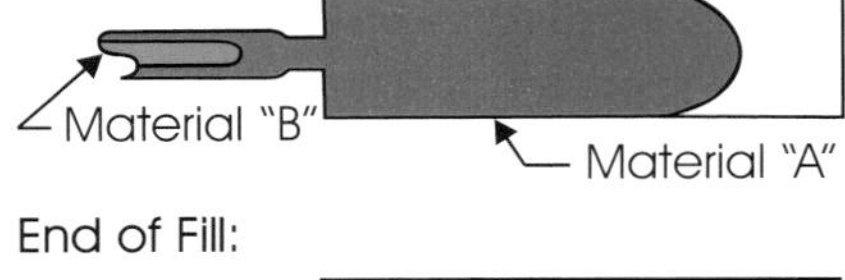

End of Fill:

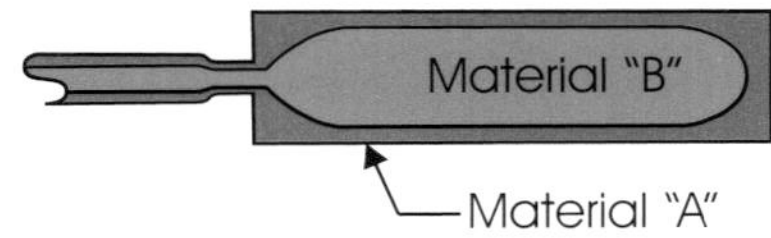

Figure 2.2 Laminar and fountain flow characteristics provide for the co-injection process, allowing for a part to be formed with a skin of one material and a core of a second material

2.3 Factors Affecting Viscosity

The laminar flow conditions during injection molding results in the melt remaining in distinct layers throughout mold filling. The relative motion of polymer layers to each other and of the polymer mass to the flow channel walls can be quite high. The rate of this change is described by the following equation referred to as shear rate.

$$Shear\ rate = \dot{\gamma}\frac{v}{h} = \frac{x}{h}\frac{dx}{dt} \quad (2.5)$$

Here t = time, v = velocity, h is the thickness, or diameter, of the flow channel, and x is the distance along the length of the flow channel. Shear rate is normally expressed as s^{-1} or reciprocal seconds.

Figure 2.3A is the velocity profile divided into a finite number of laminates. The crosshatched sections in the velocity profile illustrate the relative velocities of each of the laminate layers. Figure 2.3B is a shear rate profile derived from the velocity profile. The highest rate of change in the velocity profile corresponds to the highest shear rate. Shear rate is zero at the mold wall and at the center of the flow channel. Maximum shear rates are generally found in the melt stream near the flow channel wall.

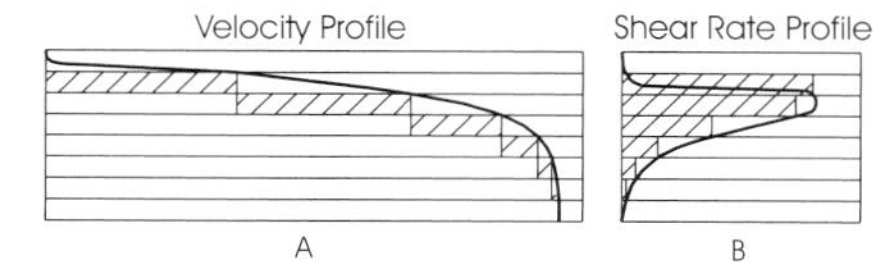

Figure 2.3 Shear rate distribution in a flow channel (B) is derived from the velocity profile (A) developed as melt flows through a channel

During flow, the relative motions of the laminate layers act on the molten polymer. The shearing force will result in an orientation of the polymer normal to the direction of the applied strain. As the polymer chains become more oriented, or aligned, there is less steric hindrance between the polymers, which allows them to move more freely relative to each other. This, in effect, reduces the viscosity of the flowing material.

Figure 2.4 shows outputs from a finite difference flow analysis program providing information on the conditions that exist within the melt through the cross section of the flow channel. These particular outputs resulted from an analysis that assumes flow is symmetrical about the middle of the flow channel. The position across the flow channel is represented by the y-axis. Clockwise from top left, the outputs include velocity, shear rate, temperature, and viscosity. Note the relationship between the high shear rate region and the high-temperature region in the temperature profile. The high shear in this region creates frictional heating, which results in the irregular temperature profile. This phenomenon is particularly pronounced for flows through high-shear regions such as runners and gates.

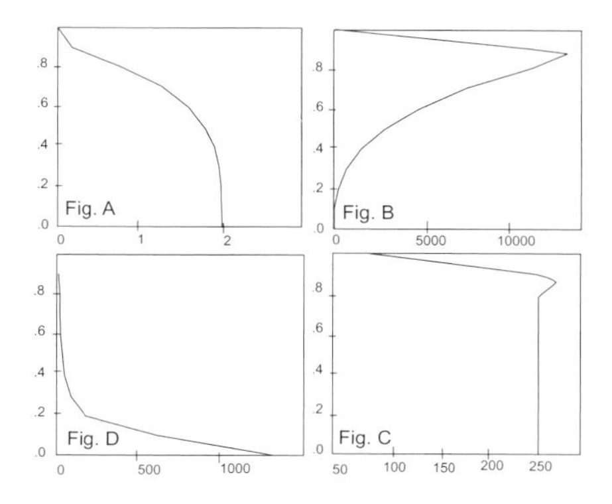

Figure 2.4 Outputs from a finite element, finite difference flow analysis program providing information on the melt conditions that exist through the cross-section of the cold runner. Y-axis is from the center line of the channel to the channel wall: (A) velocity; (B) shear rate; (C) melt temperature; (D) viscosity

2.3.1 Common Viscosity Models

A mathematical description of the effect of shear rate on the viscosity of a material is a requirement for mold filling simulations. Two of the more common models used for this are described as follows and illustrated in Figs. 2.5 and 2.6.

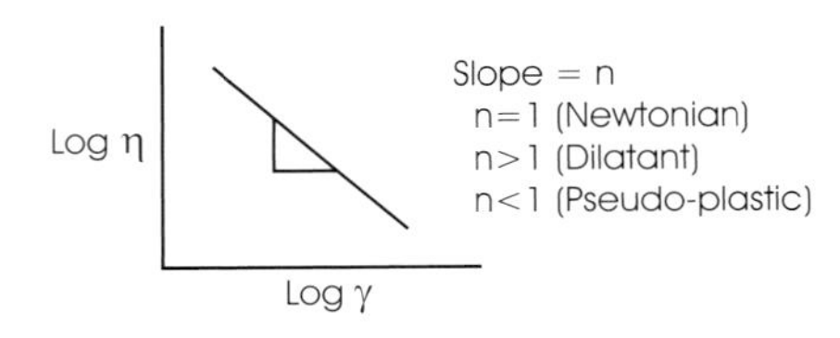

Figure 2.5 Graphical representation of a Power Law viscosity vs. shear rate model

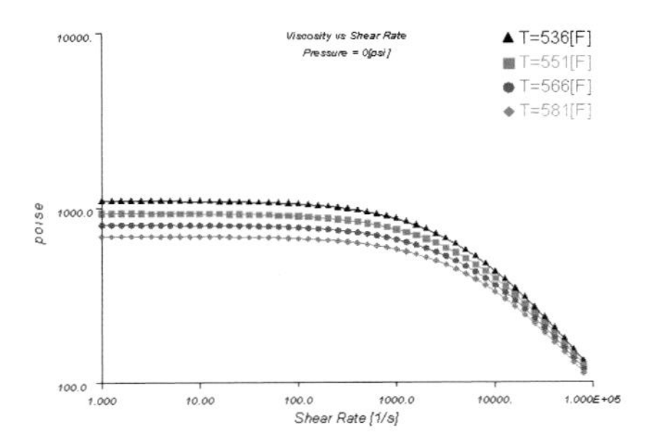

Figure 2.6 Graphical representation of viscosity vs. shear rate of a plastic material as represented by the modified Cross viscosity model

Power law and viscosity model:

$$\eta = m\dot{\gamma}^{n-1} \tag{2.6}$$

where:

η = viscosity

m = constant (consistency index)

$\dot{\gamma}$ = shear rate

n = power law index

The following viscosity model will give a viscosity versus shear curve similar to the one shown in Fig. 2.6.

Modified Cross viscosity model:

$$\eta = \frac{\eta_o}{1 + \left(\frac{\eta_o \gamma}{\tau^*}\right)^{(1-n)}} \tag{2.7}$$

$$\eta_o = D_1 e^{\left[\frac{-A_1 (T-T^*)}{A_2 + (T-T^*)}\right]} \tag{2.8}$$

where:

$T^* = D_2 + D_3 P$

P = pressure

η = viscosity

$\dot{\gamma}$ = shear rate

T = temperature in Kelvin

Unknowns are D_1, D_2, D_3, A_1, A_2, τ^*, n

The modified Cross model improves upon the power law model because it is not limited to a purely linear relationship between viscosity and shear rate. There are numerous other models, each having their own strengths and weaknesses. Certain materials will have characteristics that are best represented by specific models. The power law equation is generally recognized as the simplest model. However, it has limited capabilities of representing a polymer's viscosity over a wide range of processing conditions. This model can work reasonably well at high shear rates. It is rarely used in today's commercial mold filling analysis programs.

The ideal case is to experiment with each of the optional models to determine which one provides the best fit for a given set of raw data. The capability of using multiple viscosity models is becoming more common with today's mold filling analysis software. However, most injection molding computer-aided engineering (CAE) software suppliers favor a particular model and use that model as a standard.

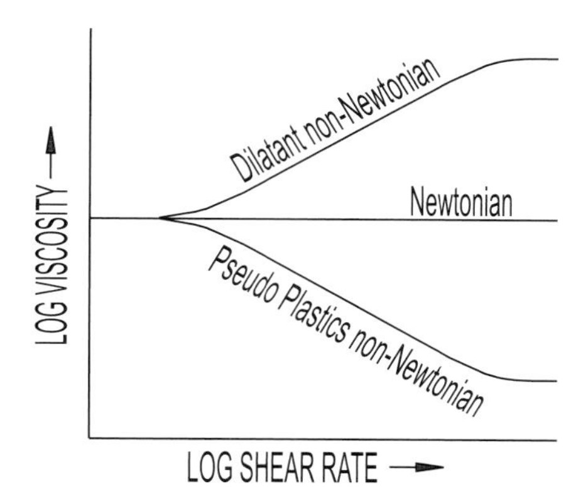

Figure 2.7 Viscosity vs. shear rate graph comparing Newtonian, dilatant non-Newtonian, and pseudo plastic non-Newtonian fluids

2.3.2 Non-Newtonian Fluids

Viscosity is defined as the resistance to flow and is determined by shear stress divided by shear rate. The lower the viscosity of a material, the lower its resistance to flow. The viscosity of many common fluids, such as oil, may be affected by temperature, but it is constant with changing flow rate, or shear rate. A fluid with a constant viscosity at changing shear rates is referred to as a Newtonian fluid. The viscosity of plastic materials is generally much more complex. At very low shear rates, the viscosity of a plastic material is essentially Newtonian. However, at the higher shear rates experienced during injection molding, the viscosity becomes non-Newtonian, taking the characteristics of a power-law fluid, with the viscosity decreasing with increasing shear rate. For most plastic materials, flow can actually be characterized as pseudo-plastic non-Newtonian. The viscosity of pseudo-plastic non-Newtonian fluids will decrease with increasing shear rates. In contrast, the viscosity of dilatant non-Newtonian fluids increases with increasing shear rate. Figure 2.7 contrasts the characteristics of Newtonian, pseudo-plastic non-Newtonian, and dilatant non-Newtonian fluids.

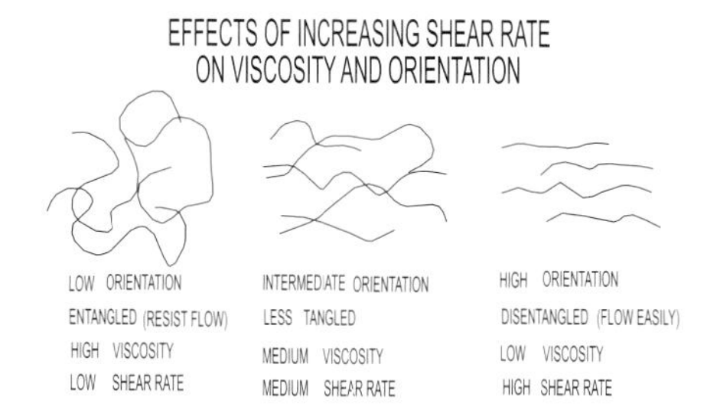

Figure 2.8 Effects of shear developed during flow on polymer orientation and viscosity

The reduction in viscosity of a pseudo-plastic non-Newtonian fluid results from shear, which orients the polymer chains. As the chains become oriented, their entanglement is reduced (see Fig. 2.8). This allows the chains to slip (or flow) past each other more easily. Due to this non-Newtonian characteristic of plastic materials, the viscosity becomes lower during injection molding the faster the material is injected into the mold. Unlike Newtonian fluids, making the polymer flow faster can decrease the pressure drop. In the case of injection molding, a shorter fill time (higher flow rate) can therefore potentially reduce the pressure to fill a mold.

The relationship of shear rate and viscosity is normally represented on a log-log graph as illustrated in Fig. 2.7. Figure 2.9 again presents the effect of shear rate on the viscosity of a plastic material without the use of a log scale. This non-log-scale graph illustrates more graphically the dramatic reduction in viscosity as a material is sheared.

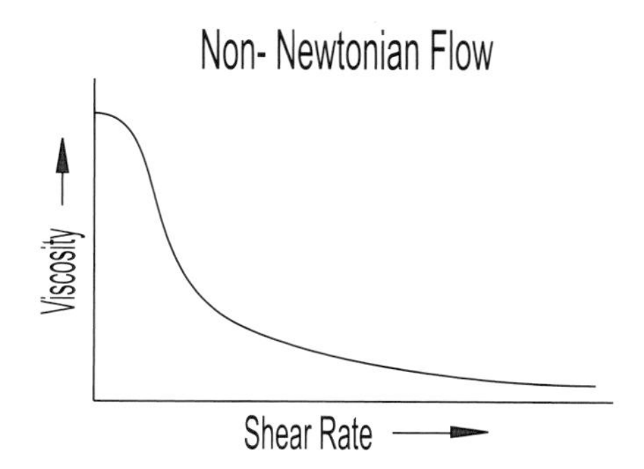

Figure 2.9 The effect of shear rate on viscosity on a non log-log scale

When the material no longer flows, the shear field disappears and the polymer will quickly return to a random high-viscosity condition. This return to a random state is driven primarily by entropy and creates an elastic effect in the polymer melt. The behavior is somewhat analogous to grabbing a pile of rubber bands and stretching them into an oriented state. As long as the stretching force remains, the rubber bands will remain oriented. However, as soon as you let go of the rubber bands, they immediately return to their low-energy random condition.

Non-Newtonian shear thinning should not be confused with the effects of frictional heating, which is developed during flow. Frictional heating is an additional factor that will reduce the viscosity, but is not the reason for the orientation-induced viscosity reduction of a non-Newtonian polymer. Because these two phenomena occur simultaneously, their effect on viscosity is commonly confused.

2.3.3 Temperature

In addition to shear rate, increasing temperature also decreases the viscosity of plastic materials. Plastics, in their molten state, consist of long molecular chains. The backbones of these chains generally consist of carbon atoms. Very strong covalent or chemical bonds hold these carbon atoms together along the chain. These separate polymer chains are subsequently held together by relatively weak van der Waals (secondary electrostatic) forces. As heat is introduced during processing, the primary bonds are weakened and the bond lengths increase. At normal processing temperatures, these weakened bonds will allow increased freedom of movement while they stay intact. If excessive heat is introduced, these bonds will break down entirely and cause a permanent degradation of the polymer.

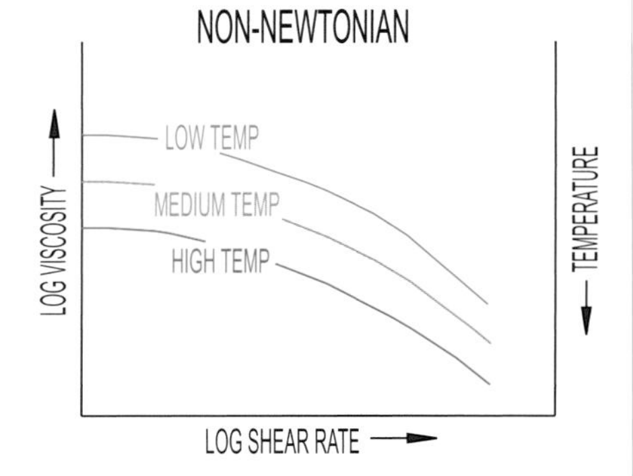

Figure 2.10 The effect of temperature and shear rate on viscosity

The secondary bonds behave quite differently under normal processing temperatures. Here, the secondary bonds are weakened considerably and provide little cohesion between the molecules. As a result, the polymer chains will separate from one another. The distance between chains will be approximately proportional to the heat introduced. The increased spacing will reduce the interaction between the polymer chains and thereby reduce the restriction to flow, creating a reduction in viscosity. Under these conditions, entropy and external forces will allow for relatively easy movement of the polymer chains. Figure 2.10 illustrates the effects of both temperature and shear rate on viscosity with a log-log graph of viscosity vs. shear rate and temperature.

2.3.4 Pressure

Pressure on a plastic melt can be quite high during both mold filling and compensation phases. During mold filling, the pressures can exceed 30,000 psi, with a maximum pressure at the injection nozzle and zero pressure at the flow front. During compensation phases, pressure is generally lower, commonly less than 10,000 psi. The pressures are also somewhat hydrostatic during compensation, with pressures at the end of fill being nearly as high as the melt pressure at the nozzle.

Under these high pressures plastics are compressible. The compression increases the interaction of the polymer chains, which in turn increases their resistance to flow, i.e., which increases their viscosity. As one might imagine, these interactions further increase the complexity of polymer melt behavior as the pressure conditions on the melt vary dramatically with time and along the melt stream throughout mold filling and compensation phases.

2.4 Melt Compressibility

Because there is a significant amount of free space as a plastic is heated, applied pressure will result in compression of the material. The effect of this compression at various temperatures can also be determined by examination of Fig. 2.11, which shows the PVT characteristic of polystyrene. At a process temperature of 250 °C (482 °F), the material is compressed from a specific volume of approximately 1.06 cm^3/g (1.83 in^3/oz) at atmospheric pressure to 0.96 cm^3/g (1.66 in^3/oz) at 160 MPa (23,206 psi). These pressures correspond to the potential range of pressure experienced during the injection molding cycle on a typical injection molding machine. The combination of thermal contraction and pressure yields potential variances of 10% in volume during molding in this particular case.

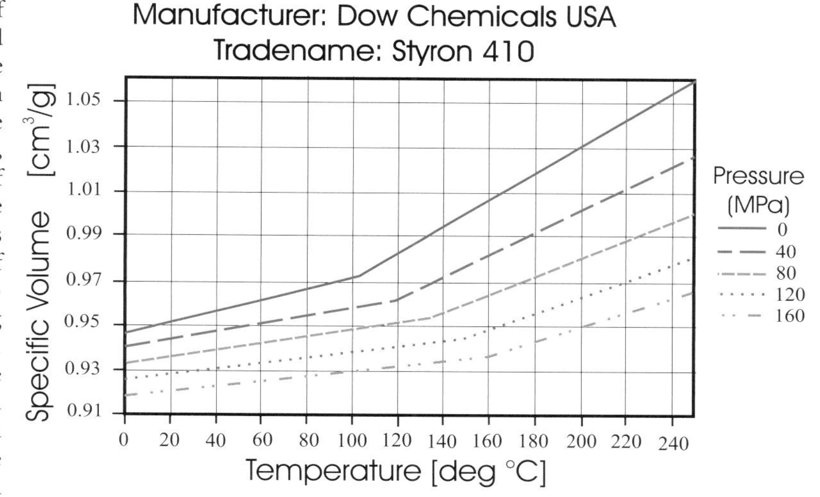

Figure 2.11 PVT graph of a polystyrene; characteristic PVT behavior of an amorphous plastic material

2.5 Melt Flow Characterization

The most common test regularly performed on a plastic resin is the determination of the melt flow index. Most resin suppliers will include this information on a material data sheet. However, the melt flow index only provides an indication of a material's flow characteristics at a single, low shear rate. During the injection molding process, the plastic melt will experience shear rates far exceeding the rate at which the melt flow index was determined. In addition, shear rates during molding will vary significantly throughout the melt flow channel (nozzle, sprue, runner, gate, and cavity) and during the injection and compensation processes. Therefore, it is important to know what the flow characteristics will be throughout the molding process. This is particularly important for the characterization of a material for use with injection molding simulation software.

Viscosity can be more accurately measured in a capillary rheometer, which is basically an extruder with a small-diameter die, through which the

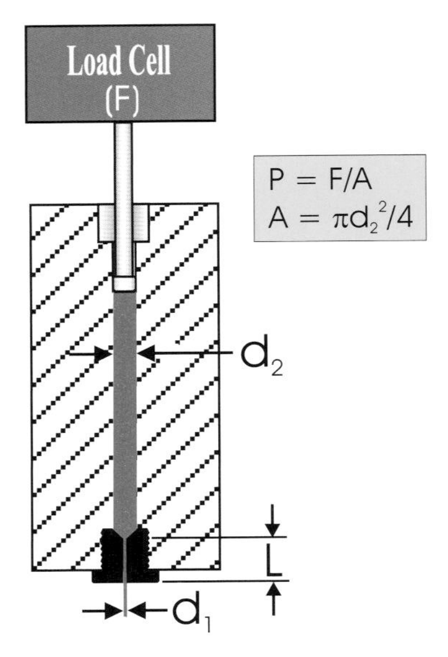

Figure 2.12 Cross-section of a typical capillary

plastic is forced by a plunger. To ensure the material is extruded at a specific shear rate, the plastic is forced through a capillary of a given length, *L*, and diameter, *d*, at a set flow rate, *Q*, after being heated to the appropriate temperature. The pressure to drive the melt through the capillary is either measured directly in the melt chamber or derived by knowing the force on, and the area of, the piston driving the material through the capillary.

Measurements are usually taken at a minimum of three temperatures – two in the operating range and one near the transition temperature for the material. Measurements should also be made at a minimum of three shear rates for each of these three temperatures. The shear rates should range from 10 to 10,000 per second. It is preferred for measurements to be taken at eight to ten shear rates at each of the chosen temperatures. The preferred length to diameter ratio of the die is twenty to one. The evaluation of the temperature sensitivity of the material's viscosity will be more accurately depicted as the range of temperatures at which measurements were taken expands. Figure 2.12 shows a schematic of a typical capillary rheometer used to determine viscosity.

The shear rate is calculated as in Eq. 2.4:

$$\gamma = 32Q/\pi d_1^3$$

where Q is the flow rate and d_1 is the diameter of the die of the capillary rheometer.

The shear stress is calculated as:

$$\tau = \Delta p d_1/4L \tag{2.9}$$

where Δp is the pressure drop across the capillary die, d_1 is the diameter of the die, and L is the length of the die.

The viscosity is calculated as:

$$\eta = \tau/\gamma \tag{2.10}$$

which can be shown as:

$$\eta = \frac{\left(\dfrac{\Delta p d_1}{4L}\right)}{\left(\dfrac{32Q}{\pi d_1^3}\right)} \tag{2.11}$$

The above assumes a Newtonian fluid where a parabolic velocity profile exists during flow. The high shear near the boundaries of a flow channel causes a viscosity variation across the flow channel of a non-Newtonian fluid. The lower viscosity material near the boundary causes the velocity profile to be more plug-like, which results in a higher shear rate. A more accurate representation of viscosity is achieved by using the Rabinowitsch correction and can be seen in the following equation.

$$\eta = \left[\frac{\left(\frac{\Delta p d_1}{4L} \right)}{\left(\frac{32Q}{\pi d_1^{\,3}} \right) \left(\frac{3n+1}{4n} \right)} \right] \tag{2.12}$$

where n is the power law index for the material.

The power law index is determined from the slope of the log viscosity vs. log shear stress curve of the material. A Newtonian fluid has a power law index of 1. When substituted for n in the above equation, the Rabinowitsch correction cancels out. When n has a value less than one, the material is non-Newtonian. The lower the value of n, the more non-Newtonian the fluid, and the more significant the effect on viscosity.

3 Filling and Packing Effects on Material and Molded Part

The flow of thermoplastics through an injection mold and its relationship to the molded part is quite complex. This chapter focuses on the development of melt conditions within a part-forming cavity and their relationship to the molded part, which will help the reader to establish an optimum gating and molding strategy.

3.1 Process Effects on Material Viscosity

In Chapter 2, the basic behavior of thermoplastic materials was discussed and the relationship between a thermoplastic's viscosity, temperature, and shear rate were explained in detail. The initial viscosity of the melt entering a mold is determined by the melt temperature, as delivered from the molding machine, and the injection rate. High melt temperatures and high injection rates result in low viscosities for the plastic melt. This combination of high temperature and flow rate can result in lower fill pressures; however, pressure can begin to increase at extreme fast or slow fill rates. High melt temperatures are normally limited by potential degradation and longer mold cooling times. It is often desirable to perform a mold filling analysis to determine the optimum balance of melt temperature, processing conditions (primarily injection rate), and runner diameter that will produce a quality product for a given part design. Use of molding techniques, such as *Scientific Molding* [1] or *Decoupled Molding* [2] can be used to determine the optimum fill rate of an existing mold. Variations of these methods are presented by Bradley Johnson in Chapter 15 of this book.

3.1.1 Melt Thermal Balance – Conductive Heat Loss vs. Shear Heating

During injection, a hot thermoplastic is forced into a relatively cold mold. As the melt travels through cold portions of the mold, heat is continually being drawn from the plastic material. Plastic immediately adjacent the cold mold walls will freeze almost immediately. The thickness of the frozen layer is dependent on the balance between heat lost to the mold through conduction and heat gained from shear. If the injection rate into a mold for a thermoplastic material is too slow, the thickness of the frozen layer builds up to a point where material can no longer be fed into the cavity and a short-shot is created.

A short-shot is the extreme outcome when the injection rate is not adequate to keep the thermoplastic melt temperature elevated enough for molding. At faster fill rates, frictional heating can overcome the heat lost through conduction and allow the material to remain molten during filling of the entire cavity. Figure 3.1 shows the result of a series of mold filling analyses of a simple rectangular plaque at three different fill rates. The plaque is 50 mm wide by 150 mm long and 2 mm thick. It is edge-gated as indicated (along the bottom edges of the figures) and molded with an ABS and a melt temperature of 255 °C. Note the change in melt temperature and frozen layer variations in each of the figures dependent on flow rate. At the fastest flow rate, it can be seen that the melt temperature at the end of fill is actually 10 °C higher than the injection temperature.

Control of frictional heating during mold filling can sometimes be difficult to achieve. With most parts, the geometry does not allow for the flow velocity of the melt to be constant without profiling the injection. Varying flow front velocities will result in a variation in the development of the frozen layer. A common example is the center gating of a disk-shaped part. At a constant injection rate from the injection molding machine, the flow front speed near the gate will be relatively high, but continually decreases as the melt progresses into the expanding cavity (see Fig. 3.2). This will cause a high amount of shear heating near the gate, but as the melt front progresses, it slows down and will begin to lose more heat to the mold than it is gaining from possible shear heat. This effect can be minimized by utilizing an injection profile with an initial slower fill rate and then gradually increasing the injection rate. However, most molding is performed without the use of profiles.

Variations in wall thickness within a part can create significant variations in flow rate and the resultant thermal balance. Thin regions will create a resistance to the flow front and cause the melt to *hesitate* as it fills other thicker regions. The hesitating melt will quickly lose heat and potentially freeze off. This is discussed in more detail in Section 4.2.10.

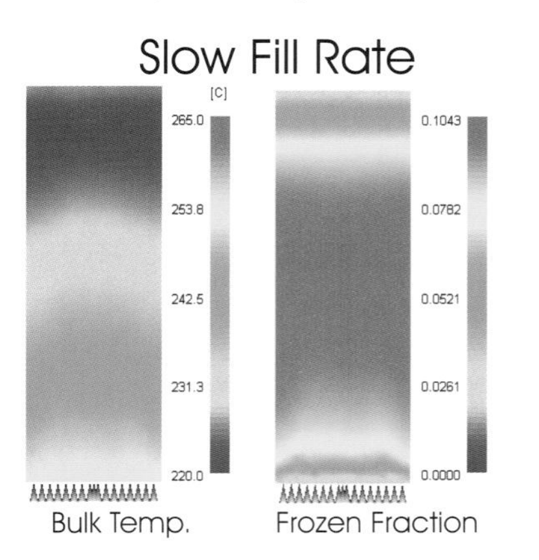

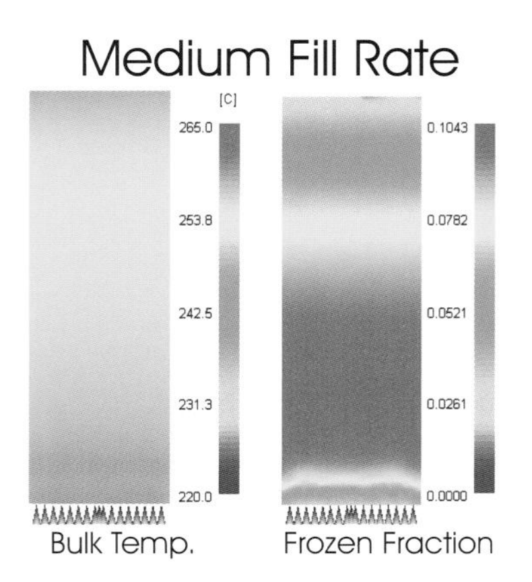

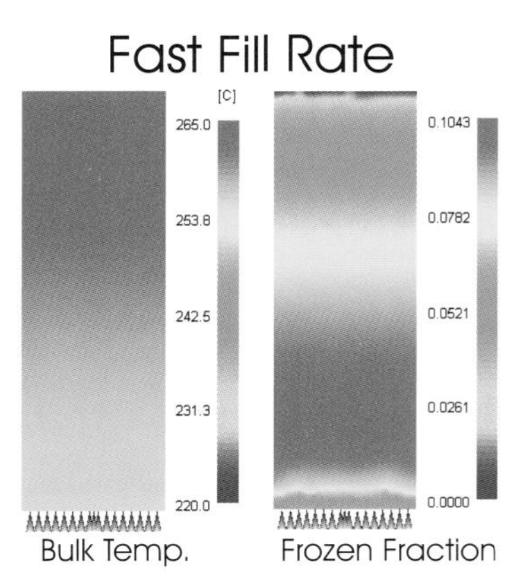

Figure 3.1 Effects of injection rate on bulk melt temperature and frozen material fraction as predicted by Moldflow's MPI

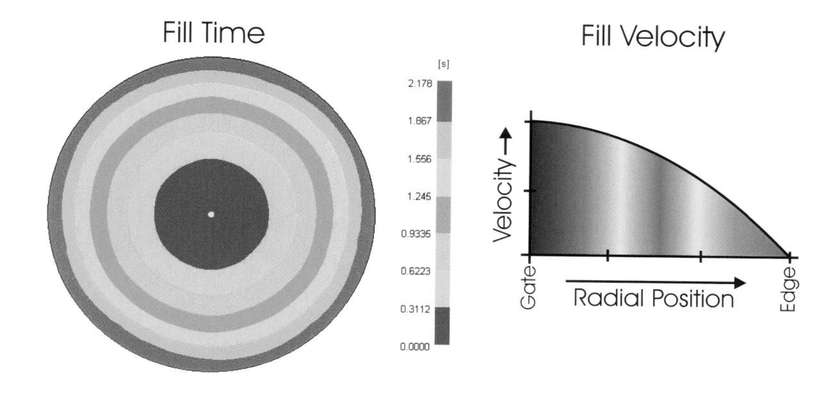

Figure 3.2 Moldflow plot of fill time and a graphical illustration of fill velocity as the melt radiates from the center to the perimeter of a center-gated disk. Note, velocity of the flow front decreases as the melt diverges from the gate

3.1.2 Development of a Frozen Boundary Layer

Both the viscosity of the plastic and the thickness of the flow channel affect the filling of a mold cavity. This is complicated by the fact that the actual thickness of the flow channel varies with the development of a frozen layer. The development of the frozen layer is affected by mold temperature and, particularly, by flow rate. Shear heating, at high flow rates, will minimize the growth of the frozen layer. This reduction in frozen layer development creates a larger flow channel cross section, which can have a significant effect on pressure drop. For example, for a cavity wall, the pressure drop at a given viscosity and flow rate is inversely related to the wall thickness of the flow channel to the third power. The effect of flow rate on the frozen layer can be seen in Fig. 3.3.

Slow Fill: Thick Frozen Layer
Smaller Flow Channel

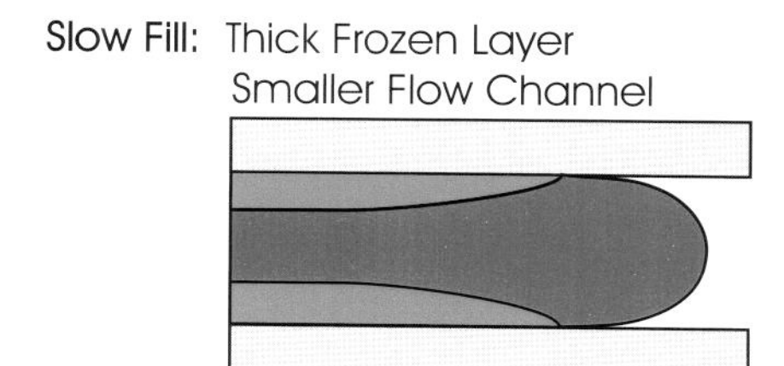

Fast Fill: Thin Frozen Layer
Larger Flow Channel

Figure 3.3 Frozen layer thickness as affected by flow rate

There are many factors that contribute to the development of the frozen layer thickness in a molded part. The primary factors are:

- The thermoplastic's thermal properties (thermal conductivity, specific heat, and no-flow temperature, or transition temperature);
- The melt and mold temperature;
- The mold material's thermal properties;
- The local flow rate; and
- The residence time of the melt.

Figure 3.4 shows a typical distribution of frozen layer thicknesses along the flow length of plastic in a cavity. The frozen layer near the gate can be very thin because of the high shear rates and the constant supply of molten thermoplastic through the near gate region of a part. The thickness of the frozen layer then builds up to a maximum between the gate and the flow front. At the flow front it again becomes relatively thin due to the short time that the melt has been in contact with the cold cavity wall.

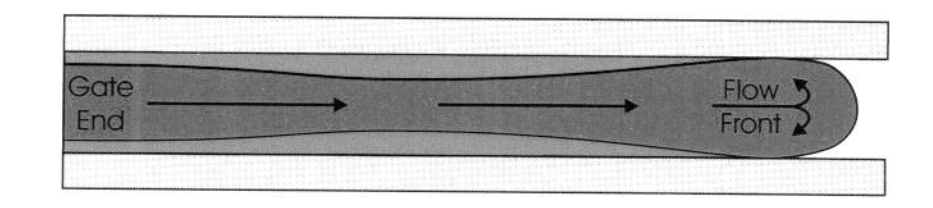

Figure 3.4 Development of frozen layer along the length of a polymer flow path

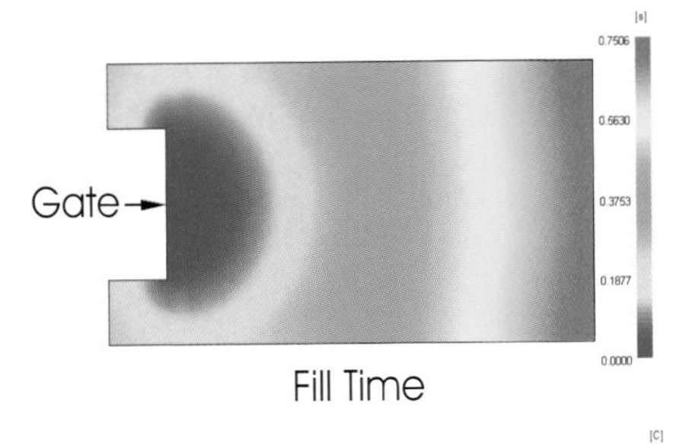

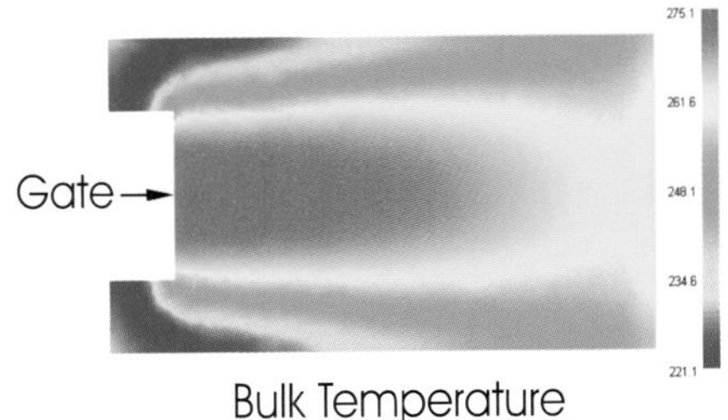

Figure 3.5 Mold filling analysis outputs of fill time and bulk temperature at the instant of fill. Note the rapid drop in melt temperature in the dead flow areas near the gate

In molds with unbalanced filling, areas that fill early develop a frozen layer much faster than areas equally distant from the gate that are still receiving continuous flow. Figure 3.5 shows an end-gated flat part with a constant wall thickness. The figure shows the fill time and bulk melt temperature distribution at he instant the part is filled. Note that the temperature distribution of the melt between the gate and the last place to fill on the right side of the part is relatively uniform as opposed to the temperature in the early filling regions near the gate. The gating location in this part causes the two tabs on the left side of the part to fill early. Once they are filled, flow into the tabs stop and the melt begins to quickly cool. Meanwhile the continuing flow in the main body of the part keeps the melt hot.

3.2 Factors Affecting Plastic Material Degradation

Degradation of a polymer will affect different polymers in different ways. Degradation of a polymer can manifest itself as a reduction in molecular weight, separation of polymer and additives, or it may result in a chemical reaction. Under high shear, a thermoplastic elastomer will experience a separation of its polymer matrix and its rubber additive. Excessive heat (time and temperature) will result in rigid PVC to go through a chemical reaction, which includes the formation of hydrochloric acid that can attack and degrade the steel components of the screw, barrel, and mold.

During the injection molding process, material degradation occurs when the plastic material reaches too high a temperature or is forced to remain at an elevated temperature too long. Additionally, a material may experience excessive shear rates. Processing conditions leading to material degradation can be caused by:

- Poor venting,
- Increasing melt temperature or injection velocity,
- Dead areas or hang-ups in hot runners, and
- Long residence times in a hot runners or the molding machine's barrel.

Over drying of hygroscopic material can also lead to degradation. With some materials, such as polyurethane, under-drying can result in degradation as a result of hydrolysis.

3.2.1 Excessive Shear

During the injection molding process, excessive shear rates and temperatures are normally associated with degradation of the plastic material. Some materials, such as rigid PVC, acetal, and polyurethane are much more shear-sensitive than others. Other materials, such as thermoplastic elastomers, will experience separation of polymer matrix and rubber additive at high shear rates. Degradation of a polymer caused by high shear rates is theorized to result from molecules being physically torn apart due to the velocity variation across the flow channel.

Highest shear rates are commonly expected at a restrictive gate; however, they may also occur at the restrictive orifice of the molding machine's nozzle. Of particular concern are tunnel gates, cashew gates, pinpoint gates on three plate cold runner molds, and restrictive gates commonly used with hot runners. The nozzle tip of the injection molding machine is another location of potential excessive shear rate. This location is not commonly recognized as a threat as it is much larger than most restrictive gates. However, in a 32-cavity mold, the flow rate will be 32 times higher through the nozzle tip than at each of the 32 gates.

Shear rate limits of most materials are poorly defined, if not unknown. Many companies and individuals have adopted shear rate guidelines established by Colin Austin in the early stages of Moldflow Inc. However, these guidelines had little scientific foundation. Numerous studies over the years have attempted to define shear rate limits by studying molded parts and materials following their exposure to high shear. Most of these data are questionable because they do not clearly separate the effects of other influencing factors experienced when the high shear rate is experienced. A part being evaluated for shear rate degradation by molding it through various gate diameters or injection rates will also experience variation in shear stress, molecular orientation, and packing. However, all studies to date appear to indicate that most materials can experience far higher shear rates than the limits established by Austin and Moldflow.

A credible ANTEC proceeding in May of 2003 from Astor and Cleveland provided some insight on ultra high shear rates and their actual effects on the mechanical and melt properties of injection molded parts [3]. This study involved a molding process, which introduced very high shear rates (on the order of 950,000 reciprocal seconds) to the material during injection. Melt flow rate and tensile tests were then performed on the highly sheared material and the parts molded from the material. The study found that these ultra high shear rates had very little effect on the actual mechanical properties of the injection molded parts. As shear rates approach extremely high levels, the mechanical properties tended to remain constant. It is theorized that these shear rates are affecting a very small percentage of the overall material near the perimeter of the flow channel, leaving a majority of the polymer's original properties intact. The study did find that viscosity was somewhat affected. Melt flow rates for the highly sheared material increased by up to 19% from the original virgin resin. This indicates some loss in molecular weight.

3.2.2 Excessive Temperature

Excessive temperature will also degrade polymeric materials. Temperature related degradation can occur prior to molding during material drying, from excessive heat in the injection unit, from excessive heat experienced in a hot runner, and potentially excessive frictional heating developed during flow in a runner.

Thermal degradation of a polymer is a function of time and temperature. The longer a polymer is exposed to an elevated temperature, the less tolerant of the temperature it is. Drying temperatures are quite low and will show no negative effects if the duration of exposure at the temperature is not excessive. However, commonly a material may be left in a dryer for days. This prolonged exposure to heat can degrade the polymer, affecting processing characteristics and mechanical properties. Spiking the temperature of the melt through a high shear runner is not expected to thermally degrade a material as the material is exposed to the high temperature only for a matter of seconds.

Some commonly used materials that exhibit degradation when exposed to excessive temperatures, or left at elevated temperatures for extended periods of time, are PVC, acetal, polyurethane, and polystyrene. Materials such as nylon, polycarbonate, and polyurethane may also experience degradation simply from over-drying.

An over-sized injection unit, or excessively long cycle times, can also cause the degradation of a polymer during molding, because they will extend the residence time of the material in the injection barrel.

Hot runner systems extend the thermal history of a polymer beyond that of the dryer and the injection unit. Ideally, the hot runner system will contain a minimum of material, which is purged with every shot. However, concerns regarding excessive pressure and shear, particularly when small thin-walled parts are molded, will result in designs with relatively large channels. These large channels may take numerous cycles to flush with fresh polymer. In addition, it should be realized that even a runner channel volume that is the same as, or less than, the cavity volume will not actually be purged every cycle. The non-plug flow conditions, resulting from the pressure driven flow and the zero flow conditions at the wall, will result in melt near the perimeter of the flow channel experiencing extended residence time. Additionally, hot runners can contain dead or low flow regions where material can collect and degrade. This is highly dependent on the design and manufacture of the system. A full round flow channel with well-aligned components throughout the flow channel will minimize the thermal degradation concerns.

During injection molding, significant heat from friction can be developed in a polymer. This heat is generated in thin laminates near the outer wall of the flow channel where shear rates are highest. Close analysis of this phenomenon with computer flow analysis programs have predicted the temperature in this area to rise by over 200 °C at very high injection speeds and with very small runner cross sections in some cold runners. However, this heat is rarely considered a problem as it is usually generated and lost to the cold mold in less than a few seconds.

3.3 Effects of Mold Fill Rate on Fill Pressure

The mold filling rate of the polymer is the most recognized direct contributor to the pressure required to fill any particular mold. The pressure required to push a polymer melt is directly proportional to the local velocity of the polymer in the melt channel and its viscosity. The following relationship of fill pressure to the melt flow velocity is shown in the common slab flow and is based on Hagen Poiseuille's Law:

$$\Delta P = \frac{12Q\eta l}{wh^3} \tag{3.1}$$

Where:

P = Pressure	Q = Flow rate
η = Viscosity	l = Flow length
w = Flow width	h = Flow channel thickness

Without close review, this basic relationship can mislead a molder to assume that reducing flow rate will decrease fill pressure. However, the results can be completely opposite.

Figure 3.6 contrasts the expected behavior of a common Newtonian fluid with that of a thermoplastic during injection molding. With a common Newtonian fluid, the pressure required to flow through a flow channel will continually decrease with decreasing flow rate. In contrast, a thermoplastic material's pressure profile will initially decrease but will then increase. The result is the characteristic "U" shaped curve seen here.

Initially, as flow is slowed, there is a corresponding reduction in pressure as experienced by a Newtonian fluid. However, as the flow rate continues to slow, the polymer viscosity begins to increase as it loses the benefits of non-Newtonian shear thinning. In addition, some of the frictional heating gained at the high flow rates is beginning to be offset by heat lost through conduction into the cold mold. This loss of frictional heating will both increase the viscosity and increase the frozen layer thickness. At a certain flow rate, the resulting increased resistance to flow reverses the trend of decreasing pressure with decreasing flow rate. Pressure begins to rise with decreasing flow rate as the material becomes more and more viscous and the flow channel becomes smaller and smaller with the increasing frozen layer thickness. Rather than pressure approaching zero at low flow rates, pressure will approach infinity as the polymer freezes completely.

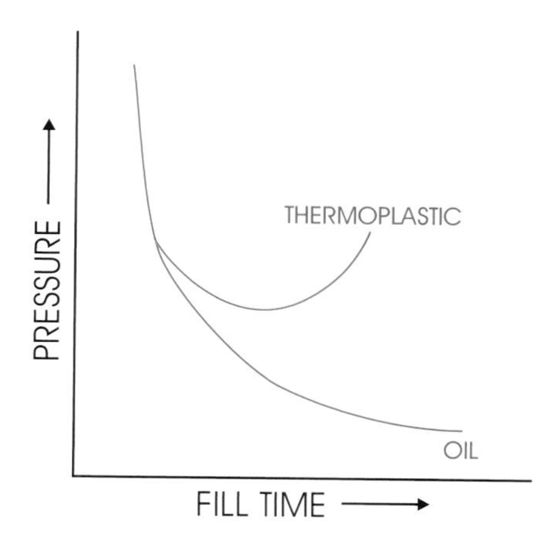

Figure 3.6 Comparison of effect of fill time on pressure development of a non-Newtonian plastic material and a Newtonian fluid

In summary, during injection molding three factors act to decrease fill pressure despite increasing fill rate.

- The non-Newtonian molten plastic will shear thin in response to the resultant increased shear rate.
- The high shear rates will increase the viscous dissipation and can offset heat being lost to the relatively cold steel of the flow channel. The net effect being a temperature increase of the flowing plastic, thereby further reducing viscosity.
- An increase in flow channel cross section will occur due to the reduction of the frozen layer.

3.4 Post Filling or Packing Phase

3.4.1 Thermal Shrinkage as Plastic Cools

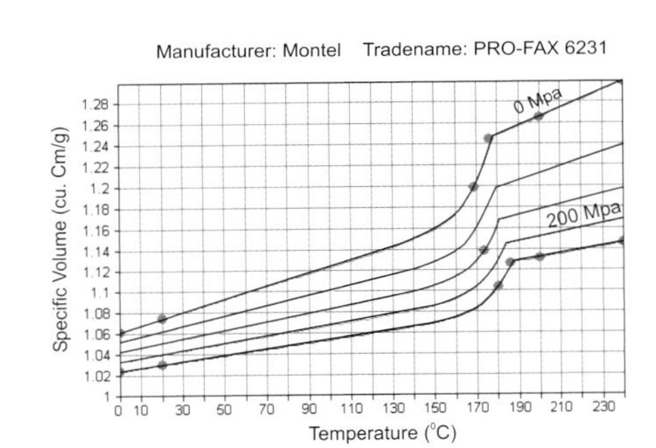

Figure 3.7 PVT data of a polypropylene (Courtesy: Moldflow Inc.)

Review of the pressure/volume/temperature (PVT) relationship of a polymer, as presented in Chapter 2 will show that the volume change of the polymer as it cools is very significant. The volumetric shrinkage at zero pressure can be as much as 20 to 30%. An example of the PVT relationship of a polypropylene is shown in Fig. 3.7. As the material is cooled from a melt temperature of 240 °C to room temperature (23 °C), the specific volume of the material at 0 MPa pressure is reduced by nearly 20%. This shrinkage should not be confused with the linear in-mold shrinkage values used in mold construction. Figure 3.7 also shows that at the maximum pressure of 200 MPa, the specific volume of the material (at 240 °C) is reduced to approximately 88% of its original volume. The PVT relationship characterizes the primary volumetric shrinkage behavior of a polymer as it cools under varying pressure throughout the molding cycle.

The shrinkage of the plastic in the mold can cause many problems in addition to its effects on warpage, residual stresses, and predicting part size. Shrinkage is also the reason for high ejection forces required to remove the part from the core of a mold. The shrinkage over any core of a mold leaves a residual tension stress in the part, which applies a normal load to the side surfaces of the core section. This normal force increases the friction experienced by the part when it is ejected, sometimes causing excessive ejection forces, which can damage a part.

Another problem introduced by polymer shrinkage is the tendency to pull away from the mold's cavity wall, thereby losing intimate contact, which will increase the cooling time for the part. The general tendency on a simple core/cavity part would be for the polymer to shrink away from the cavity wall but maintain good contact with the core wall. However, if adequate pressure is maintained on the part as the frozen layer thickness builds during the packing phase, good contact with the cavity wall can be improved, which will help uniformity of cooling and minimize cooling time.

3.4.2 Compensation Flow to Offset Volumetric Shrinkage

Because of the high volumetric shrinkage of plastic as it cools, it is critical that pressure be maintained on the melt as it solidifies to prevent many problems associated with material shrinkage. For any given polymer, the volumetric shrinkage will vary based on both the local part thickness as well as the pressure history that the polymer has experienced throughout the cycle. The basic volumetric shrinkage behavior as a function of the part thickness, gating configuration, and pressure history is critical to the success of the part design and the part processing.

As the plastic in a mold cools it will shrink. If the material is allowed to freely shrink, the resulting molded part will be poorly shaped and have poor surface finish. As the material shrinks, additional material must be fed into the cavity to compensate for the shrinking material. This requires that not only pressure is maintained by the molding machine but that all flow paths to all shrinking regions within a part remain open to the compensating flow. Cold runners and their gates must be large enough so that they do not freeze before the part. Hot runner gate tips must remain open and unfrozen. Gating into the cavity must be in a location that allows material to continually flow, during the compensation phase, through the gate region of the part to all extremities as they cool and freeze. This dictates that a part with varying wall thicknesses be gated into its thicker regions. In cases where this is not possible, it should be anticipated that the thicker regions will experience some free shrinkage. This will generally result in sinks and/or voids in the thicker regions and the increased potential for warpage. Figure 3.8 shows a center-gated disk with a thickened flange. Note the relative volumetric shrinkage prediction from a mold fill/pack analysis and the resulting warpage. The volumetric shrinkage in a majority of the thinner (1 mm thick) center region is about 2.5% whereas shrinkage in the 2 mm thick flange is more than 5%. Also, note the predicted warped bowl shape resulting from the excessive shrinkage in the outer flange.

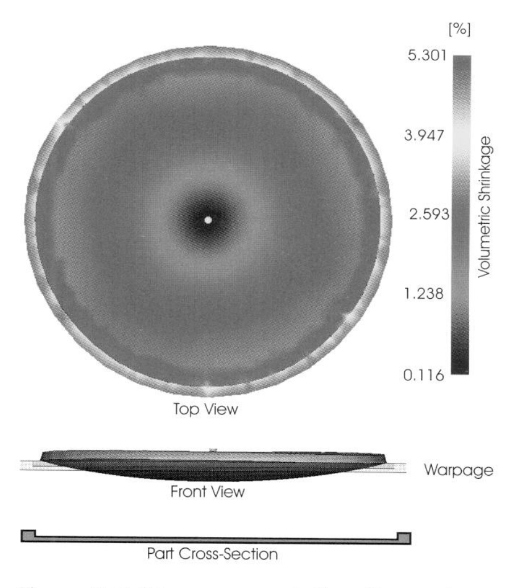

Figure 3.8 Warpage prediction, illustrating classic deflection caused by variations in wall thickness

3.4.3 Pressure Distribution During the Post Filling Phase

During mold filling, the pressure distribution across a part ranges from a maximum at the gate to zero at the free surface of the polymer flow front. Once the cavity is filled, the pressure distribution within the cavity becomes somewhat hydrostatic, with the pressure distribution becoming much more uniform. The molten polymer in the cavity acts like a fluid in a vessel. However, conditions are not quite hydrostatic as compensating flow continues. During the post filling pressurization and compensation phases, the pressure on the melt will remain highest at the gate region and decrease moving further from the gate. Figure 3.9 includes the results of a mold

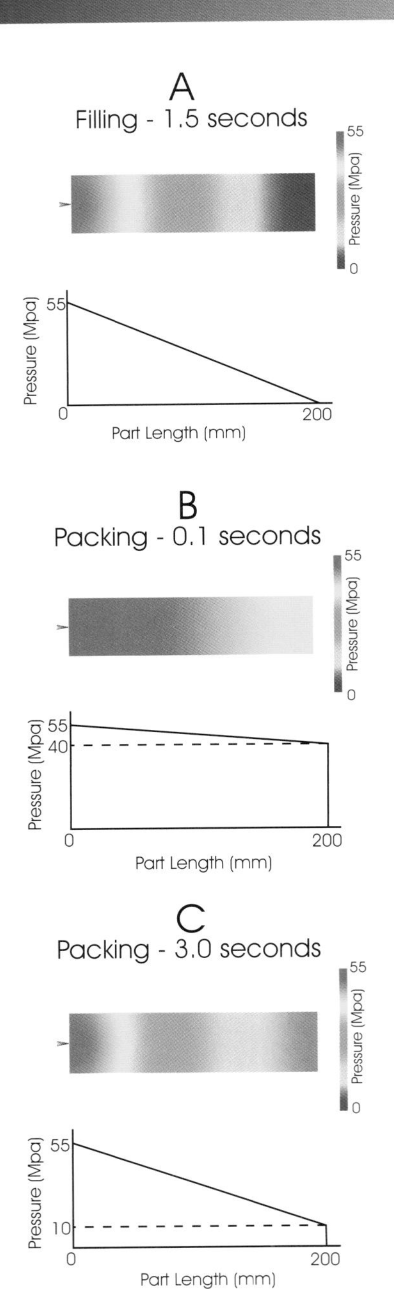

Figure 3.9 Predicted cavity pressure distribution during mold filling and mold packing. Note the decrease in pressure away from the gate with time during pack stage

filling and packing analysis of a flat rectangular strip. Pack pressure was kept the same as the predicted filling pressure. During filling, pressure graduates from 55 MPa at the gate to zero at the flow front. Immediately after filling (Fig. 3.9B), the melt pressure distribution becomes approximately uniform. Figure 3.9C shows the melt pressure distribution 3 seconds after packing. Note that the pressure near the gate remains at the inlet pack pressure, while pressure on the opposite side of the part has decreased as the material between these regions continues to cool and freeze. The amount of pressure drop seen both during the filling and post-filling phase is dependent on the frozen layer thickness development and the viscosity of the material. The frozen layer thickness in turn is dependent on the specific material's thermal properties, the part geometry, and the processing conditions under which the part is being molded.

3.4.4 Gate Freeze-Off

Designing a gate for proper freeze-off is one of the more important factors when assuring that a part is packed properly and its related effect on control of the part's quality. An under-packed part can display poor surface finish, sinks, voids, uncontrolled shrinkage, and warpage. The most common technique for determining when a gate is frozen is relatively simple. With a reasonable pack pressure set, gradually increase the packing time and weigh each part (without runner system attached). When the part weight stops increasing, the gate has frozen-off and is preventing any additional material to be fed to the part. This gate freeze-off time will probably only vary slightly with process variations.

Gate freeze time is primarily controlled by gate size. However, additional influencing factors include localized heating of the gate by the flowing melt. In addition, gate freeze can be affected by increasing or decreasing the thermal conductivity of a gate insert. Pack pressure can also have an effect. Generally, a high pack pressure is expected to improve packing. However, sometimes it can be found that a lower pack pressure increases the time for compensating flow and keeps the gate open longer. A high pack pressure can compress the material in the cavity and thereby require less compensating flow. The resulting low flow rate during the compensation phase can cause the gate to freeze quickly and eliminate further compensating flow.

Too short a gate freeze time will result in under packing and loss of pressure control in the cavity as the part cools and shrinks. Too long a gate freeze time can result in over packing of the part, particularly in the gate region. This is more of a potential problem with hot gates vs. cold gates. In addition, if pressure is relieved before the gate is sealed, material can back flow through the open gate resulting in high shrinkage in the immediate gate region.

3.5 Melt Flow Effects on Material and Molded Parts

3.5.1 Shrinkage

The complex shrinkage behavior of plastic materials is probably the most dominant underlying problem associated with the successful design and manufacture of injection molded plastic parts. During the injection molding process, plastic materials undergo a reduction in specific volume of as much as 35% as they are cooled from their molten state to a solid state. If this did not present enough of a design and molding problem, flow of plastic through the mold induces molecular orientation, which introduces a directional component to this shrinkage. The direction and magnitude of the shrinkage varies significantly from part to part and within a given part. The direction and magnitude of the resulting shrinkage affects the final size, shape, and mechanical properties of the finished product. It becomes imperative to understand these shrinkage phenomena in order to develop and design molding strategies that will minimize their potential negative impact.

Thermoplastics are comprised of long organic molecular chains that are made of numerous repeating carbon based units (monomers). The atoms within these chains are held together by relatively strong covalent bonds. Entanglement and relatively weak electrostatic forces hold the individual polymer chains together. These weaker secondary forces have only a fraction of the strength of the primary bonds holding together the individual atoms within the polymer chain.

During the injection molding process, a polymer mass is heated to a point at which it melts. In the molten phase, the material can flow and is forced into a cold mold where it will take the shape of the cavity. After cooling, the material can be removed from the mold and it will retain the general shape of the cavity from which it was removed. However, the part will experience shrinkage and potentially distortion, or warpage, resulting from variations in the shrinkage.

3.5.1.1 Volumetric Shrinkage

Thermal contraction and expansion is a well-known and understood phenomenon that occurs with almost all known matter. With few exceptions it is expected that as a material is heated it will expand, and as it is cooled it will contract.

When relatively low levels of heat are introduced to a polymer, this external energy increases molecular motion and weakens the cohesive bond energies of both the primary (atomic) and secondary (molecular) bonds. The result is an increase in the specific volume of the polymer mass as the atoms and molecules move apart. The amount of expansion will be proportional to the increase in heat input to the body. As the secondary bonds are weaker, the introduction of heat will have a more dramatic effect on them. During subsequent cooling, if no external forces are applied, the polymer would experience uniform, orthotropic contraction (shrinkage).

Ordered polymers, such as HDPE, can form crystals in which the polymer chains fold back upon themselves, resulting in densely packed parallel chains. The effect of this can be seen in Fig. 3.7, which is a characteristic PVT graph of a semi-crystalline polymer. At the elevated temperatures at which the polymer is molten, its structure is amorphous. As the material is cooled, initially there is a gradual reduction is specific volume. At the material's crystallization temperature, there is a sudden decrease in specific volume as the polymer transforms from a molten amorphous to a solid semi-crystalline structure. Once the material has solidified, the reduction in specific volume with change in temperature again becomes more gradual and linear.

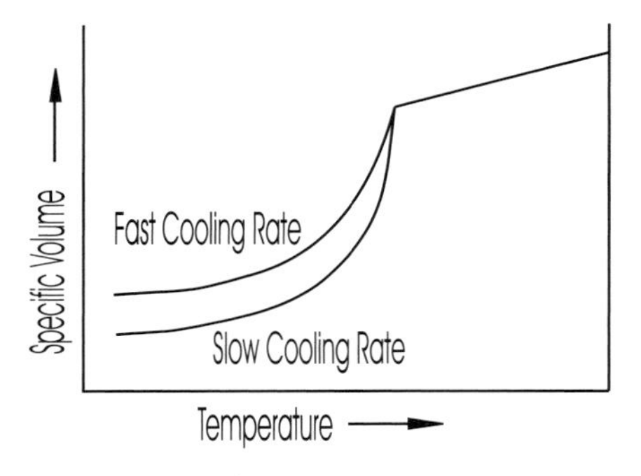

Figure 3.10 Effect of cooling rate on the shrinkage of a semi-crystalline material

The formation of the crystalline structure requires that the molecules must organize themselves from the random state in which they are while molten. As this physical structuring of molecules requires freedom of movement during the time the crystals are forming, the rate at which the polymer is cooled will affect how much of this structuring occurs, that is, what percentage of the polymer will be crystalline vs. amorphous. It can also affect the size of the forming crystals. As the crystals are denser than the amorphous regions of the polymer, it can be expected that the rate of cooling will affect how much the polymer will shrink. Therefore, unlike the density of amorphous materials, the density of a semi-crystalline material will be affected by the cooling rate. Figure 3.10 illustrates the relationship of cooling rate and shrinkage of a semi-crystalline material at atmospheric pressure. The differences in specific volume at room temperature result from changing the cooling rate of the molten semi-crystalline polymer. Also note the abrupt change that occurs in specific volume between the material's melt temperature and room temperature. This is the temperature region in which the polymer's crystals are forming during cooling (dissociating during heating).

Because of the high volumetric shrinkage of plastic as it cools, it is important to maintain adequate pressure on the melt as it solidifies to minimize many of the problems associated with material shrinkage. For any given polymer, the volumetric shrinkage will vary based on both the local part thickness as well as on the pressure history that the polymer has experienced throughout the cycle. The basic volumetric shrinkage behavior as a function of the part thickness, gating configuration, and pressure history are critical to the success of the part design and the part processing.

Figure 3.11 shows the results of a 2D thermal analysis, which illustrates the basic problem that can be found in part designs that include variations in wall thickness. If the gate(s) is not located in the thicker areas of the part, as is illustrated in the "poor design," the thinner section between the gate and the thicker region will freeze off first, and packing pressure from the gate cannot be maintained on the thicker section as it continues to cool and

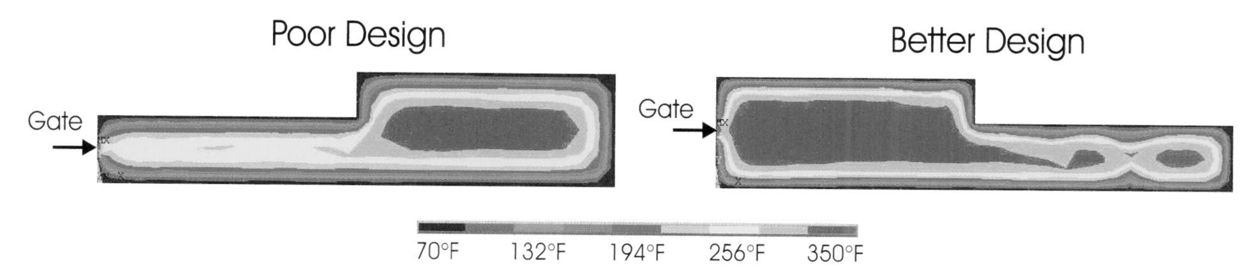

Figure 3.11 FEA thermal analysis of a part with varying wall thickness. Part should be gated into the thicker wall section whenever possible to assure the ability to feed the last regions to freeze during compensation/packing stage

shrink. The inability to adequately pack the thicker sections of the part will lead to higher shrinkage in those areas. The differential shrinkage between the different thickness regions of the part can cause warpage and residual stress problems.

Particular effort should be made to design a part and its mold such that the situation in the "better design" illustration in Fig. 3.11 is achieved. This will allow better packing of the part to produce more uniform shrinkage and less potential warpage and residual stress problems. Owing to part complexity, the guideline of "gating and filling from thick to thin areas" cannot always be maintained. Care should be taken to at least be knowledgeable of the potential for packing problems to arise when this guideline is not observed. Another consideration is the pressure difference between the gate and the end of the cavity, especially when molding viscous materials. The pressure lost from transmitting pack pressure through viscous polymers could give rise to variations in shrinkage between near and away gate locations. This will result in residual stresses and the potential for warpage.

3.5.1.2 Orientation-Induced Shrinkage

Orientation of polymer and fillers, developed during the flowing of the material, can create anisotropic shrinkages in a molded part. These orientation-induced shrinkage variations are dependent on material, process, gating, and part design. Orientation of a polymer and any asymmetric filler in the polymer is directly related to the direction and magnitude of the applied stress developed during the mold filling and compensation/packing phase. The cooling rate will also effect the molded-in orientation of the polymer but not the filler. A fast cooling rate will freeze more of the stress induced polymer orientation into the part, whereas a slow cooling rate will allow for more relaxation. An asymmetric shaped rigid additive, like a glass fiber, cannot relax and therefore will maintain whatever orientation was established during flow.

Shear stress and extensional flow effects create orientation of both polymer and asymmetric fillers. This orientation will create both anisotropic shrinkages that are a major source of warp and residual stress in molded parts.

The direction of the applied stress is dependent on flow type. The stress can either be induced by the shear between the flowing melt and the stationary cavity wall or by extensional flow effects. Shear induced stress and strains result from the resistance to flow as the material drags against the stationary walls of mold cavity. This shear, or drag, stretches the polymer in

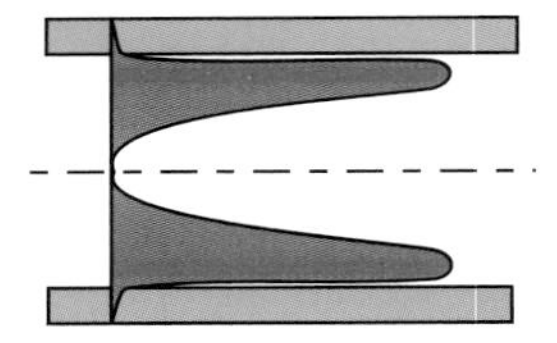

Figure 3.12 Profile of the shear rate through a cold runner

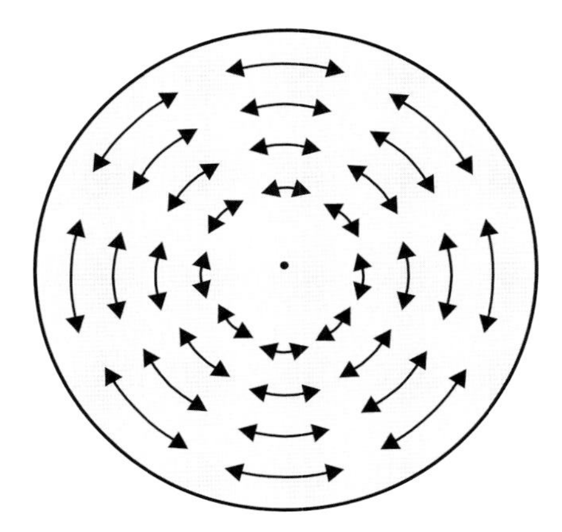

Figure 3.13 Extensional flow effects will cause polymer chains and asymmetric additives to be oriented transverse to flow in the mid laminates of a center gated disk

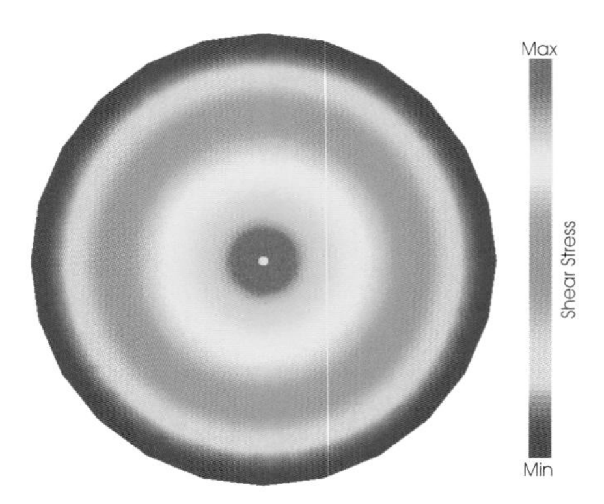

Figure 3.14 The shear stress in a center gated disk is greatest near the gate and decreases as the flow front velocity decreases as it radiates outward

the direction the material is flowing. The magnitude of orientation is closely related to the shear rate distribution across the flow channel, which is highest near the stationary channel wall and zero at the center of the flow channel (see Fig. 3.12).

Extensional flow develops from an expanding flow front, such as in a center-gated disk (see Fig. 3.13) or in the initial phases of flow from a small gate onto the edge of a part. With an expanding flow front, there is an extensional force acting on the polymer, which is perpendicular to the direction of flow. As the melt in a center-gated disk progresses across the radial length of the cavity, the polymer is also being stretched in the circumferential direction, similar to the surface of an expanding balloon. The extensional flow will cause the polymer molecules, and its fillers, to become oriented in the transverse-to-flow direction.

The net orientation of the polymer and filler in an expanding flow front depends on where the shear-induced or the extensional flow-induced stresses are dominant. Near the cavity wall it is expected that shear stresses are dominant. As shear stresses decrease to zero in the center of the flow channel, extensional flow forces will become increasingly influential. In addition, proximity to the gate will affect the balance of orientation. In a center-gated disk, flow rates and resultant shear forces are highest near the gate. Therefore, orientation in the direction of flow will be dominant in the near-gate region of the part. As flow slows down as it advances away from the gate, shear stresses decrease and extensional flow forces will have increasing influence. Figure 3.14 shows a predicted shear stress distribution in a center-gated disk Highest shear stresses near the gate region are shown in red and progressively decrease to a minimum value around the perimeter, shown in blue. Similar effects can be expected with changes in wall thickness or material viscosity. Thin walls and high viscosity material will increases shear stresses and contribute to their dominance of orientation.

The effect of flow-induced orientation on shrinkage is somewhat inconsistent. With neat materials, shrinkage is not always in the direction of the dominant orientation. However, it is common that increased shear stresses will increase the tendency for most materials to shrink in the direction of the dominant applied stress.

While the directional shrinkage of neat materials is not always consistent relative to the direction of the applied stress, the direction of shrinkage of fiber-filled materials is always transverse to the direction of the orientation established by the stress developed during mold filling. For glass fiber filled materials, shrinkage in the orientation direction is commonly only 20% of the shrinkage in the cross direction. This difference is significant enough that anisotropic shrinkage should be taken into account when cutting the cavity for parts that are intended for high-precision molding. In some cases of highly linear flow, shrinkage in the flow direction with glass-filled materials has been found to be less than 6% of the cross flow shrinkage. [4]

Table 3.1 summarizes a review of shrinkage studies performed by Moldflow Inc. to characterize materials for their shrinkage and warpage analysis

Table 3.1: Material Shrinkage and Warpage

Material	**Thickness Effect Induced Shrinkage**		**Average linear shrinkage**
	Thick (3 mm +)	**Thin (< 3 mm)**	
General materials:			
Amorphous (unfilled)	Isotropic	Parallel*	0.0045
Semi-crystalline (unfilled)	Parallel*	Parallel**	0.0182
Filled polymer (amorphous or semi)	Perpendicular**	Perpendicular***	0.0062
Specific materials:			
ABS	Isotropic	Parallel*	0.0043
PC	Isotropic	Parallel*	0.0057
PS	Parallel*	Parallel**	0.0035
PVC	Perpendicular*	Perpendicular*	0.0044
HDPE	Parallel*	Parallel**	0.0234
PP	Isotropic	Parallel*	0.0133
Acetal	Isotropic	Not dependent	0.0200
PET	Perpendicular*	Perpendicular**	0.0149
Nylon 6	Perpendicular*	Parallel**	0.0109
Nylon 6–6	Perpendicular*	Parallel**	0.0162
PET glass filled	Perpendicular*	Perpendicular**	0.0053
Nylon 6-6 30–35% glass filled	Perpendicular*	Perpendicular***	0.0070

Note: Observations based on experimental data, use for reference purposes only.
Definitions: Parallel – Polymer shrinks more parallel to direction of flow
Perpendicular – Polymer shrinks more perpendicular to direction of flow
Isotropic – Linear shrinkage is independent of flow direction.
* Low level of orientation effects
** Medium level of orientation effects
*** High level of orientation effects

programs. The characterization studies evaluated the shrinkage of a material in the flow and in the transverse-to-flow direction in an end-gated rectangular plaque. The fan gate at the short edge of the rectangular plaque helped to develop a simple linear flow path across the cavity. Materials were molded at various thicknesses, each under a matrix of process conditions. For each of the materials listed below, several like materials were selected from Moldflow's database and their shrinkage data averaged. Shrinkage for each material, at two different thickness ranges (thick and thin), is then identified as being predominately "parallel" to flow; "perpendicular" to flow or "isotropic". Average linear shrinkage for each material is then identified in the last column.

From the data in Table 3.1 it can be seen that shrinkage with the three neat amorphous materials tended to be either isotropic or to shrink more in the direction of flow at the 2 mm thickness. Also notable is the fact that higher shear, resulting from thinner walls, tended to increase the anisotropic shrinkage in both the polycarbonate and polystyrene but not in the ABS.

The semi-crystalline materials not only experienced much higher shrinkage, but also were less predictable in their behavior. The addition polymers (polyethylenes and polypropylenes) behaved similar to the amorphous materials. The condensation semi-crystalline polymers acted oppositely. The PET's and the polyamides shrank more transverse to flow rather than in the flow direction. The glass filled nylon, as expected, demonstrated extreme anisotropic characteristics.

In most cases, linear shrinkage variations were found. All of the amorphous material, aside from PVC, showed either a non-dependence or parallel with flow dominant shrinkage in thicker sections. This pattern was amplified as orientation increased with the thinner sections making the amorphous material shrink parallel to flow again, except for PVC. Semi-crystalline materials are expected to have similar anisotropic shrinkage characteristics to amorphous, just to a higher extent, which may be up to ten times as much. This trend can also be seen by the semi-crystalline materials in the previous chart. Fillers commonly oppositely affect the anisotropic shrinkage of the polymer. Because the fibers are oriented parallel to the flow direction and do not shrink with the polymer, the dominant shrinkage becomes perpendicular to the flow direction.

3.5.2 Development of Residual Stresses and Warpage

Variations in shrinkage in a molded part will result in conflicting strains, which will develop residual stresses and warpage. If the part is molded of a rigid material and has a sufficient rigidity in its shape, it will better resist warpage from residual stresses. However, if the residual stress is sufficient, warpage can be expected. This can be observed with many glass fiber-filled nylon parts. The modulus of a glass fiber-filled material is relatively high and is often used in structural applications utilizing rigid shapes. Despite the high modulus and rigid part geometry, designers and molders continually struggle with the warpage of parts molded of fiber-filled materials.

If a part is heated, the modulus of the plastic will decrease. With decreasing rigidity, the residual stresses will more easily distort the plastic part. This distortion will relieve the residual stresses. If the part is fixed so that it cannot distort, the stresses will still be at least partially relieved as a result of increased molecular mobility.

Any stress developed from differential shrinkage will experience some relaxation owing to the visco-elastic properties of the plastic. The amount of relaxation is dependent on time and temperature. Residual stresses can contribute to premature failure of a part. Areas of high residual stress, such as the gate region, are particularly prone to failure.

3.5.2.1 Warpage and Residual Stress from Side-to-Side Shrinkage Variations

Side-to-side shrinkage variations can occur when differential cooling is present in the molded part; that is, when one side of the part cools more quickly than the other, for instance, when the mold surfaces experience non-uniform cooling. Imagine a flat rectangular plastic part being molded between mold halves of different temperatures. The side of the part in contact with the cooler side of the mold will freeze sooner and experience less shrinkage than the side in contact with the warmer half of the mold. The side-to-side shrinkage variations will create a bending moment causing the part to warp (see Fig. 3.15).

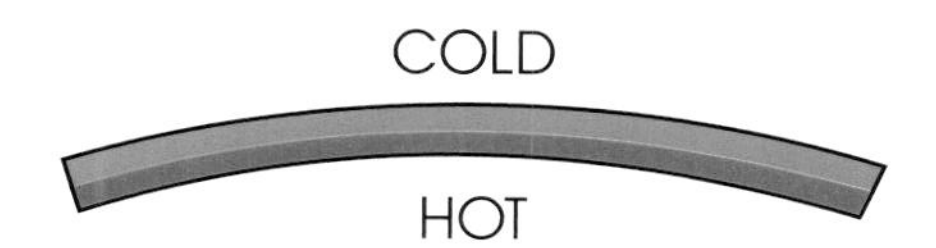

Figure 3.15 Variations in mold temperature will create a bending moment in the part, which will cause it to warp as shown

This variation in side-to-side cooling is the major contributor to the classic hour glass warp seen in most box shaped parts. During mold cooling, the inside corner of a box tends to be much warmer than the outside corner. The heat from the inside corner must be extracted from the limited geometry of the associate corner of the mold core. The heat is more easily extracted from the increased surface area on the cavity half (see Fig. 3.16). The result is that the increased shrinkage on the inside corner pulls the sidewalls of the part/box inward, creating the hourglass shape shown.

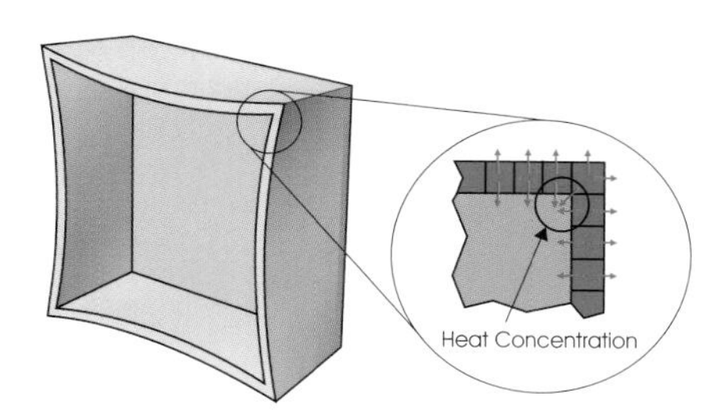

Figure 3.16 Difficulty in removing heat from the inside of corners results in box-shaped parts warping as illustrated

3.5.2.2 Warpage and Residual Stress from Global/Regional Shrinkage Variations

During mold filling and packing there will be a pressure gradient across the part. Highest pressures will be near the gate compared to away from the gate. Distance, wall thickness, material viscosity, and fill rate will affect the amount of pressure variation. Areas of high pressure will be packed out better than areas of low pressure. This will cause these regions to shrink at a different rate. Normally, it can be expected that the gate region of the part will shrink less than far regions of the molded part.

This becomes evident when molding parts such as a flat center-gated disk. Here it can be expected that the part will undergo less volumetric shrinkage

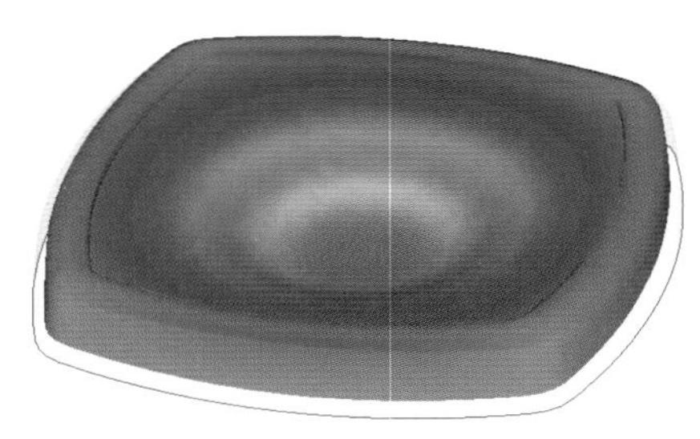

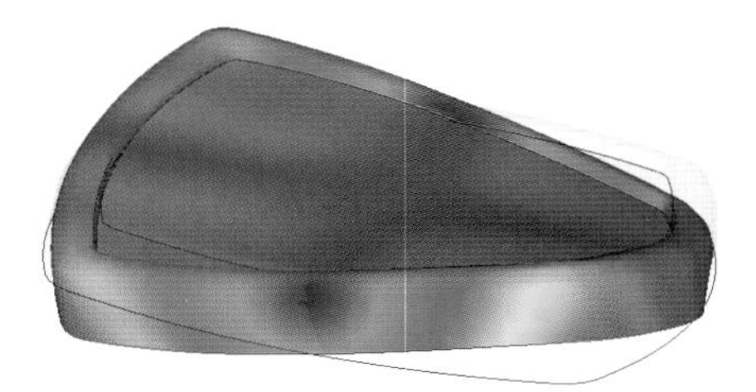

Figure 3.17 Effect of wall thickness on part warpage. A relatively thick perimeter will result in a "bowl" shape; a relatively thin perimeter will cause a "saddle" shaped warp

in the center region of the disk vs. at the perimeter as a result of higher pack pressures in the center. As the perimeter shrinks, it will force the part into the shape of a bowl. This same condition can develop in square or rectangular shaped parts (see Fig. 3.17A).

This type of warpage will be most dramatic if the perimeter of the part has a greater thickness than the interior. One method to try to combat this problem is to increase cooling near the perimeter of the part relative to the center region.

Potato chip warpage, also known as saddle warpage, occurs when the central region of a molded part is shrinking more than the perimeter. Again, this type of warp can occur with many differently shaped parts. The predicted saddle warp, as seen in Fig. 3.17B, was a result of the part being altered to have a thinner wall around its perimeter.

Two design modifications can be made to address saddle warp:

- Thicken the perimeter of the part and/or
- Improve the cooling near the center region of the part relative to the perimeter.

3.5.2.3 Warpage and Residual Stress from Orientation-Induced Shrinkage Variations

Warpage resulting from orientation-induced shrinkage may be one of the most difficult to overcome problems. This is particularly true with fiber-filled materials, which experience extreme anisotropic shrinkages resulting from flow. As discussed earlier in this chapter, the development of orientation in a molded part is extremely complex.

Using a center-gated disk as an example, under these expanding flow conditions, if the dominant shrinkage is in the radial direction, the shrinkage will cause the part to warp in a twisted saddle or potato chip shape. If the dominant shrinkage is circumferential, the part will warp into a bowl shape. A thin-walled part molded with a fiber-filled material is most likely to warp into a bowl shape. A neat thin-walled molded part is most likely to warp into a twisted potato chip shape. However, this warpage is dependent on material properties and the counter effect of volumetric shrinkage as discussed earlier in Section 3.5.5.2.

3.5.3 Physical Properties as Effected by Orientation

The flow of material during injection molding can have a significant effect on many of the physical properties of molded parts. These include strength, rigidity, ductility, and electrical conductivity. Many of these anisotropic characteristics are compounded when a fiber is added to the material as reinforcement.

When applying a flexural load, a part molded of a neat material will normally be stronger and more ductile when loaded perpendicular to the direction of orientation (see Fig. 3.18). When applying a tensile load, the molded part exhibits greater strength parallel to the direction of orientation.

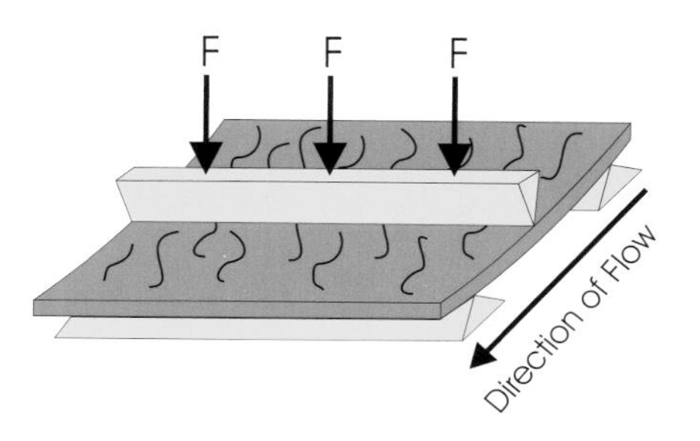

Figure 3.18 Flexural loads should be applied perpendicular to the flow direction of the polymer for maximum strength

3.6 Annealing a Molded Part

During annealing, a molded plastic part is heated enough to permit sufficient movement of the polymer molecules, thus allowing relaxation. This reduces the stress within the part. Annealing can also be performed on warped parts to correct their shape. A molded part can be placed in a fixture forcing it into the desired shape. As the part is heated and then cooled, the stressed polymer molecules reposition themselves to a relaxed position within its corrected shape. Though this is a costly operation, it is often required to achieve the desired shape of a troublesome molded part.

Note: Other molding influences on material properties and defects are presented in Chapter 15 – Troubleshooting Guide. This includes issues such as weld lines, sinks, voids, jetting, flow marks, air traps, splay, and burning.

3.7 Summary

The following summarizes the major causes for shrinkage variations, which will result in the development of residual stresses and warpage.

Orientation:

- Magnitude of polymer and polymer additive orientation
  Orientation creates a variation in shrinkage relative to the orientation. The higher the magnitude of orientation, the more variation in shrinkage will result. High shear stress, seen with mold filling analysis, is the direct cause of high orientation. The higher the shear stress, the higher the orientation, the higher the residual stresses, and the higher the potential for warpage. Highest shear stresses are most commonly found near the gate and can also be found at the end of fill. A warmer mold will allow some of the orientation to relax during cooling.
- Direction of orientation
  Variations in flow direction create conflicting shrinkage vectors. Consider a center-gated disk. Here the flow radiates out from the gate in 360°. These conflicting flow directions will result in material wanting to shrink in an infinitely variable direction around the gate. This is in contrast to a long narrow part, which is end-gated with a fan gate. Here the flow would be purely linear with no variation in flow direction. No conflicting strains will develop.

- Type of flow
  The flow front pattern indicates the direction of orientation on the outer laminates of a molded part. The orientation in the mid-layer laminates will be effected by flow type and transient flow. Flow type can be linear, expanding, or converging. Linear and converging flows will cause the mid laminates to be oriented in the direction of the primary flow direction. Expanding, or diverging, flow fronts create an extensional flow force, which acts transverse to the direction of flow. This will cause the mid laminate to become oriented transverse to flow. The resultant variation in shrinkage through the cross section of the part will cause localized residual stresses.
- Transient flow
  Transient flow is a condition where the flow direction changes through the thickness of the flow channel during mold filling. This will occur due to unbalanced filling in a cavity. Once a given region in a cavity fills, the molten material under the outer frozen layers will change direction toward unfilled regions of the cavity. As with the extensional flow types, the resultant variation in shrinkage through the cross section of the part will again cause localized residual stresses

Regional Volumetric Shrinkage:

- Variations in volumetric shrinkage can occur due to variations in wall thicknesses, pressures, and cooling. Regions of a part, which are thinner or formed under high pressure or low temperatures, can normally be expected to shrink less than thicker regions of a part or regions formed under low pressure or high temperature. These variations in shrinkage from one region of the part to another will create residual stresses. This will cause a center-gated disk with a thick flange around its perimeter to warp in a bowl shape.

Cooling:

- Effect on Orientation
  A warm mold will allow some of the orientation set up by flow to relax. A cold mold will lock in more of the orientation and the resultant residual stress. This will not only affect stresses between adjoining global regions of a part, but it will affect stresses developed through the thickness of the part. In general, it can be expected that a warmer mold will reduce polymer orientation and the resultant stresses.
- Effect on Regional Volumetric Shrinkage
  Temperature affects thermal contraction and with semi-crystalline material, the degree of crystallization. A region of a part molded with high temperatures will shrink more than a region formed at low temperatures.

- Side-to-Side Cooling Variations
  Side-to-side cooling variations will affect orientation and volumetric shrinkage from one side to the other of a part. The side-to-side shrinkage variation will develop a bending moment, which will tend to cause a part to warp towards the warmer side.
- Pressure
  During filling and packing, regions under high pressure will shrink less than regions under low pressure. Highest pressures will exist near the gate of a part and will decrease along the flow path. In the case of unbalanced filling, early-filling regions will become nearly hydrostatic at a higher pressure than the late filling regions of the cavity.

References

[1] Injection Molding Solutions, 1019 Balfour Street, Midland, MI 48640 Tel. (989) 832-2424; Fax (989)832-8743; http://www.scientificmolding.com/

[2] RJG, Inc. 3111 Park Drive Traverse City, MI 49686 Tel. (231) 947-3111; Fax (231)947-6403; http://www.rjginc.com

[3] K. Astor, S. R. Cleveland, "Ultra High Shear Rates and their Effects on the Physical and Melt Properties of an Injection Molded Part," SPE Annual Technical Conference, Nashville (2003), p. 3378

[4] J. A. Doolittle, "Effects of Glass Fibers on Shrinkage of Molded Parts," SPE Annual Technical Conference, Dallas (1990), p. 1986

4 Gate Positioning and Molding Strategies

Gate positioning is critical to the successful injection molding of plastic parts. In addition, it is important to establish a basic strategy related not only to gate position but to part design, mold design, and processing. This chapter first outlines some important considerations in the positioning of gates. This is followed by 14 design and process strategies, which will help assure a successful molding operation.

4.1 Gate Positioning Considerations

The following is a checklist of factors to consider when selecting a gating location for your cavity. For those readers not familiar with some of the gate types mentioned below, please refer to Section 8.4.

- Ability to reach gating location with the desired runner system/mold type: A gating location on the perimeter of the mold cavity can be fed by most runners including a standard two-plate cold runner system. A gate location inside the perimeter is more restrictive and will require a three-plate cold runner system or a hot runner system. A cashew gate may be used to gate inside the perimeter of a part, but it will typically extend at most 0.5 in. beyond the perimeter. A jump gate can extend up to 0.75 in. inside the perimeter of the part with a two-plate cold runner, but both these gate options provide limited access to the part interior.
- Gate types available for the selected location, mold type, and material: The use of a diaphragm gate would require a three-plate cold runner or hot runner mold. Cashew gates are limited to more ductile materials or amorphous materials with gate removal occurring when the gate is still hot and ductile.
- Consideration of wall thickness variations in molded parts: If there are variations in wall thickness, the gate should be located in the thickest wall section. Gating into thinner wall sections will restrict the control of the packing of the thicker regions. This can result in excessive shrinkage, warpage, sinks, and voids.
- Effect of gating location(s) on the creation and position of weld lines: This can be both a cosmetic and a structural concern.
- Potential cosmetic and structural concern at the gate location on the part: The gate region of a part may have problems with gate blush and gate vestige, and may contain high-residual stresses.
- Effect on flow pattern and its influence on shrinkage: Anisotropic, or directional, shrinkage results from flow-induced orientation and is one

of the most significant contributors to warpage and residual stresses in molded parts. Gating resulting in a pure linear flow across a part will generally minimize the potential for conflicting shrinkages. This is particularly true for fiber-filled materials.

- For long narrow parts, a gate at one end of the part will best ensure that the flow-induced orientation is structured within the part such that potential of flow induced warpage, and residual stresses are minimized.
- Parts with dimensions that are relatively symmetrical about a centriod generally do not lend themselves to gating along one end where they could develop a linear flow pattern.
  - Three-dimensional shapes (cup, bowl, or box) generally prefer a gate location at the centroid. This will result in a radial flow pattern developing from the gate. Though this radial flow pattern will result in residual conflicting strains, the symmetry of these strains, the resulting stresses, and the structural rigidity provided by the part's shape will generally result in the most acceptable molding conditions.
  - Two-dimensional shapes (flat disk or square) will also often prefer a gate location at the centroid. However, conflicts in shrinkage can result in bowl- or saddle-shaped warpage with square on rectangular, shaped parts use of flow leaders, which can improve fill balance from a centroid gate, can reduce warp. In cases where flatness is critical, one should attempt to gate along the edge of the square part using either a wide fan gate or numerous small gates having balanced flow so that they develop a linear flow pattern similar to a wide fan gate.

- Length of flow through the required runner and part cavity: Flow length will affect the pressure to fill the mold and potentially influence gate placement, runner diameter and part thickness.
- Effect on clamp tonnage: Gate location on a part can increase clamp tonnage with increasing flow length, by gating into a region with a large projected area versus a region with a smaller projected area, and by creating unbalanced filling within the cavity.
- Effect on core deflections: Unbalanced filling around a core can cause the core to deflect.
- Effect on venting: The flow pattern developed from a gate will either allow for venting along the parting line or it will require special venting. Also consider the potential for gas traps.
- Jetting: Jetting is caused by restricted gates and poor placement of gates in a mold. The problem results from the inertial effect of the melt entering a cavity at a high velocity. If the melt does not impinge on something as it enters the mold, it will jet across the cavity in a stream (see page 245).

4.2 Design and Process Strategies for Injection Molding

This section provides a series of 16 design and process strategies based on what was discussed in the previous three chapters. These strategies will summarize information and focus on design and process issues that will help minimize injection molding problems. Both injection molding analyses and actual case studies of moldings are utilized to explain these strategies. The optimum application of many of these strategies requires the use of mold filling simulation.

Figure 4.1 Part showing deformation caused by variation in wall thickness

4.2.1 Maintain Uniform Wall Thicknesses in a Part

Uniform wall thickness in an injection molded plastic part is critical to minimize both shrinkage and mold filling related problems. Non-uniform wall thicknesses will result in both volumetric and directional shrinkage variations. It is these variations that are often at the heart of warpage problems and other defects, such as sinks and voids. In addition, non-uniform wall thickness will disrupt filling patterns, potentially causing race tracking and hesitation related problems.

Figure 4.2 Gas traps resulting from the addition of flow leaders to the thin center region of the part

Example 1

The center region of the defective part shown in Figs. 4.1 and 4.2 has a wall thickness of 1 mm, while the perimeter was designed with a wall, which is approximately 2 mm thick. Eight radial flow leaders (thickened regions radiating from the center of the part) were added to the thin center region in an attempt to help with production problems. This is a low production part, to be molded in a single-cavity three-plate cold runner mold. The part shown is molded with eight gates, each feeding into one of the eight radially positioned flow leaders.

Resulting problem and explanations:

1. The thick perimeter is shrinking more than the thinner center region. This is causing the center region of the part to pop up. The radial flow leaders disrupt this effect, causing some irregularity in the warpage. Figure 4.1 is a side view of the part showing the warpage resulting from the variations in wall thickness.

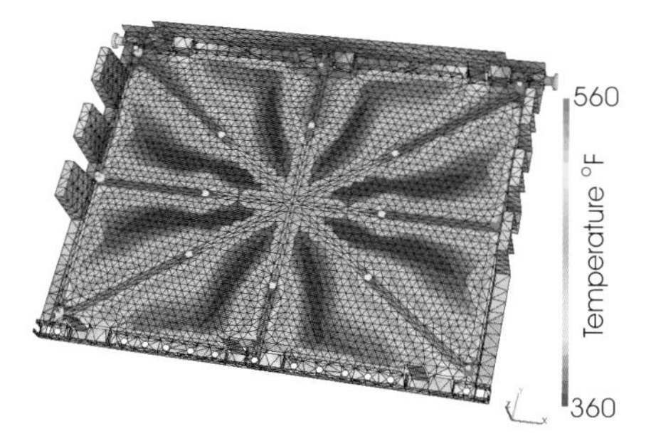

Figure 4.3 Melt front temperatures from a mold filling simulation, showing effects of the melt hesitating in the thin region of the cavity

2. The thin center region not only creates a problem of differential shrinkage between the center and the perimeter, but is also very difficult to fill. The thickened radial spokes help the melt get to the perimeter, but they also create gas traps as the melt races down the spokes and around the thicker perimeter, while hesitating in the thinner center regions (see Fig. 4.2). Figure 4.3 shows the result of a

mold filling simulation for this part showing the melt flow front temperature. Note, how the melt cools as it hesitates in the thin region, while maintaining a relatively high temperature in the ribs and at the perimeter, where it is race tracking around the thinner region.

There is no good solution to molding this part other than to provide a uniform wall thickness. With the current design, a single center gate would result in even worse gas traps in the same places as with the current gating. Without the radial flow leaders, the warping problem would be more extreme. Gating into the perimeter would result in the material racing around the entire perimeter, while experiencing severe hesitation near the gates, resulting in little chance of filling the part at all.

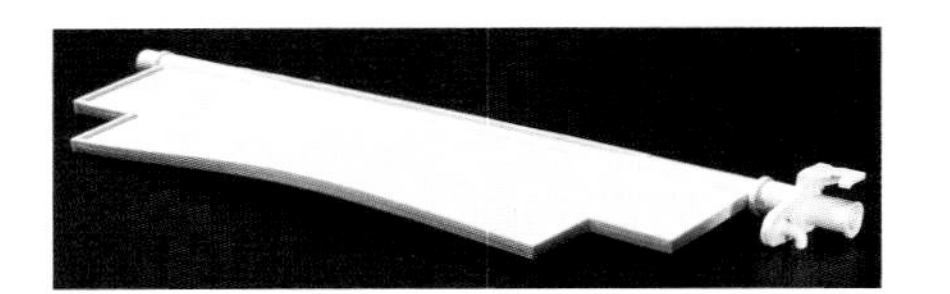

Figure 4.4 The thick cylinder along the thin rectangular region of the part initiates warpage because of differential shrinkage

Example 2

The part in Fig. 4.4 has a thick cylindrical region, which runs along a thinner rectangular region. Regardless of gating, the thicker region will want to shrink more than the thinner rectangular region. As the thinner region freezes first, it is forced to warp as the thicker cylindrical region continues to shrink inward. In this case, there is a possibility to reduce the problem by coring the thicker cylindrical region.

Figure 4.5 Non-foamed injection molded plastic pallet (Skidmarx®) developed with the aid of solids modeling, injection molding simulation and structural analysis

Example 3

The Port Erie plastic pallet – Skidmarx® shown in Fig. 4.5 is the first known pallet to be fully injection molded without the use of structural foam. The part was initially designed with a solids modeling program. A mid plane mesh was created and a series of injection molding, mold cooling, and structural analyses were performed. The part has a near constant wall thickness of about 2.8 mm. For ejection, the part incorporates alternating angled ribs crossed with alternating tapered ribs with wall thicknesses averaging 2.8 mm. The alternating ribs facilitate ejection. As a result of utilizing CAE to help optimize the design and evaluate manufacturability prior to mold build, this part started up with virtually no molding problem. Good parts were produced within 10 shots with a cycle time of just over a minute. With a weight of only 6.8 kg (15 pounds) when molded of HDPE, this part is shown holding 1,361 kg (3,000 pounds) in Fig. 4.6.

Figure 4.6 The lightweight Skidmarx® plastic pallet shown in industrial application carrying over 2000 lbs

4.2.2 Use Common Design Guidelines for Injection Molded Plastic Parts with Caution

There are many sources offering general guidelines for designing plastic parts. Using these can often avoid problems that might otherwise develop. However, it should be realized that many of these are very general and they commonly focus on local problems, while ignoring negative global effects on the part. These general guidelines also often do not consider the characteristics of the specific material being molded.

One should not only understand the benefits of using a particular design guideline, but should understand the potential negative consequences of applying the guideline.

Of particular concern is the common practice of thinning ribs that are attached to the primary wall of a molded part. Thinned ribs, which will reduce the potential for local sinks and voids at their intersection with the primary wall, can result in warpage, residual stresses, and numerous mold-filling problems. If cosmetics are a concern at the location of the rib, a thinned rib may have to be used. Voids at this location are usually not a real concern as the highest stress in a rib under a flexural load is normally at its tip rather than at its intersection with the primary wall.

The part shown in Fig. 4.7 is from a test mold and is molded from a neat nylon. It has a primary wall of 3 mm with a 1.5 mm rib, which is 6 mm tall. This design is based on common design guidelines for injection molding of high-shrink materials. The mold cavity is fed by a wide fan gate, which creates a near ideal linear flow. Note how the variations in wall thickness, resulting from following the guideline, are causing the part to warp. The thinner rib freezes first, while the thicker primary wall will continue to cool and shrink causing the part to warp as shown.

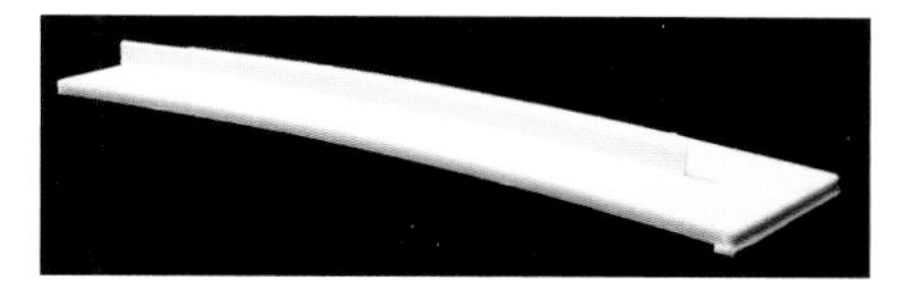

Figure 4.7 Molded test plaque showing warpage resulting from the addition of a rib that was designed according to common design guidelines for ribs

4.2.3 Avoid Flowing from Thin to Thick

Whenever a part must have variations in wall thickness, it is highly desirable to gate into the thicker region so that plastic flows from thick to thin. This will minimize the potential for sinks and voids in the thicker region as well as the risk of uncontrolled shrinkage.

Whether gating from thick to thin or from thin to thick makes little difference during initial mold filling. However, during the compensation phase (packing phase), a thin wall can be expected to freeze-off prior to a thicker wall. If the part is gated into the thin wall, which will freeze during the compensation phase, all flow to the thicker region will be blocked. The thicker region will continue cooling and shrinking without any compensating flow. The molder will have lost control of the shrinkage of this portion of the part. The resulting high shrinkage in the thicker region could result in sinks, voids, and a stress, relative to the thinner region. This residual stress is what leads to warpage in plastic parts. Figure 4.8 illustrates the progression of a developing frozen layer during the compensation phase and the resulting loss of shrinkage control in the thicker region.

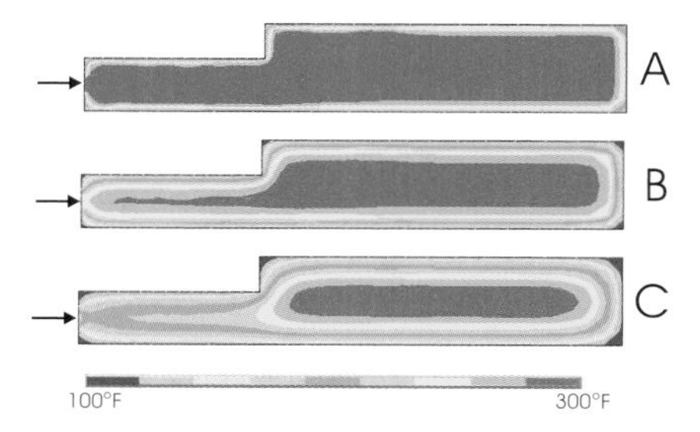

Figure 4.8 FEA analysis shows the thin region of a part freezing off prior to the thick region. Gating in the thin region will result in a loss of shrinkage control in the thicker region

4.2.4 Establish a Simple Strategic Flow Pattern within a Cavity

As plastic is forced through a mold cavity under high pressure, shear and extensional flow forces act on the polymer and any fillers or reinforcements it is carrying. These forces cause the polymer molecules, and any asymmetric additive, to become oriented in the direction of the principle strains developed from these forces. This orientation in turn results in anisotropic residual strains, shrinkage, and changes in mechanical properties. The effect is most dramatic with fiber-filled materials. Therefore, when positioning the gates of a mold a designer should consider the following:

- In parts with elongated shapes, it is generally preferred for the plastic to flow in one direction when filling the cavity. This will result in fewer conflicts in residual strains and shrinkage and thereby reduce stresses and tendencies for the part to warp. Therefore it is generally preferred that a designer attempt to gate from one end of a cavity resulting in the material flowing across its length. This simple flow pattern is particularly important with fiber-filled materials, where warp developed from anisotropic shrinkages can be severe. The ideal position for a fan gate is along one edge of a part so that the plastic melt is flowing across its length.
  It should be noted that gating from one end of a part also creates the biggest variation in melt pressure across the part during filling and packing. Without use of packing profiles the regions near the gate will be packed out better than regions away from the gate. However, the reduction of conflicts in orientation-induced strains normally has a more significant impact on reducing warpage than the effects of differential pressures.
- A circular shaped part should be gated in its center, thus providing a uniform flow in all directions throughout filling. Circular geometries are not conducive to a linear flow path. Gating from one side to develop a linear flow would cause the part to become more oval in shape. However, if the part includes regions that are not linear, another gating location might be considered. A preferred position might again be from one end or centrally positioned to provide a balanced fill between the two most extreme locations of the part.
- If gating from one end is impractical, multiple gates might be preferred to develop a simple but balanced flow (see Section 4.2.7 – Balance Filling Throughout a Mold).
- As mechanical properties are enhanced along the direction of flow, flow should be directed perpendicular to expected flexural loads and parallel to tensile loads. Again, this is particularly important for fiber-filled materials.

Example 4

The part shown in Fig. 4.9 is molded of a 30% glass-filled nylon 66. The part has a 100 mm radius, is 2 mm thick, is tab gated, and develops a radial flow pattern as the melt expands out from the gate. Note the characteristic warpage due to the variation in shrinkage from the gate to the perimeter. A second part with the same 2 mm wall thickness is a rectangular 50 × 200 mm plaque, which is gated with a fan gate at one end and establishes a nearly ideal linear flow across the entire part during mold filling. Regardless of process variations, this part is very stable and highly resistant to any significant warpage from molding.

Figure 4.9 Warpage resulting from radial flow pattern of a fiber-filled material

Example 5

The part shown in Fig. 4.10 is a wire harness molded of fiber-filled nylon. This part has a complex shape that is assembled with a second part. Despite the 600 mm flow length and an average wall thickness of only 2.3 mm, the cavity is gated from one end with a fan gate positioned on one end. As a result of the simple linear flow path, the part is produced with a minimum of distortion problems.

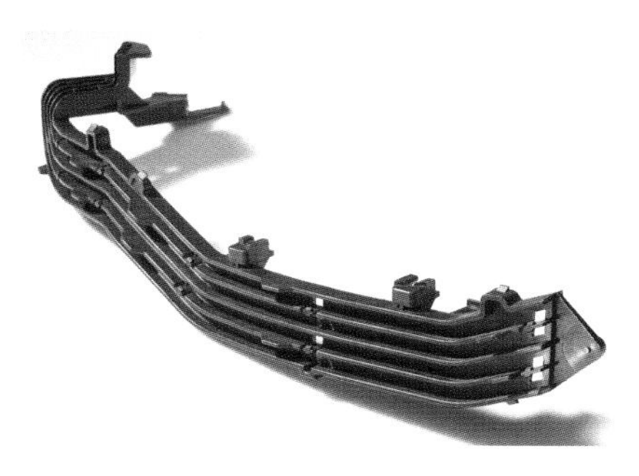

Figure 4.10 Wire harness with fan gate resulting in a linear flow pattern, which minimizes the potential for warpage

Example 6

The rectangular part at the top of Fig. 4.11 is molded of a neat polycarbonate and is gated with a tab gate along the edge as shown. In a secondary operation a small metal plaque is mounted to the part with a thermally activated adhesive. During application it turned out that the metal plaques were loosening and falling off. Placing the part in an oven at an elevated temperature revealed the orientation induced residual stresses that were developed in the gate region. The effect of the high shear stress and radial flow pattern on orientation becomes obvious in this case. At the elevated temperatures the part loses its stiffness and can no longer resist the residual stress developed by the highly strained polymers in the gate region. A second mold, which was end gated eliminated this problem.

Figure 4.11 Rectangular plaque as molded (bottom) and after being exposed to heat (top). Note the distortion from residual stresses acting on the part as the polymer chains are allowed to relax at the elevated temperature

Example 7

Radiator end tanks are designed with a thick flange along their perimeter for mounting purposes. Gates are normally positioned near the top center of these parts to minimize race tracking and gas traps developed as a result of the thickened flange. Without considering orientation-induced shrinkages it would be expected that the position of the thickened flange would cause the ends of the part to warp/bow downward. However, these parts nearly always warp in the opposite

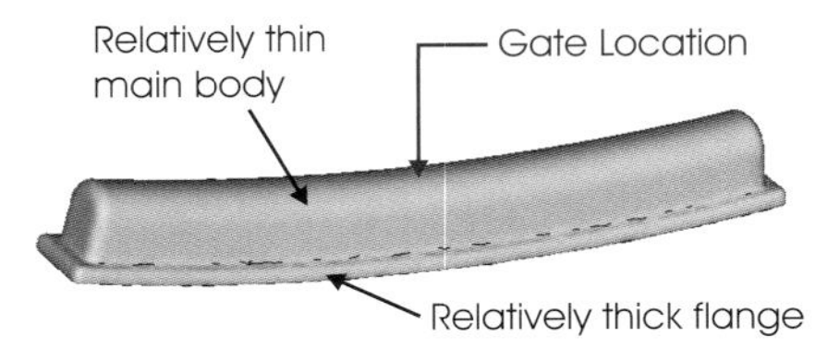

Figure 4.12 Typical warpage in most radiator end tanks. Warpage, relative to the thick flange, is opposite to what is expected because of orientation of glass fibers as developed during mold filling

Figure 4.13 Gas trap development in a radiator end tank that is center-gated on its top. During mold filling, material reaches the thick flange near the gate and races down the length of the flange creating a gas trap at the end as shown

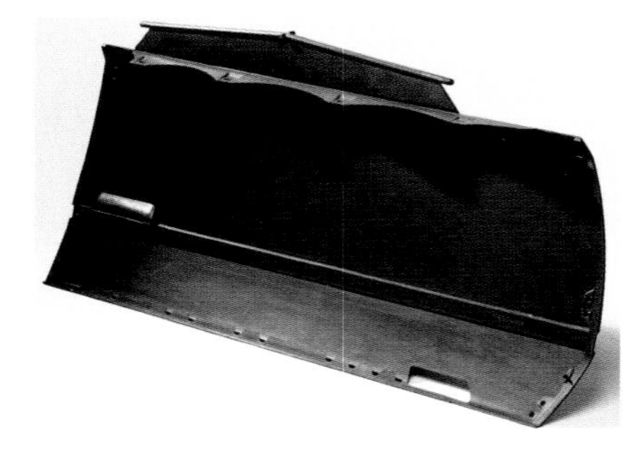

Figure 4.14 Automotive body panel molded with specially designed fan gate to eliminate weld line and minimize potential for warpage and residual stresses

direction (see Fig. 4.12). This warp results from the filling pattern that is developed in these parts, combined with the use of a fiber-filled material. The radial flow in the gate region creates both shear and extensional induced fiber orientation. As the melt reaches the flange, it begins to race down the flange and causes nearly pure linear flow in this region. This linear flow orients the fibers along the length of the flange and significantly reduces its ability to shrink along its length. Flow along the upper body is more complex, combining both radial and linear regions. The resulting reduction in fiber orientation along the length of the upper body will allow it to shrink more along its length relative to the thicker flange. This will force the ends of the part to warp upward. To address this warpage, most molders must either build the molds with a counter warp or anneal the parts, while in a fixture, after molding. Flow leaders are also often applied to the top length. These are used primarily to avoid gas traps, which are created due to the race track effect created by the thicker flanges (Fig. 4.13).

Example 8

Figure 4.14 shows an exterior automotive body panel with a specially designed fan gate (724 mm wide) attached. The part was molded originally from multiple hot drops placed along where a decorative strip would be placed (the strip would be able to cover up any local cosmetic flaws at the gate locations). The part required dimensional stability and minimal stresses so as to help withstand elevated temperatures experienced during post-mold painting operations. The painted part had to have a nearly flawless Class A surface. The original multiple gates resulted in numerous cosmetic and dimensional problems. A single, specially designed, fan gate was able solve all these problems. The fan gate encouraged the development of a more linear flow path, thereby providing a much more dimensionally stable part. In addition, this approach eliminated weld lines and other related distortions.

4.2.5 Avoid Picture Framing

There are two different sets of conditions under which *"picture framing"* can be developed. These are presented in the following examples.

Example 1

Figure 4.15 shows the results of a mold filling analysis of a part that has a thinner center wall region surrounded by a thicker perimeter. This "picture frame" design would be typical of a flat part with a thicker flange. This is primarily a part design problem, for which there is generally no good solution other than to modify the part design so as to core out the flange. If gated into the thicker region of the part, as shown in the analysis output, the melt will race around the thicker section while it hesitates trying to fill across the thinner section, thus creating the picture frame effect. The melt racing around the perimeter will meet, trapping air in the center thinner-walled region, thus creating a gas trap. If gated in the center region, the melt will flow from thin to thick creating uncontrolled shrinkage in the thicker perimeter. Figures 4.1 and 4.2 are examples of a part with a thin center region and thick perimeter. The part is molded with eight gates each feeding midway along one of the spoke-patterned flow leaders. The result is a twisted bowl shape as well as formations of gas traps. The bowl shape is evident in Figure 4.1 and is causing the three vertical features on the left and right sides to bow outward.

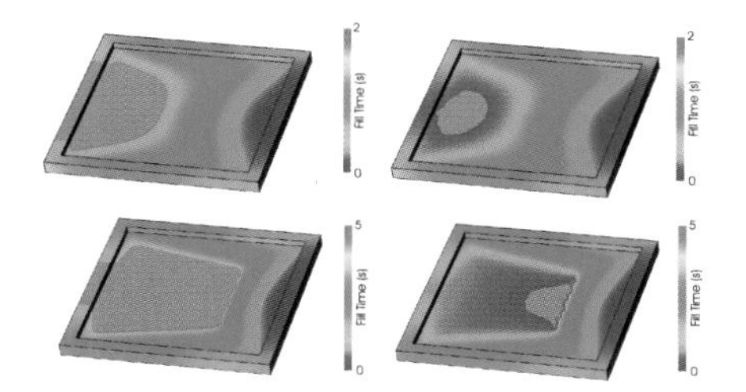

Figure 4.15 Injection molding analysis showing the development of "picture framing" resulting from a part with a thick perimeter and a thin center region

Example 2

Another variation of picture framing occurs in a part with a cored section that is gated along its perimeter. Figure 4.16 is an edge-gated cup. The flow path from the gate around the perimeter is a shorter distance than across the top. This results in a gas trap.

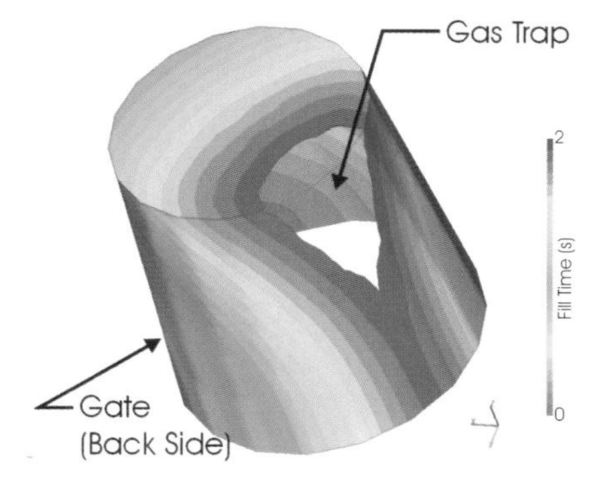

Figure 4.16 Mold filling analysis showing the development of an air trap resulting from poor gate positioning relative to the part geometry. Melt flows around the perimeter before it can move over the top of the mold core

4.2.6 Integral Hinges

Integral hinges create a particular problem in injection molding. Most commonly formed from polypropylene, these hinges commonly span the width of a part connecting a container and its lid. The hinges are much thinner than the adjoining walls, commonly only about 0.25 mm thick. They get their characteristic high strength and excellent flexural strength from the orientation of the plastic melt as it is flows across the hinge. To avoid flow hesitation at the hinge, the gate(s) should be placed away from the hinge.

Example 1

Figure 4.17 shows a mold filling analysis of two equal halves of a rectangular part separated by a thin integral hinge. A poorly placed gate is positioned along one edge near the hinge. Here the melt hesitates as it hits the restrictive hinge. Some of the melt crosses the hinge near the gate into the left side of the part and slowly moves up the left side. Meanwhile, the melt on the right side quickly travels up the length of the part. The melt traveling along the right side hesitates along the hinge, welding with material meeting it from the left side. When the hot material racing along the right side reaches the far end, it blows across the top hinge region and begins to back-fill while meeting up with the slower moving melt front on the left side of the hinge. Except at the two

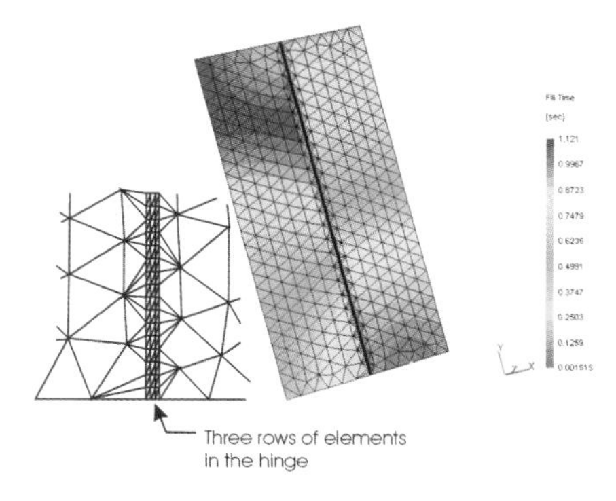

Figure 4.17 Mold filling analysis showing the poor flow conditions across an integral hinge that result from a poorly placed gate

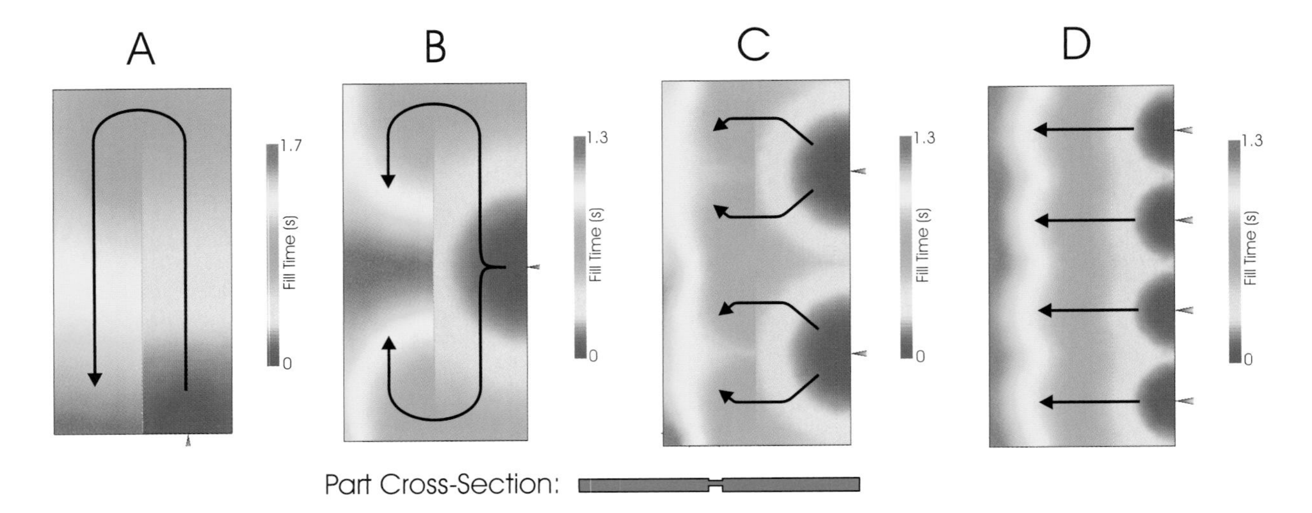

Figure 4.18 Effect of gating locations on mold filling across a thin integral hinge

extreme points, the hinge has been formed as a weld, with very little flow across it. This will result in a very weak hinge.

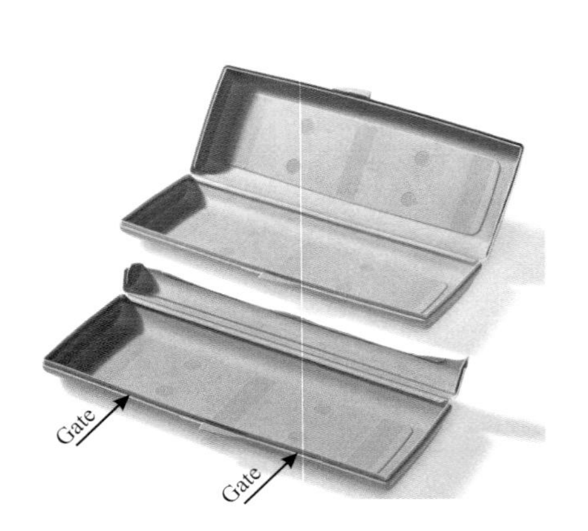

Figure 4.19 Baby wipe container with optimal filling pattern across its integral hinge

Example 2

It is desirable to gate into a part in such a way that there is a minimum of hesitation at the hinge. Ideally the gating would develop a broad flow front, which hits the width of the hinge at the same time after all other extremities of the part have been filled. This is not always practical, but conditions close to this can be achieved. Figure 4.18 illustrates mold filling analysis results of four different gating options. Option A is the least desirable as it repeats the conditions discussed in Example 1 earlier. Option B improves on Option A but still results in hesitation and poor hinge strength at the center region of the hinge closest to the gate. Option C further improves flow across the hinge by using two gates at balanced locations along the edge of the part. Here hesitation is significantly reduced at the hinge. Option D shows the near optimum condition where four gates provide for a broad flow front to be developed prior to reaching the hinge. This virtually eliminates any hesitation at the hinge, improving orientation and maximizing hinge strength. The small shallow box in Fig. 4.19 was originally gated near the hinge (as in Fig. 4.18B), which resulted in premature hinge failure. Two gates were then placed along the edge of the part as shown. Note the uniform filling pattern on the opposite side of the part from the hinge that resulted from this new gating location.

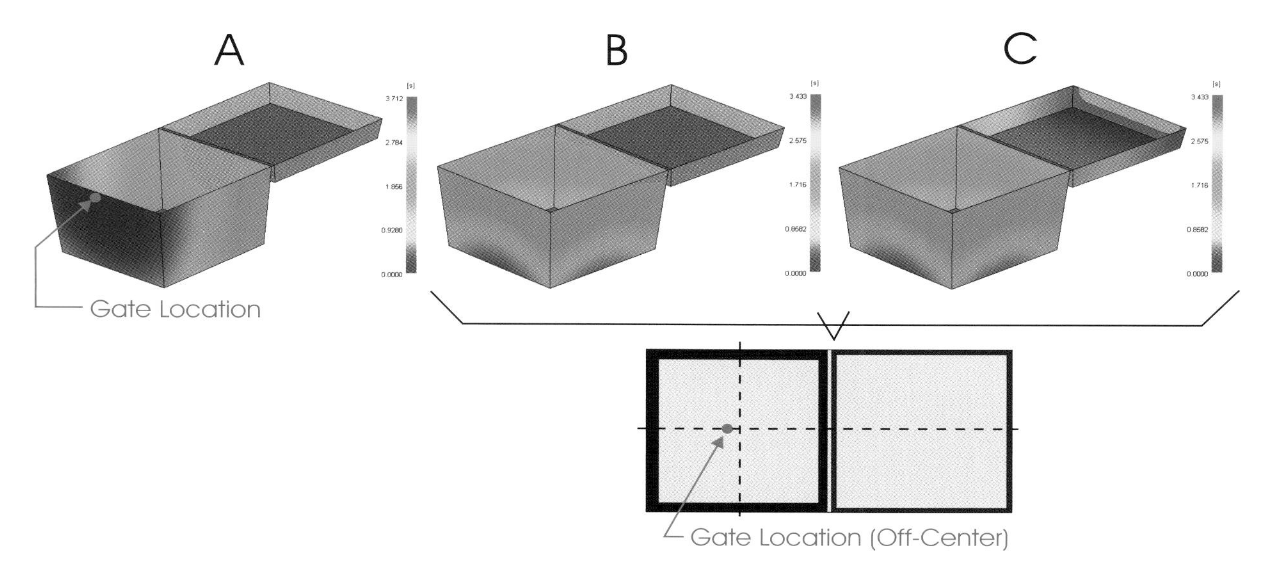

Figure 4.20 Moldflow filling output for a tall box with undesirable gating location on the front edge opposite the hinge (A) and proper gating location (B and C) at the base of the box. Gating on the base, slightly offset away from the hinge, results in a uniform melt front reaching the hinge

This approach cannot be used in a part such as the one shown in Fig. 4.20, because the depth to the box would corrupt the uniform flow front as the melt wrapped around the side walls resulting in a severe weld or gas trap (see Fig. 4.20A). The preferred approach in a case like this is to gate the part approximately in the center of the main body, i.e., in the center of the base of the taller box region, so that all extremities of the box, including the hinge, fill at about the same time (see Fig. 4.20B and C). In some cases it is desirable to position the gate slightly off balance so that the side opposite the hinge fills first. This will accelerate the melt approaching the hinge and minimize hesitation.

Special Note: In Example 2, as the melt flows across the thin hinges into the thicker lid portions, one of the general guidelines is violated – *Avoid flowing from thick to thin*. In this case, the hinge will quickly freeze and prevent complete packing of the lid. This creates a challenge in properly sizing the two halves, box and lid that would normally require assembly (lid closing onto the container portion of the part). To make matters worse, a warpage issue will arise because the two regions shrink differently. This is less of a problem in thinner parts. With a thicker part wall, there is a higher risk for a problem to develop.

Gating on either side of the hinge can help control packing, but will result in a weld line. The position of this weld can be difficult to control, as the melt on each flow front will tend to hesitate at the hinge. This will result in an extremely weak hinge. The problem can be addressed through careful modeling and sizing of the runner branches feeding these two halve using injection molding simulation software. It can also be addressed using sequential hot valve gating. One gate could be positioned on the main container portion and opened to fill the entire part. A second gate, located on the lid, would open after filling from the first gate simply to assure packing of the lid portion.

4.2.7 Balanced Filling Throughout a Mold

The volumetric shrinkage of a plastic material can vary by over 20% during injection molding. This variation will occur over the range of temperatures and pressures the plastic material experiences from its molten to its solidified states. Temperature variations of over 250 °C and pressure variations of 140 MPa are not unusual.

Variations in pressure will influence how much material is fed into the cavity as the material is shrinking during the packing, or compensation, phase of the molding cycle. The principle of balancing of pressures is not only important to provide consistency between cavities of multi-cavity molds, but also to provide balanced conditions within a given cavity to minimize the potential for flashing, residual stresses, and warp. Balanced pressures will increase the process window for making higher quality and more consistent parts.

4.2.7.1 Gating Position(s) Within a Cavity

Balancing pressures within a cavity should be attempted first by varying gating location and then by using either flow leaders or deflectors. Unbalanced filling of a cavity develops near hydrostatic pressure regions and transient flow. Hydrostatic pressure regions are developed in locations of a cavity that fill early while other locations are continuing to fill. As soon as a location within a cavity is filled, pressure will dramatically increase between this location and the gate and will approach the high pressure at the gate location. In contrast, the pressure at the continuing flow front is zero with a constantly increasing pressure gradient back to the gate. There are a number of problems created by this phenomenon:

- The rapid and high pressure development, in the near hydrostatic early filling regions, will unnecessarily increase the force opening the mold and potentially allow the melt to flash into the parting line.
- Due to the stoppage, or decreased flow, in the early filling region, the melt will begin to cool there as other regions are still filling. The melt will also cool under higher pressures causing it to be packed out better and shrink less than the later-filled regions. These variations in shrinkage will develop residual stresses, which can cause a part to warp.
- If the early-filled region is not a dead end, the melt will change direction under the frozen layer and be directed toward the continuing flow front. This will continue through the packing, or compensation phase. This transient flow will create variations in orientation through the cross section of the early filled region, which will develop complex strains resulting in localized stresses and contribute to warpage of the part.

- As early filling regions are filled, the flow front velocity at the continuing flow front will accelerate as flow is diverted to it. The larger the early filled region, the faster the acceleration of the flow front. This will create variations in the conditions under which the melt was flowing and forming in early filled vs. later-filled regions of the part. The early filled region is likely to have been developed from material exposed to less shear and lower melt temperature. The acceleration of the melt front in the later-filled regions will increase the shear and thereby the shear-induced orientation and frictional heating of the melt. Both of these regions will be formed under contrasting conditions, causing them to shrink differently, which is the fundamental cause of residual stresses and warpage.

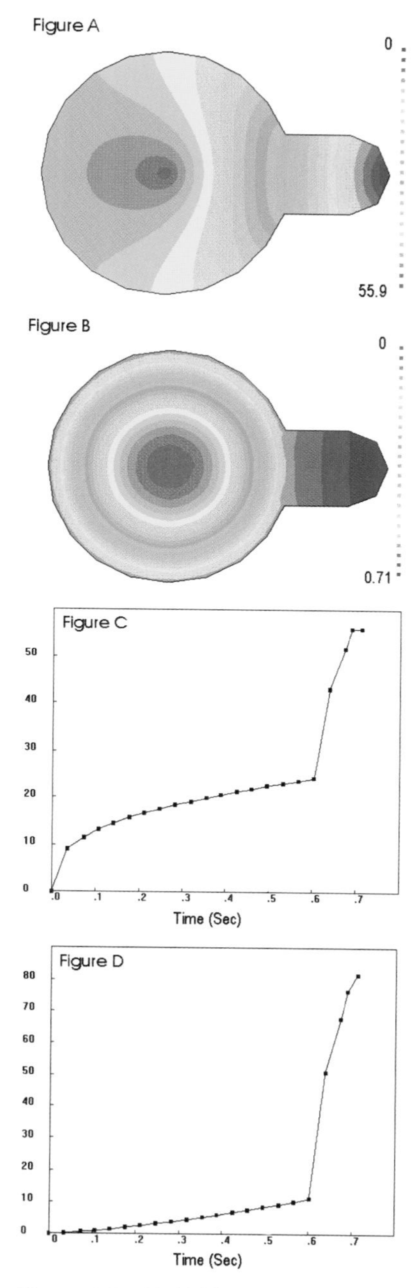

Figure 4.21 Mold-filling analysis results for a non-symmetric part (see text for explanation)

Example 1

In many parts it is not feasible to avoid filling imbalances; however, the imbalances can often be minimized. The part presented in Fig. 4.21 develops many of the conditions discussed earlier. The gate in this part is positioned in the center of the main cylindrical body. This would be ideal if it were not for the appendage on the right side of the part. The appendage creates a situation where the melt reaching the perimeter of the main cylindrical body approaches a hydrostatic condition, redirecting the melt to the continuing flow front in the appendage. Following clockwise from the top left:

a) At the instant the main body is filled, the early filling regions approach a hydrostatic condition (note the wide spacing in the isobars in this region indicating a more constant pressure with a minimum pressure gradient)

b) The flow front accelerates into the appendage (notice the dramatic increase in the flow front indicated by the wide spacing in lines)

c) The acceleration into the handle causes a pressure spike (plot of Pressure vs. % Fill). This pressure spike increases fill pressure from 23 MPa to 57 MPa (2½ times increase) during the last 12% of fill.

d) The pressure spike compounds the high pressure condition in the hydrostatic regions causing the force within the cavity to increase dramatically – which in turn will increase tonnage requirements (plot of Clamp Tonnage vs. % Fill). Clamp tonnage increases by over 600% during the last 12% of fill.

Further evaluation of the effect can be seen in Fig. 4.22. Again, clockwise from the top left:

a) The increased velocity and shear causes the flow front temperature to spike as the melt enters the appendage: A sudden increase of over 18 °C can be observed. This is in contrast to the reduction in temperature occurring just prior to entering the appendage.

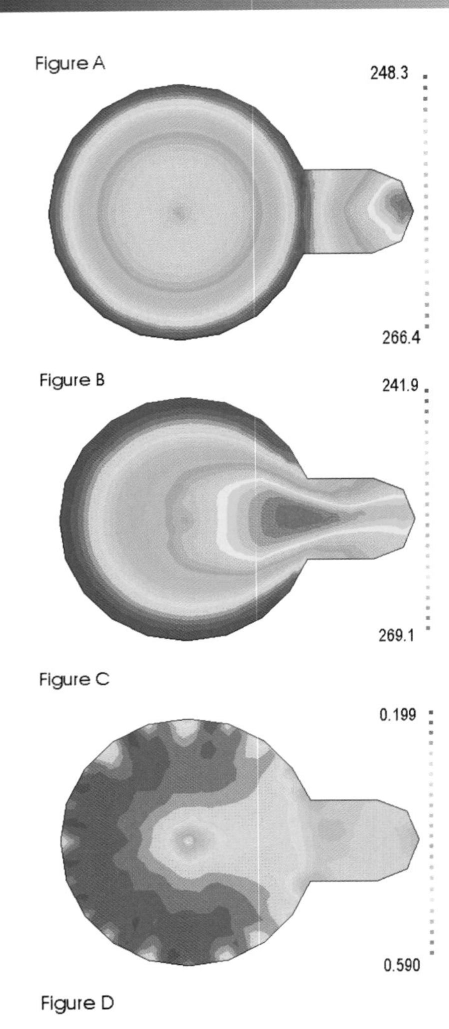

Figure 4.22 Mold-filling analysis results for a non-symmetric part (see text for explanation)

b) The effect on the melt temperature distribution is even more dramatic at the instant the mold fills. The material in the early fill regions has had a chance to cool a bit more, while the entire path between the gate and end of fill can be seen to spike in temperature, reheating some of the material that had previously been losing heat. Temperature variations within this part reach almost 30 °C.

c) The increase in velocity in the handle causes a high shear stress to develop between the gate and the appendage. Note that the shear stress spikes to about 0.400 MPa as the material first enters the appendage and then drops to about 0.300 MPa as the melt temperature spikes.

d) Significant transient flow, or changes in flow direction, is developed in the early-filled regions. This can be seen by contrasting the filling pattern indicated by the flow angle lines in this figure to the original fill pattern shown in Fig. 4.21. Also note the changes occurring in the early fill regions as opposed to the direct line of flow between the gate and the later-filling appendage.

Reflecting on the factors affecting the development of residual strains and shrinkage of a plastic part, it is apparent that the imbalanced filling will create extremely complex strain patterns through the cross section of this molded part, as well as between different regions of the part. This part would be very sensitive to warping. When considering alternate gating, one would realize that there is no ideal position in a part with this non-symmetrical geometry. However, there may be better positions. An edge gate on the right side would provide a more uniform flow path across the part and hydrostatic regions would be virtually eliminated (see Fig. 4.23). More appropriate fill rates would provide more uniform melt temperature and shear stress conditions throughout the part. Another alternate gating location would be in a balanced position between the left and right extremities of the part. This would reduce some of the imbalances, though they could not eliminate the remaining imbalance between the vertical and horizontal directions in the part.

Example 2

Figure 4.24 contrasts some alternate gating locations within a long rectangular part. The designs are presented in order of their resistance to warpage.

- Fan gating this part along one of its ends would develop a nearly ideal flow condition and minimize the potential for warping. This gating is commonly used with long electrical connectors. A compromise would be a tab or tunnel gate positioned at the same location. In some cases, multiple tunnel gates could be positioned along the edge to help to more quickly develop a linear flow front.

- Multiple gates symmetrically placed and balanced along the centerline of the part also provide well-balanced filling and will be resistant to warpage. This arrangement will reduce the pressure variation across the length of the part but will increase orientation variations and introduce weld lines.
  - This gating arrangement provides a reasonable and symmetrical balance. Though not as ideal as "b", it reduces the number of welds.
  - This gating arrangement is sometimes seen but has no positive attributes. This gating generally results from a misunderstanding of balanced filling. The gates are positioned such that the gates and the extremities of the part are equidistant. The problem with this design is that the melt between each of the gates will meet while they are only halfway to the left and right extremities of the part. This will immediately create hydrostatic conditions in the center region, transient flows, melt front acceleration at the continuing flow fronts, pressure, and clamp tonnage spikes.
  - If edge gating is required and the part cannot be gated from one end, the multiple well-balanced arrangements shown here might be considered. The less the spacing between gates, the quicker a linear flow front may be developed flowing across the short length. The obvious drawbacks are the resultant weld lines and the difficulty to provide balanced filling to all the gates.
  - The center edge gate in this elongated part might be used in a two-plate cold runner mold where welds are considered unacceptable. This gating promotes a poor filling balance and an increased potential of warping.

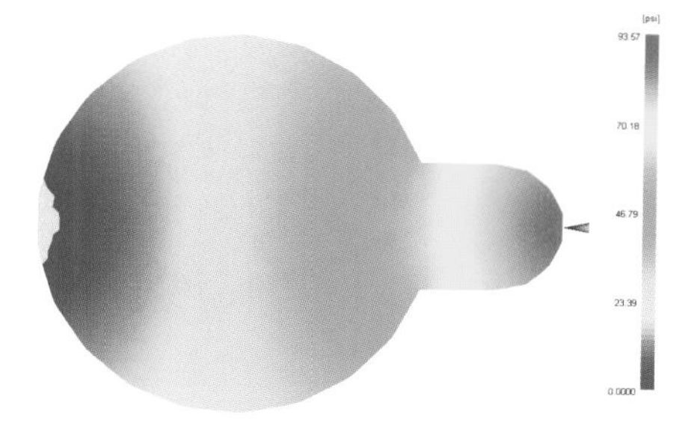

Figure 4.23 Mold-filling analysis of non-symmetric part

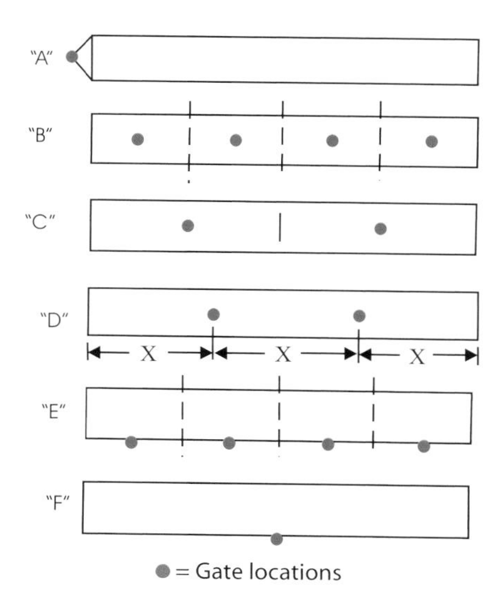

Figure 4.24 Six alternative gating locations on a rectangular part. Dots indicate gate locations; dashed lines are sub-regions fed by each gate

4.2.7.2 Multi-Cavity Molds

Balanced flow to each of the cavities in a multi cavity mold will maximize the potential of producing parts within specification and provide the largest process window. It can generally be stated that the more parts produced to specification in a single molding cycle, the higher the profit.

The objective of a multi-cavity mold is to produce multiple identical parts in one shot within a mold. This is much simpler said than done, even if we assume that steel dimensions and cooling between cavities are identical. Successful multiple-cavity molding requires that the melt conditions introduced to each of the cavities be the same. As will be discussed in Chapter 6, even the current industry standard geometrically balanced runners result in significant variations in melt flow, temperature, and material balances.

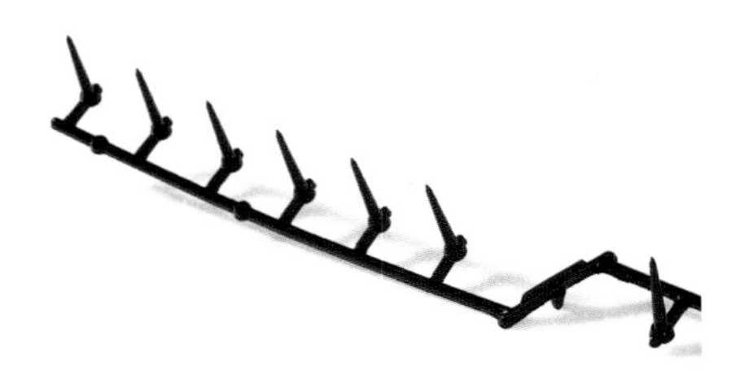

Figure 4.25 Typical fishbone runner layout for 3-plate cold runner mold

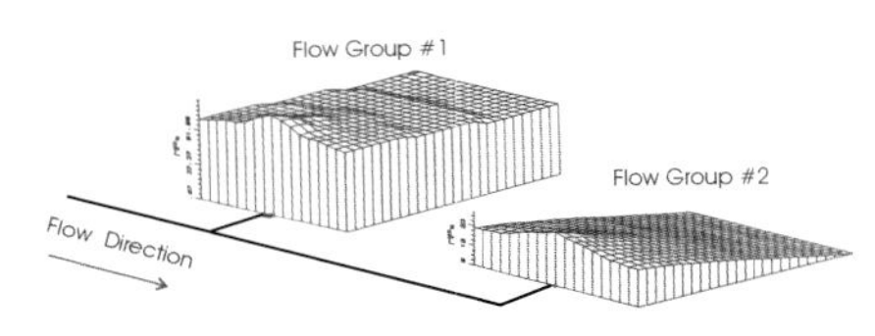

Figure 4.26 3D plot of a mold filling simulation showing the variation in pressure distribution developed from unbalanced mold filling

Consider the design shown in Fig. 4.25. This tree or fishbone, style runner layout will use less material than most runners but will result in the most imbalanced filling of cavities in multi-cavity molds. Though there are applications where this runner type provides some advantage, it cannot be recommended for precision molding. As melt enters this runner system, it will progressively reach each of the gates and cavities along its length. At a relatively fast fill rate, in a mold with non-restrictive gates or cavity wall thicknesses, the cavities closest to the sprue will fill first and then progressively fill the remaining cavities along its length. In doing this, each cavity will be filled at a different pressure, flow rate, and melt temperature.

The instant the cavities closest the sprue in an eight-cavity mold fill (Flow #1 cavities), the flow front velocity feeding the remaining cavities (Flow #2 cavities) will nearly double, as flow is no longer split into eight cavities. As the velocity increases, there will be a sudden corresponding pressure spike. The pressure inside the Flow #1 cavities will immediately become near hydrostatic and the total pressure compounded by the pressure spike from the increased flow velocity. At the instant before the Flow #2 cavities fill, the pressure at the flow front will still be zero, while pressure in the Flow #1 cavities will be significantly higher (see Fig. 4.26).

A detailed examination of the conditions experienced by the melt that formed the parts in these two sets of cavities would reveal that the parts formed in the Flow #2 cavities saw less pressure, a higher and longer shear history, a higher shear rate through the cavities, and a higher melt temperature due to the increased shear. Pressure, temperature, and shear-induced orientation are the three fundamental factors that control how a part shrinks and warps. In addition, considering the increased clamp tonnage requirements resulting from the unnecessary pressure spikes and high hydrostatic pressure buildups within the Flow #1 cavities, this will reduce the process window in which good parts can be produced.

The above discussion was based on a relatively fast fill rate, in a mold with non-restrictive gates or cavity wall thicknesses. If the flow rate is slowed down, or gates or cavity channels are downsized, a reversal in the filling sequence could occur due to hesitation of the melt at the Flow #1 cavities.

Similar effects found in the fishbone runner layouts are now known to exist even in conventional geometrically balanced runners (see Chapter 6). The only true runner balances achievable in molds with more than four cavities require the use of non-branching radial patterned runners, stacked manifolds in hot runner molds, or the use of melt rotation technology. Some less practical means include the use of static mixers in hot runner manifolds. The amount of imbalance found in various other runners depends on the length and diameters of the runner channels. There are many products that are successfully produced with conventional runners. However, the tighter the tolerance, and the greater the number of cavities, the greater are the problems.

Balancing mold filling through changing gate diameters is typically not a successful strategy. Though this approach is used, it provides a very unstable solution with many compromises. Minimum changes in process or material will offset the balance. In addition, even if a balanced fill is obtained during a given shot, the variation in gate diameter will cause the gates to freeze off at different times during the packing/compensation phase. If a fishbone layout is required, one can attempt to artificially balance the filling using the branching runners (see Fig. 4.27). Though this is better than balancing with gates, it will also result in a relatively small process window for which pressure can be balanced. However, it should be realized that even if a filling balance is achieved, the melt and packing conditions to each cavity would be different. This design should only be considered with relatively low-tolerance parts.

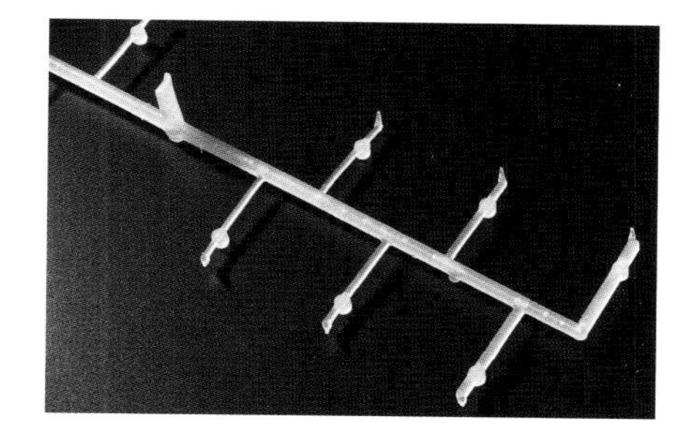

Figure 4.27 Artificially balanced runner attempting to address the expected unbalanced filling in a fishbone/ladder type runner

4.2.8 Provide for Uniform Temperatures (Mold and Melt)

The rate at which plastic is cooled in a cavity will directly affect the degree of anisotropy and, with semi-crystalline material, its density. In all cases, variations in temperature will result in variations in shrinkage and contribute to the development of residual stresses and warp.

Uniform cooling should be established both through the mold's cooling design and the melt temperature as determined by the mold filling rate.

Variation in cooling across a part can result in many of the same effects of variations in wall thickness presented in Section 4.2.1. Poor cooling in the perimeter of a disk, combined with good cooling in the center region, will result in the perimeter shrinking more than the center and cause the disk to warp in a bowl shape. The part shown in Fig. 4.28 was molded under the opposite conditions – the center core was poorly cooled, while the perimeter cooling was more favorable. This combined with the high stresses developed during molding, the relatively thin part, and the low modulus high-shrink polyethylene, caused the part to twist as the center region shrank more than the better-cooled perimeter.

Figure 4.28 Round polyethylene part was molded with better cooling around the perimeter of the cavity than in the center. The resulting higher shrinkage in the center region causes the saddle-warped shape

The same effect can be created by the fill rate. At a relatively slow fill rate, the flowing melt will not develop sufficient shear heating to offset the heat loss to the mold by conduction. As a result, the melt near the gate will be hotter than at the end of fill. These two regions will want to shrink differently, which can induce warping. Alternately, too fast a fill rate will result in melt temperature being higher near the end of fill than near the gate. This is potentially worse, as melt near the gate is likely to freeze first and block additional material being fed to the extremities during the compensation stage.

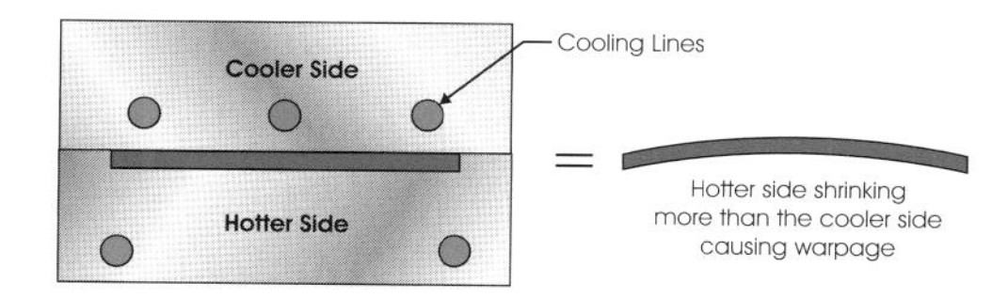

Figure 4.29 Characteristic warpage resulting from non-uniform cooling. The hotter side will normally shrink more and cause the part to warp as shown

Side to side variations will create a bending moment; with the ends of the part bowing toward the warmer side (see Fig. 4.29). This is also the primary cause of the hourglass warp, which develops in box-shaped molded parts. The problem develops from the limited ability of the mold's core, forming the inside corners of a box, to remove the heat at the same rate as on the cavity side of the same corner. The result is that the plastic on the inside surface of the corner will be warmer than on the outside surface. The resulting increased shrinkage on the inside surface causes it to pull in the side of the box. In a six-sided box, there are eight such corner edges, all pulling the sides in (the four corners formed by the intersection of the four side walls and the intersection of the four side walls with the bottom of the box). This is a very difficult problem to address. Techniques to minimize this problem include maximizing the radius of all eight edges/corners or the use of beryllium copper inserts to improve cooling on the inside corners. Often molders are forced to build the molds with a counter bow in the steel in order for the part to warp into a straight edge.

4.2.9 Eliminate, Strategically Place, or Condition Welds

Welds result from the merging of two melt fronts within a cavity. The resulting joint is weak relative to the surrounding material. The joint also will commonly result in a visual flaw on the surface of the part, which is often cosmetically unacceptable. The loss in part strength at a weld is dependent on material and the condition under which the weld was formed. Loss of strength results from the combination of a small notch/line created at the wall surface and the reduction of orientation across the weld region.

Welds can form when two flow fronts meet straight on (180° to one another) or at some smaller angle. Welds formed with melt fronts approaching each other at 180° are the worst case and decreasing angles of approach will improve the strength. A weak weld will also result, if the flow conditions of the two joining melt fronts are perfectly balanced such that there is no flow across the weld during the filling and compensation/packing stages. Without any continued flow across the weld, there is little chance of establishing any orientation of molecules or reinforcement across the initial weld from which some strength could be gained.

Fiber-filled materials are particularly prone to loss of strength at a weld. Though there may be bonding between polymers across a weld and some migration of polymer chains, fibers have no means of contributing to the strength across a weld unless they are physically positioned there by polymer flow.

Table 4.1 shows the results of a study contrasting tensile bars molded with and without a weld line. The samples without weld lines were molded with a single gate on one end of the bar. Those with a weld line had a gate on either end resulting in a butt weld midway between the two ends of the bars. Neat materials retained 80 to 100% of their strength when molded with a weld. However, the same materials reinforced with fiber were significantly affected, losing as much as 66% of their strength.

Table 4.1 Tensile Strength with and without Weld Line

Material	Filler	Tensile strength (psi)		Percent retained (%)
		One gate	Two gates	
Nylon 66	None	11,560	11,170	97
Nylon 66	10% Glass	13,980	13,060	93
Nylon 66	40% Glass	28,830	14,990	52
Nylon 66	40% Mineral	14,500	11,500	76
Polypropylene	None	5,400	4,650	86
Polypropylene	30% Glass	9,800	3,330	34
Styrene acrylonitrile	None	11,300	9,625	80
Styrene acrylonitrile	30% Glass	16,180	6,470	40
Polysulfone	None	9,600	9,600	100
Polysulfone	30% Glass	16,800	10,400	62

Ref: [1]

Through the strategic placement of gates and the use of flow leaders and diverters, welds can be placed in non-sensitive areas of the part. When welds are unavoidable, attempt to join the melt fronts at a minimum angle and consider developing some transient flow across the initial weld. Push-pull techniques, such as British Technology Group's Scorim™ process, develop continued movement of melt flow across an initial weld throughout the compensation phase. With the Scorim process, melt is introduced through multiple gates into a cavity by independent high pressure feed pistons. These pistons alternately reciprocate; driving melt back and forth through the cavity under very high pressure as the plastic is cooling. This approach has shown to virtually eliminate any flaws expected at a weld.

4.2.10 Avoiding Flow Hesitation

Molten plastic being forced into a mold will follow the path of least resistance. As plastic reaches a thin section (region of high resistance) it will hesitate if it has an alternate thicker region to flow through. The viscosity of the melt, which is hesitating in the thin region, will quickly increase as it loses its non-Newtonian shear thinning, it will generate less shear heat and is quickly cooled by the surrounding steel. This will compound the

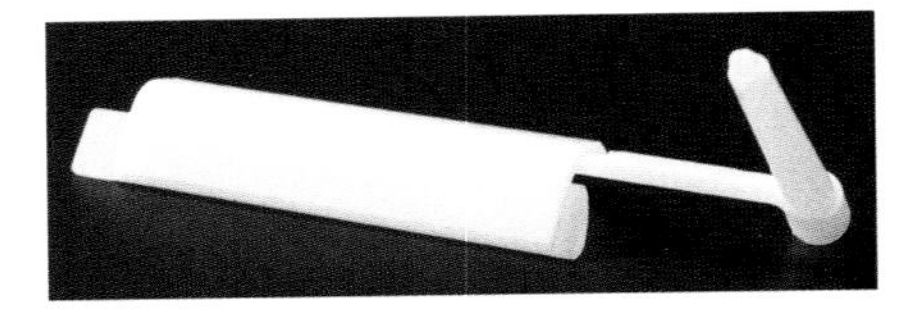

Figure 4.30 Symmetrical end-gated part with thin features on either end. Hesitation effects cause melt entering the thin feature adjacent to the gate to freeze before filling, while the same feature on the opposite end fills without a problem

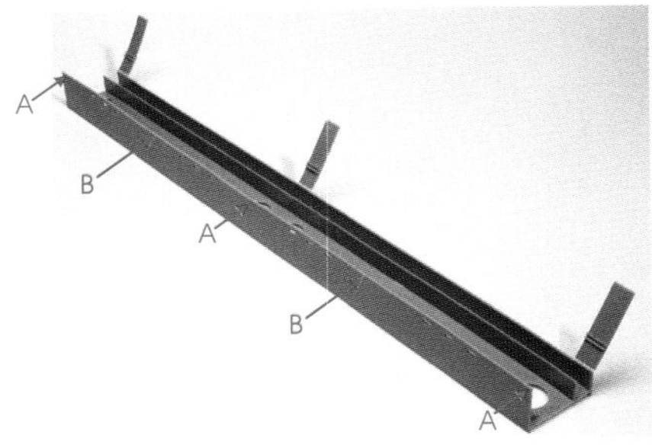

Figure 4.31 Wire harness with incorrect (A) and correct (B) gating locations. Correct gating locations avoid hesitation at the thin integral hinges connecting the three flaps to the main body (Courtesy: CP Plastics Group [2])

difference in flow velocity between the thin and thick regions, which can result in no-fills, gas traps, and complex filling patterns. Flow front advancement in a part with variations in thickness is much more difficult to anticipate without the use of mold filling analysis.

Variations in wall thickness promoting hesitation should be avoided in the design of the part. If variations exist, position the gate as far away from the thinner regions as possible in order to minimize hesitation problems. It is this hesitation effect that often creates problems when molding parts with integral hinges (see Section 4.2.6 for discussion of integral hinges).

Figure 4.30 shows a part molded of HIPS. The gate is positioned on the primary wall of the part having a thickness of 3 mm (0.120 in.). Immediately adjacent to the gate is a small tab, 1 mm (0.040 in.) thick. There is an identical thin tab on the opposite side of the part from the gate. Notice that the plastic was not able to fill the thin tab next to the gate but has filled the same feature opposite the gate.

Figure 4.31 is a part used as a wire channel for office furniture. The part contains three integral hinges connecting three hold down flaps. In application the flaps are bent over, along the integral hinges, to restrain the wires in the main wire channel. During initial molding gates were incorrectly placed near the hinge (locations marked as “A”). The thought process was that as the thin hinges were expected to create a resistance to flow, the gates should be near them in order to drive the melt across the hinge. This is a common mistake as it overlooks the effects of hesitation. During initial molding, the melt would hesitate resulting in no fills and/or weak hinges. The gates were relocated as indicated in the figure (locations “B”). The new gate locations cause the hinge regions to be the last to fill. Hesitation was eliminated and the parts filled with no problems.

If hesitation is occurring during molding causing no-fills, a faster injection rate is the best means to potentially overcome this problem.

4.2.11 Managing Frictional Heating of the Melt

As plastic flows through a mold, heat is generated as a result of the shear created from the relative movement of the plastics laminates and their movement relative to the stationary wall of the mold channels. This heat, and the orientation of the polymer from the shear field, will reduce the viscosity of the polymer melt. Runners with a small cross section will induce frictional heating of the material during filling. This controlled “spiking” of temperature can reduce the plastics viscosity without degradation. This in turn can reduce stresses within the molded part, because the material entering the cavity will have a lower viscosity and thereby lower shear stresses. In some cases, this way of reducing the material’s viscosity may be beneficial. For materials, such as acrylics, a small increase in temperature can have a dramatic effect on viscosity, while having a minimum effect on cycle time. However, one should be more cautious of using this method with thermally unstable materials like PVC. These materials can quickly degrade if shear heating in the runners is excessive.

With thermoset materials this approach of spiking the temperature of the melt through the runner can reduce cure/cycle time. The hotter material will require less time to be heated through conduction by the hotter mold.

4.2.12 Minimize Runner Volume in Cold Runners

Use of higher molding pressure will allow for the runner size (cross sections) to be reduced. Runner size can have a significant effect on part cost if the runners cannot be recycled. Reground runners can be sold at only a fraction of the cost of the original virgin thermoplastic. Even if the runners can be fed back into the process, there are numerous drawbacks – some of which are less obvious than others.

Most reground material will loose some of its original properties. Each time the material is processed it experiences an additional shear and thermal history, which can reduce the molecular weight of the material. This will not only affect the properties of the molded part, but also the viscosity of the material, thereby affecting further processing. The higher the percentage of regrind fed back into the process, the greater the amount of material that will have been processed again and again. If the percentage of regrind is kept relatively small and is metered back into the virgin material through a controlled metering devise, variations can be minimized.

Oversized runners can also limit the cycle time of a molded part. The part shown in Fig. 4.32 has a wall thickness of about 0.5 mm (0.020 in). The original runner has a trapezoidal shape with an equivalent cross section of a 6.25 mm (0.25 in.) diameter runner. The large diameter was more than what was required to fill the parts and was significantly increasing cycle time as well as material regrind. A mold filling analysis found that the runner diameter could be significantly reduced. The new runner is now less than 25% the weight of the original and cools in a fraction of the time.

Additional drawbacks of oversized runners include:

- The potential of material contamination with the additional handling required
- The need to inventory and track regrind (or dispose of it) at the end of a run.

Figure 4.32 Cold runner reduced to less than 25% of original runner volume through the use of mold filling analysis, thus significantly reducing mold cooling time and filling time

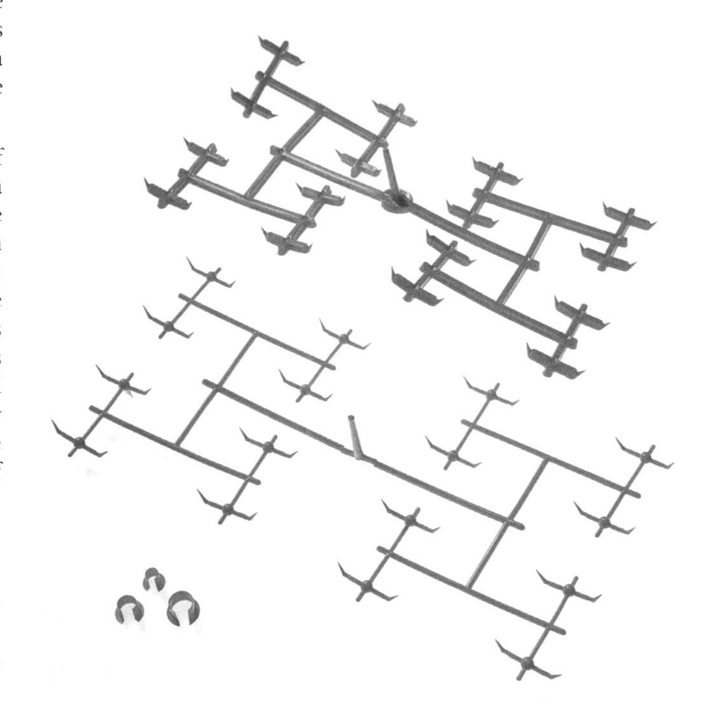

Caution should be taken in downsizing runners too much. Shear sensitive materials like RPVC can degrade if the resulting shear is excessive. Mold filling analysts should be careful to assure that a smaller runner, which is capable of filling a mold cavity, will also remain open long enough for compensation during the packing stage. This normally requires that the runner be at least 1.5 times larger in cross section than the primary wall of the part.

4.2.13 Avoid Excessive Shear Rates

The shear rate is a measure of the relative speed of material laminates flowing through a flow channel and can be easily calculated.

Shear rate through a round channel: $\dot{\gamma} = \dfrac{4Q}{\pi r^3}$ or $\dot{\gamma} = \dfrac{32Q}{\pi d^3}$ (4.1); (4.2)

Shear rate through a rectangular flow channel: $\dot{\gamma} = \dfrac{6Q}{wh^2}$ (4.3)

Where $\dot{\gamma}$ is shear rate, Q is flow rate, r is radius of a round channel, d is the diameter of a round channel, w is width of a rectangular channel, and h is the height (or thickness) of the rectangular channel.

It is expected that excessive shear rates will result in degradation of a flowing polymer. This degradation is not well understood, nor are the shear rates known at which this happens. It is expected that at some high value, the relative velocity may be so high that the polymer molecules are torn apart. This is compounded by the local development of frictional heating. This can often be seen when molding clear thermally sensitive materials like PVC. Black streaking can be seen to develop in outer laminates as the PVC flows along a runner under high shear conditions (see Fig. 4.33). Note that this phenomenon is developed along the length of the runner and not simply at high shear points like the gate and corners. The degradation was developed with sustained shear. Degradation from sustained shear heating may be limited to some of the most thermally sensitive materials.

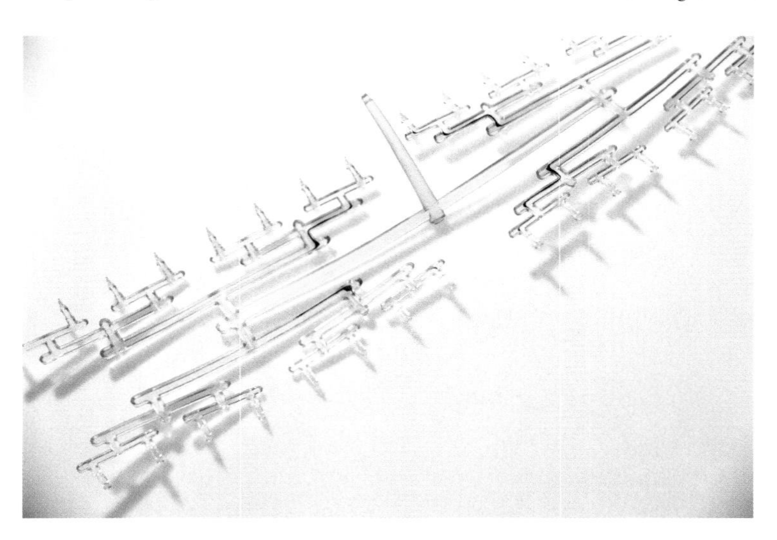

Figure 4.33 Visible decomposition of PVC material resulting from shear development along the length of the runner

Though guidelines for shear rate limits have been available from developers of mold filling analysis software, and in some cases from material suppliers, it should be recognized that none of these have been developed by any well-defined scientific method. Some of these existing guidelines were based on a ratio of the mechanical properties of the molded material. Though there is no real relationship between these values, they have been broadly adopted due to the lack of an alternative. Some of the problems of developing these data are discussed in Chapter 3.

It is the experience of the author that common guidelines for most non-thermally sensitive materials can be exceeded. However, using these values as guidelines can generally be expected to provide a conservative safe design. Table 4.2 below provides both shear rate and shear stress limits offered by Moldflow Inc.

Analysts who regularly use simulation programs often try to develop their own guidelines. This is done by empirically developing correlations between suspected shear related molding defects to the predicted shear rates.

Table 4.2: Shear Rate and Shear Stress Limits

Material type	Description	Shear stress limit		Shear rate limit
		psi	MPa	1/s
ABS	Acrylonitrile butadiene styrene	43.5	0.30	50,000
GPPS	Polystyrene (general purpose)	36.3	0.25	40,000
HIPS	High impact polystyrene	43.5	0.30	40,000
LDPE	Low density polyethylene	14.5	0.10	40,000
HDPE	High density polyethylene	29.0	0.20	40,000
PA6	Nylon 6	72.5	0.50	60,000
PA66	Nylon 66	72.5	0.50	60,000
PBT	Polybutylene terephthalate	58.0	0.40	50,000
PC	Polycarbonate	72.5	0.50	40,000
PET	Polyethylene terephthalate	72.5	0.50	Unknown
PMMA	Polymethyl methacrylate (acrylic)	58.0	0.40	40,000
PP	Polypropylene	36.3	0.25	100,000
PVC	Flexible polyvinyl chloride	21.8	0.15	20,000
RPVC	Rigid polyvinyl chloride	29.0	0.20	20,000
SAN	Styrene acrylonitrile	43.5	0.30	40,000
PSU	Polysulphone	72.5	0.50	50,000

Ref: [4]

4.2.14 Avoid Excessive, and Provide for Uniform, Shear Stresses

Mold filling analysts are provided information on shear stresses developed in the melt during mold filling. Shear stress is a measure of the stress on a flowing polymer and is directly related to the velocity and viscosity of the melt. A relatively cold material is likely to develop high shear stresses. Analysts use shear stress information as their most direct indicator of the magnitude of orientation developed by a polymer molecule during flow. Unlike shear rate, shear stress is not normally used to interpret potential degradation of the polymer.

High shear stresses in runners and gates are expected but are not normally considered a concern. The stress developed in the polymer while in the runner is a localized effect and is independent of the shear stress, which is developed in the cavity. Again, a stressed runner is not a concern.

Shear stress limits have been provided by some developers of mold filling analysis software and typically have no scientific basis. Existing shear stress limits are even less regularly used by analysts than the shear rate limits. Essentially, the shear stress provides an indicator of the force acting on the polymer chain during flow. If the plastic is cooled quickly, the effects of the high shear stresses are more likely to be trapped in the molded parts. This normally means higher residual strains, stresses, and a potential for related problems. If the part is cooled very slowly, many of these highly oriented, or stressed, molecules will have an opportunity to relax.

Basically, high shear stresses in a cavity are an indicator of potential stress in the molded part. An analyst must use his/her judgment in regards to whether a high shear stress might be a problem. This should be weighted against the type of flow existing in the high shear stress region. High shear stress in a linear flow region is less likely to create a problem than in a radial flow region. High shear stresses experienced through an integral hinge actually provide a benefit. High shear stress in the radial flow region of a cavity at the gate can develop significant stresses, which can result in stress cracking. Thickened shear circles are often added in parts at the gate to reduce the shear stress and extend the cooling time in order to allow the highly stressed material to relax. The shear circle may be radiused or of constant cross section. The diameter should range from 6 to 10 times the thickness of the primary wall of the part, and the thickness of the shear circle should be approximately 1.5 times the primary wall thickness (see Figure 4.34).

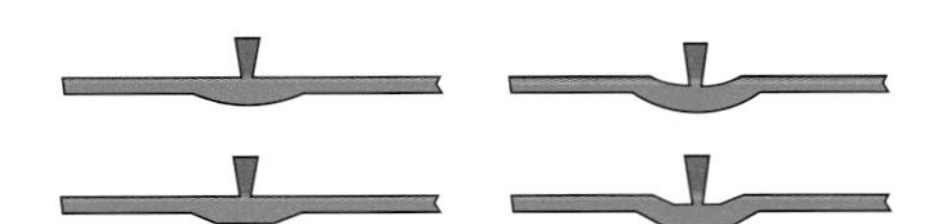

Figure 4.34 Variations in design of shear circles. The two designs on the right also help hide any gate vestige

References

[1] LNP Clouw, McDowell & Gerakaris, *Plastics Technology*, Aug. 1976

[2] CP Plastics Group, 1890 Lyndon Blvd.; Falconer, NY 14733; Tel. (716)664-4444; Fax:(716)664-2556; http://www.cpplasticsgroup.com

[3] Cinpres Gas Injection, Ltd. Head Office and AIM Centre, Prosperity Court, Prosperity Way, Middlewich, Cheshire CW10 0GD, United Kingdom; Tel: +44 (0) 1606 839800; Fax: +44 (0) 1606 839801; http://www.gasinjection.com/map/uk.htm

[4] Moldflow Corporation Headquarters, 430 Boston Post Road, Wayland, MA 01778 USA; Tel. (508)358 5848; Fax (508)358 5868; http://www.moldflow.com

5 The Melt Delivery System

5.1 Runner Design Fundamentals

The following list itemizes the basic design objectives for a runner system:

- Deliver melt to desired gating location(s) on the part
- Minimize excess material
  - Cold runners: reduce regrind, scrap and handling
  - Hot Runners: reduce residence time, increase wall flush and minimize melt stagnation
- Accommodate filling pressure limitations
- Provide for cavity packing
- Provide desired melt conditions to the cavities
- Provide same melt conditions at the same process to all cavities in multi-cavity molds
- Cold Runners:
  - Minimize diameter to avoid negative effect on cycle time. Consider this with the sprue design and also consider use of a hot sprue.
  - Provide for ejection that does not interfere with melt flow.
  - Provide cold slug at intersection of the sprue with the primary runner. Additional cold slug wells at runners branches may be desirable, but are less critical.
  - Use full round or correctly designed parabolic runner.
  - If full round runner is used, tolerance the match-up of the two runner halves to be +/– 0.050 mm (+/– 0.002 in.).
- Hot Runners:
  - Be sure to select a hot runner manufacturer that has a good history, has good references, and can provide you good engineering and service.
  - Be sure the plastic material to be molded is known and specified to the hot runner designer.
  - Critique the runner design based on the checklists provided in Chapter 14 and the criteria established in this book.

Do not be penny wise and a dollar short. The short-term cost of a well designed and manufactured runner is minimal relative to its long-term impact on your molding.

5.2 Overview of Runner/Melt Delivery System

The melt delivery system begins with the injection molding machine nozzle and continues into the mold, progressing through the sprue, runner, gate, and the part forming cavity. Each of these components, or regions, has an impact on molding, which can be significant. These components cannot only affect the process, but also the molded part. Process effects include the ability to fill and pack the part, the injection fill rate, the clamp tonnage, and the cycle time. Effects on the part include size, weight, and mechanical properties of the part and variations in these characteristics between parts formed in various cavities within a multi-cavity mold. Despite the significant influence of the melt delivery system, its various components are generally poorly designed relative to the time, effort, and cost put into the other components of a mold or the molding machine.

Evidence of the lack of understanding of runners includes the fact that the impact of shear-induced flow imbalances in runners was not documented, or clearly explained, until 1997 [1]. For the first time, it became widely publicized that the industry standard "naturally balanced" runners were creating significant imbalances. Imbalances ranging from 65:35 to as high as 95:5 were discovered to develop in a standard "H–patterned" geometrically balanced runner used in a four-cavity test mold with only two branches. This phenomenon was being overlooked by nearly the entire molding industry for both cold and hot runner molds. In addition, the industry's leading state of the art mold filling simulation programs had been developed without the realization of the shear-induced imbalance. As a result, these programs did not predict the imbalance and thereby left the analyst with a false impression of a given design.

Runners are designed as either cold runners or hot runners. Hot runners are most often made of standardized components, developed by companies that specialize exclusively in hot runners. A company can purchase these components and include them in its own design or contract to have the hot runner company design and build a complete system for them. An entire assembled "hot half" can be purchased, which would include a complete assembly of hot runner components in a frame with all wiring in place. This can then be directly assembled to the mold with little additional work. In some cases, the supplier of the hot runner systems can be contracted to develop the entire mold. Owing to the complexity of the hot runner system and its potential problems, most mold builders will purchase standard components and assemblies in order to minimize their startup and maintenance problems. In addition, if problems develop, the faults can be isolated more easily.

5.2.1 Machine Nozzle

The machine nozzle is a tubular-shaped component that mounts on the face of the injection barrel. The machine nozzle screws into the end cap of the injection barrel. It is intended to bridge the injection unit to the mold, providing passage for the melt through the stationary platen to the mold's sprue. It commonly consists of a main body, which screws into the end cap of the injection barrel, and a separate nozzle tip, which screws into the main body of the nozzle. The main nozzle body extends from the barrel to the sprue and is usually independently heated using heater bands. In some cases, heat pipes may be used, which will draw their heat from the screw barrel. Heat pipes have the advantage of avoiding exposed heater bands and wires, which can be damaged. However, they do not have the advantage of independent temperature control. The nozzle tip provides the final transition from the larger diameter of the nozzle body to the sprue inlet (see Fig. 5.1).

The nozzle can be of various lengths depending on how far it might need to reach to get through the stationary platen and mold. Some molds, such as three-plate cold runners, commonly eliminate their cold sprue. Here, a molder may either use a hot sprue or an extended machine nozzle. The extended nozzle requires that it reaches through both the stationary machine platen and through a portion of the mold to connect with the runner.

The nozzle tip makes the final transition between the molding machine and the mold. With a cold sprue, the outlet orifice of the nozzle tip is slightly smaller than the inlet orifice of the sprue. This ensures that the frozen sprue can be ejected. With hot runners, it is desirable that the orifice of the nozzle tip and the sprue inlet be aligned and of the same size which will provide a more stream lined flow channel and eliminate places where material could get hung up. Nozzle and sprue alignment is assisted by matching the radius on the nozzle tip face and on the sprue base (see Section 8.1).

Shut-off devices such as needles, springs, sliding balls, or a combination of these can be incorporated in the nozzle designs. Often their purpose is to shut off the flow of low viscosity plastic for a material such as nylon that tends to drool from standard nozzles. Also these nozzles can be used to extend plastification time in applications where sprue break is desirable and the normal plastification following fill and pack will limit the cycle. Common cases are stack molds, where the sprue bar is retracted from the nozzle during mold opening. The nozzle shut-off will allow plastification while the nozzle is separated from the mold.

A removable nozzle tip is often desirable to provide for changing molds with different sprues and molding different plastic materials. The orifice of the nozzle tip can also be designed for various materials and processes (see Fig. 5.1).

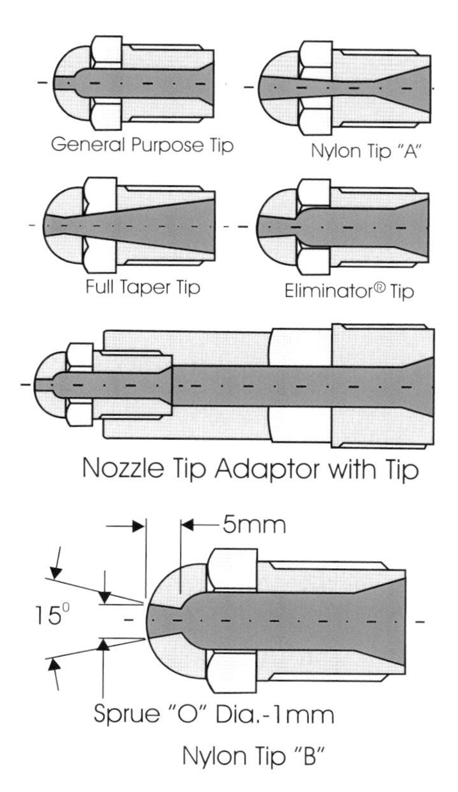

Figure 5.1 Typical Nozzles and nozzle tips used in injection molding

5.2.1.1 Nozzle Filter

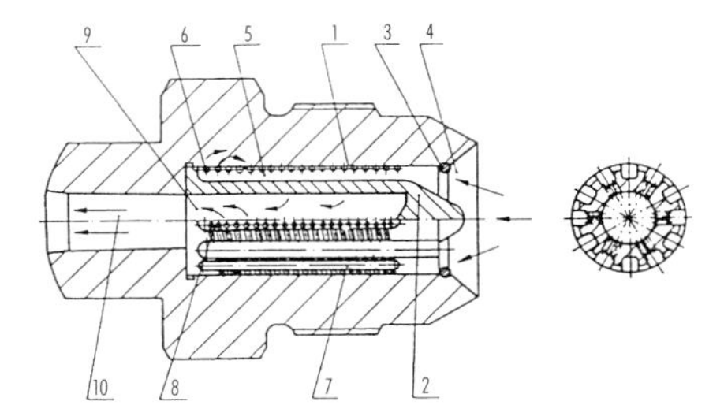

Figure 5.2 Typical nozzle filter that removes contaminants in the melt;
1: Location holes, 2: Filter inserts, 3: Locking ring, 4: Transition to nozzle of injection moldes machine, 5: Feed channel,
6: Tangential filter groove, 7: Intermediate channel, 8: Radial filter holes, 9: Collecting channel, 10: Die orifice

The purpose of a nozzle filter (Fig. 5.2) is to remove possible contaminants from the plastic melt during injection molding. Examples of such contaminants are any un-melted plastic pellets, scrap metal, or virtually any object that was accidentally placed in the hopper. Nozzle filters are particularly useful in hot runner systems to prevent blockage in the gate tips from any foreign objects. Instead, the objects, which would normally be trapped in the hot runners gate tip, are trapped in the machine nozzle. The nozzle is much easier to clean out, which saves time and money.

The disadvantage of nozzle filters is both the high pressure drop through the filter and the changing pressure drop experienced during a production run as the filter clogs. The changing pressure drop may be the bigger issue as it can result in changes in the molded part as the filter progressively clogs. As long as the pressure limits of the molding machine are not exceeded as the filter clogs, flow rate to the cavity during mold filling should not be affected if injecting under velocity control. However, the increasingly restrictive flow channels can potentially increase shear and shear heating of the plastic. In addition, pack pressure could be affected unless it is controlled through cavity pressure. Monitoring the pressure buildup during injection can minimize these negative effects of clogging nozzles.

5.2.1.2 Static Mixers

Static mixers are occasionally used in the injection nozzle to improve the homogeneity of the melt delivered from the screw. The practice of using static mixers after the plasticating screw is much more common in extrusion. The most notable reason for using static mixers in injection molding is to improve color dispersion when color concentrates are being used. Generally, static mixers are avoided due to the significant pressure drop required to get good mixing (see Section 7.1 for more information on static mixers).

5.2.2 Sprue

The sprue provides a transition from the machine nozzle to the runner system or directly into the cavity in the case of a single cavity sprue-gated part. There are both hot and cold sprues. The sprue is typically the component of the melt delivery system with the largest diameter. Cold sprues are tapered to facilitate ejection during each cycle. Hot sprues can be used to feed a cold runner or may be integral to a hot runner manifold. They may also have a slight taper to facilitate removal when cooled for cleaning or inspection.

5.2.3 Runner

The term runner refers to the melt delivery channel between the injection molding machine nozzle and the part-forming cavity in the mold. Though it most directly refers to the portion of the melt delivery system between the point where the melt exits from the sprue to the point where it enters the gate, it can include both of these. The runner has the potential to influence the size, shape, and mechanical properties of the molded parts. The primary objective of the runner in a multicavity mold is to supply the same melt condition to each cavity with the same process conditions to ensure part-to-part consistency. Runners are classified into two main categories – hot and cold runners.

5.2.4 Gate

The gate is the final component of the melt delivery system. It is a passage for the plastic melt to travel through between the runner and the part-forming cavity. Gates should be located in areas of minimal end use stress in the part, where cosmetics are not a concern and where a desirable cavity flow results. Gate placement and number of gates on a part are critical as they can affect numerous mechanical, dimensional, and cosmetic factors of a part. In addition, they can significantly affect the process including pressure to fill, clamp tonnage, and the ability to fill at a desired rate.

> When in doubt of proper gate size, it is best to be "steel safe." Start with a gate size smaller than you think will do the job, and increase the size until proper packing of the cavities is achieved consistently. In multi-cavity molds, it is critical that this gate adjustment is well controlled so that the gates are of the same size for each cavity.

5.3 Melt Flow Through the Melt Delivery System

5.3.1 Melt Preparation – The Injection Molding Machine

Unlike the screws used in most continuous extrusion processes, injection molding screws are normally of a "general purpose" design. That is to say, regardless of the material being run, the same screw is used for plasticating the material. With most continuous extrusion processes, a different screw may be used for nearly every material run. The practice of using the same screw for all materials is common throughout the industry. It is the unusual case that a screw be designed for a particular material. Exceptions can occur when a molding machine is being purchased for a specialized

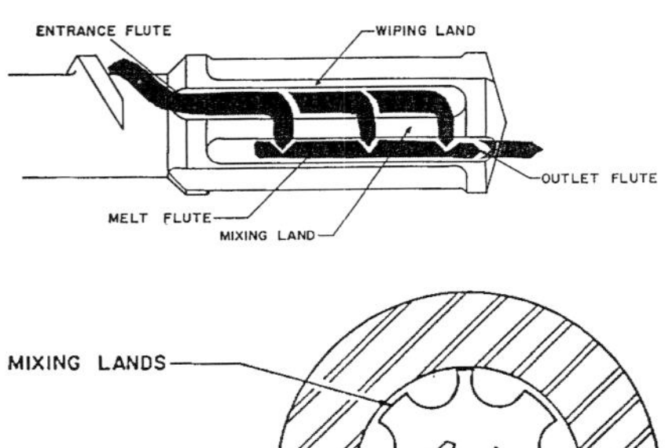

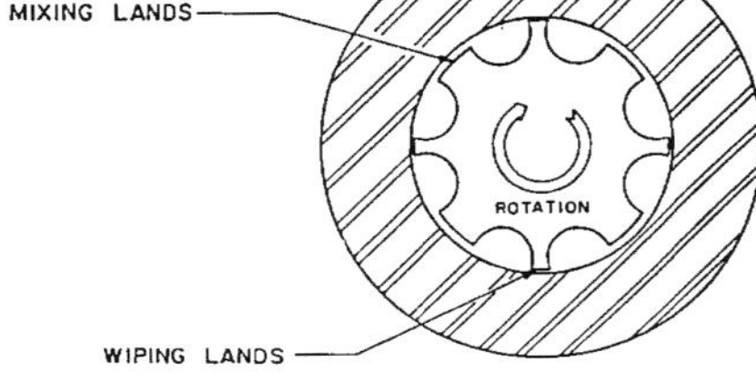

Figure 5.3 Maddock mixing device

Figure 5.4 Maddock mixing device

application, which is going to mold one product and one material, or in the processing of particularly shear sensitive materials such as rigid PVC.

Using a "general purpose" design screw for all applications may be acceptable, but it is not optimum. It is known that the plastification characteristics of plastics are quite different from each other. A specialized screw can improve plastification rate, homogeneity of the material, color mixing, mixing of regrind, and the degree of degradation of the material.

Improved mixing can reduce the let-down required if a color concentrate is to be used. Improvements in screw mixing can be provided through the design of the feed, transition, and metering zones or through the inclusion of mixing devices such as the Maddock mixing section (Figs. 5.3 and 5.4). B. H. Maddock from Union Carbide first published results of experiments with this mixing section, hence the name. [2]

It is also possible that poor mixing in a screw can create mold-filling imbalances. Owing to the laminar flow conditions through a mold, if the melt has asymmetric conditions entering the mold, these varied conditions can continue through the mold's runner where they may become separated into different branches. Though this has not normally been recognized as a major problem in injection molding, it is worth reviewing. The imbalances created from this poor mixing should be erratic and unpredictable. If a regular imbalance pattern is recognized, it is not likely a result of poor mixing from the screw.

5.3.1.1 Pressure Development from a Molding Machine

The most common method of developing the driving force to push the plastic from the injection barrel through the mold is a hydraulic system that works on the rear of the injection screw. Here, the melt injection pressure is intensified through sizing of the injection piston to the screw diameter. These intensification ratios can commonly range from 8:1 up to 15:1 in injection molding machines. A molding machine with a maximum hydraulic pressure of 2,000 psi and an intensification ratio of 10:1 will result in melt pressures of 20,000 psi. Most molding machines allow injection screws and barrels to be changed. A small diameter screw will increase available pressure on the melt versus a large diameter screw. Therefore, one must know the hydraulic pressure and the screw diameter to determine the potential melt pressure that can be developed. The intensification ratio is determined by the ratio of the area of the piston driving the screw and the area of the screw's cross section. Electric molding machines do not have the same intensification ratio because the screw is driven forward by electromechanical means, but a change in screw diameter can change the pressure available on the melt; therefore, this factor should also be taken into consideration.

The pressure on the melt is calculated as injection force divided by the area of the screw (see Fig. 5.5):

$$P = F/A \tag{5.1}$$

Where:

P = Melt pressure

F = Force driving the screw forward

A = Area of the cross section of the machines screw and barrel

During the injection process, pressure losses occur from what is supplied by the hydraulic system. These can include frictional losses of conveying the screw forward and leakage of the melt over the screw's check ring and flights. These losses can be as high as 25% during the injection phase and are affected by fill rate. Therefore, a machine that is expected to deliver a melt pressure of 20,000 psi may actually only be able to deliver a melt filling pressure of 15,000 psi. During the pack stage, it can be expected that there would be less of a difference between machine output and melt pressure, as the dynamics of the machine, and the related losses, are significantly decreased. [3]

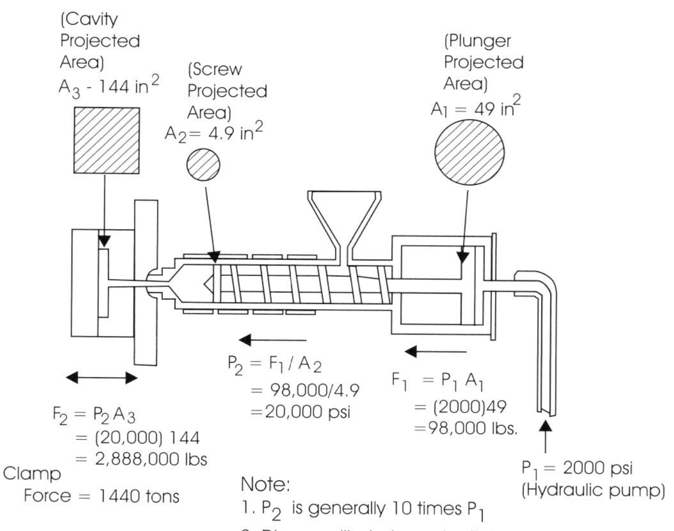

Figure 5.5 Development of melt pressure and clamp tonnage in a typical injection molding machine. The design of the machine normally provides for the melt pressure to be intensified by a factor of 10 relative to the machines hydraulic pressure

5.3.1.2 Flow Through a Runner Channel

The flow of plastic in a runner is quite complex and has a much more significant influence on successfully producing parts than thought in the past. In nearly all situations, flow in a runner is considered to be laminar and flow along the channel wall is considered to be zero. This condition results in fountain flow, where the plastic in the flow channel's center is flowing faster than the material along its perimeter. The material at the flow front is pushed forward and "fountains" to the channel wall. The result is that during mold filling, the first material to enter the runner fountains to the slower filling, or freezing, perimeter while the later filling material advances to the flow front.

The velocity across the flow channel increases dramatically from the zero flow at the wall to the laminates just inside these layers. The rapid change in velocity results in zero shear rates at the wall to very high shear rates just inside the wall. From a maximum shear rate just inside the wall, shear rates decrease more gradually towards the center of the flow channel where they again become zero. The result is a continually varying shear rate across the flow channel. As plastics are non-Newtonian, the result is dramatic variations in material viscosity across the flow channel. In addition, the high shear rate region causes significant frictional heating in those relative laminates. This further affects the viscosity and material property variations across the flow channel.

As flow progresses along a channel, it can also be expected that temperatures and effects of shear and thermal history on the material will change. The overall result is that there can be significant variations in material properties across and along a runner channel, which can include temperature, flow rate, viscosity, molecular weight, and filler/reinforcement damage.

The high shear rate region near the outer wall has a combined effect on viscosity. Viscosity in this region is reduced because of its non-Newtonian characteristics and the resultant frictional heating. This frictional heating will cause the melt in these outer laminates to be well above the temperature of the melt in the center of the runner channel. In thermoset injection or transfer molding, the frictional heating in the outer laminates is compounded by the heat gained from the heated mold.

The most dramatic effect of these variations is their impact on multi-cavity molds. As flow branches in the runner, it has been found that material from the various laminates becomes separated and flow into different cavities. This will result in parts produced with essentially different materials under different molding process conditions. Chapter 6 discusses this issue in detail.

These areas of high and low shear across a runner are observed at all flow rates and in both hot and cold runner molds. It can be expected that the shear rate will be highest in the runner and gates of a mold where the velocity of the melt is normally at its highest.

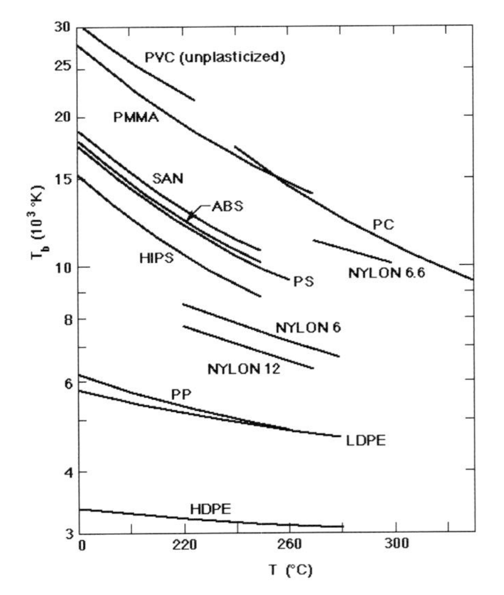

Figure 5.6 Temperature sensitivity factor (T_b) versus temperature level provides a contrast of how sensitive a material's viscosity is to temperature change (Courtesy: Moldflow Corp., all rights reserved)

5.3.2 Effect of Temperature on Flow

5.3.2.1 Melt Temperature

After the effects of non-Newtonian shear thinning, melt temperature is the greatest determining factor of the viscosity of a plastic material during injection molding.

The viscosity of some materials is influenced more by melt temperature than others. Table 5.1, and its accompanying graph (Figure 5.6), provides information on the sensitivity of a material's viscosity to temperature and change in temperature. The table lists the average value of T_b of the thirteen different materials presented in the graph. $T_{b,}$ as used in the Cross-WLF viscosity model, represents the sensitivity of a material's viscosity to change in temperature. The viscosity of an unplastisized PVC, with an average T_b value of 25,000 K, is nearly five times more sensitive to change in temperature than is that of a LDPE or a polypropylene.

Figure 5.7 contrasts a PMMA and a polypropylene. Note the viscosity of the PMMA, having an average T_b value of 19,000 K. Note that the viscosity, at low shear rates, varies from 1,000 to 10,000 Pa-s with only a 40 °C change in temperature. On the other hand, with the same 40 °C, the polypropylene's viscosity changes less than half of this amount.

Table 5.1: Viscosity Sensitivity to Temperature and Change in Temperature

Material	T_b (10^3 K)
PVC (unplastisized)	25
PMMA	20
SAN	14
ABS	13.5
PS	12.8
PC	12.8
HIPS	11.6
Nylon 6/6	11
Nylon 6	7.5
Nylon 12	7
PP	5.3
LDPE	5
HDPE	3.3

5.3.2.2 Mold Temperature

The mold temperature does not have nearly the impact on flow as the melt temperature. This is particularly true at the relatively high injection rates normally used during molding. At the usually high injection rates, the shear heat generated in the melt from friction dominates the heat loss to the mold through thermal conductivity. At slow flow rates, this may change. This is most evident in cases where a part includes thick and thin features with the thin features near a gate. As the melt reaches the thin features it will slow down, or hesitate, while quickly advancing through the thicker regions. The slow flowing melt will quickly cool due to loss of frictional heating and non-Newtonian shear thinning, and potentially freeze off.

5.3.3 Cold vs. Hot Runners

There are fundamental differences in the flow of plastic melt through cold runners compared to hot runners. Perhaps one of the most obvious of these is that the melt flowing along the heated surfaces of the hot runner will not freeze. A frozen skin forms in a cold runner as the plastic is pushed along the cool walls of the mold. This layer will have a minor

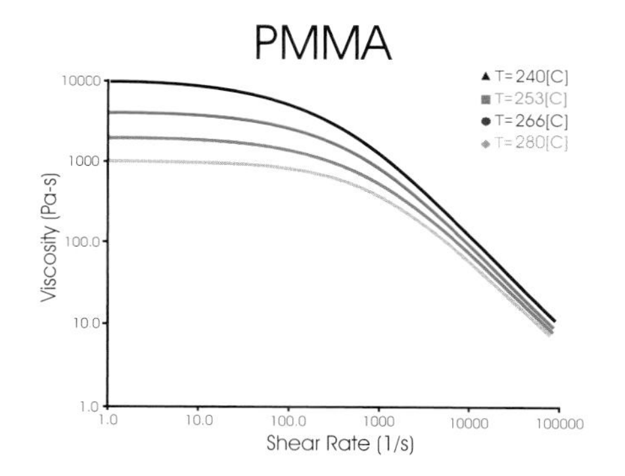

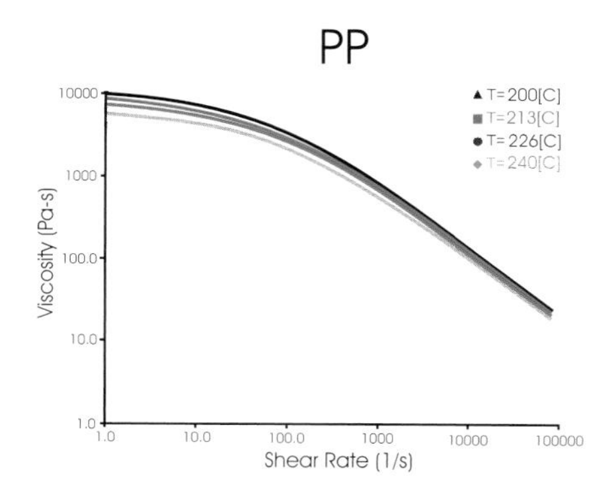

Figure 5.7 Viscosity vs. shear rate graphs of PMMA (A) and PP (B) showing the relative sensitivity of their viscosities to temperature. Note both viscosity curves have a 40 °C temperature variation

impact on pressure to fill at fast flow rates and a more significant effect at slow fill rates. The most significant effect of a developing frozen skin occurs during the compensation phases of the injection process. At the slow compensating flow rates, a runner that is too small in cross section may freeze prematurely. This will prevent the part from being properly packed out.

Often, hot runners are able to have larger diameters than their cold runner counterparts because there is no concern with waste or regrinding runners. In addition, there is no concern with the excessive cooling time of a large diameter cold runner. On the other hand, the plastic melt in the hot runner will go through an extended thermal history as it is not ejected every cycle. The larger the runner diameter, the longer the thermal history of the material and the greater is the potential for thermal degradation. For some heat-sensitive materials, the thermal degradation can affect viscosity, degradation of material properties, part quality, and in some cases result in toxic agents, which can be hazardous to both humans and to the tooling.

The larger diameter of hot runners can also be an area of concern if the molder anticipates that color changes will be performed. The larger the diameter of the hot runner, the longer it will take to flush the old color from the system.

5.3.4 Pressure Drop through the Melt Delivery System (Nozzle vs. Sprue vs. Runner vs. Gate vs. Part Forming Cavity)

The injection pressure delivered from the injection molding machine must overcome pressure losses as the melt travels through the machine nozzle, sprue, runner, gate, and part forming cavity. The impact of the machine nozzle is often underestimated, yet may exceed pressures of 30 MPa (4350 psi). Unfortunately, its contribution is commonly ignored by analysts performing mold filling analyses. This alone can result in significant errors in the analysis.

Pressure loss of a laminar flowing material of constant viscosity in a round channel is directly proportional to the length and inversely proportional to the forth power of the diameter. Though the viscosity of a plastic material is not constant along a runner channel, this relationship of geometry to pressure loss gives us some indication of the relative impact of the runner channel's diameter. That is, changing the length of a runner has much less effect on pressure relative to changing its diameter.

Pressure loss through a runner system will vary with mold and material. However, it is not unusual for the pressure loss in a runner to be 50% of the total pressure drop during mold filling. The low pressure loss that might be experienced in the cavity of a small part might allow for the use of very small diameter runners, thus reducing the negative effects of a larger runner. The filling pressure of the wire protectors shown in Fig. 4.32 was relatively low. Therefore, the cold runner feeding this part was downsized using mold filling analysis in order to minimize regrind. The result was that the cavity pressure reached only 4,200 psi while the runner pressure was 10,800 psi. As the machine was capable of 20,000 psi, the high pressure loss in the runner did not create a problem.

It should be obvious now that performing a mold filling analysis without considering the nozzle and runner system could result in significant misjudgments about the ability to fill a part.

5.4 Use of Mold Filling Analysis

Injection mold filling analysis programs by companies like Moldflow Inc. and CoreTech Systems Co. provide an excellent tool for sizing runner systems. These programs provide information on pressure, melt temperature, and shear rate at various fill rates. Though shear rate can be determined using simple hand calculations, fill pressure and melt temperature at various fill rates require much more sophisticated solution methods and detailed material data. Of particular interest to most molders is determining if their mold will fill with a given runner and gate design and a given gating location on their part. To determine this, the melt delivery system and the part forming cavity must be modeled. To size runners, a skilled analyst does not require a detailed model of the cavity. Often they can use simplified geometries that represent the volume of the cavity and a flow length and thickness representative of the most difficult flow path through the cavity [4]. Early 2-D injection molding simulation programs used this method successfully for years. The advantage of this older 2-D method is that the modeling and analysis can take as little as a half an hour for a skilled analyst. These programs used a simple 1-D beam for runners, and although they did not provide any graphical feedback, they did provide good information on pressure, temperature, shear rate, and shear stress on the melt during mold filling. The risk of this technology originated mostly in poor application by the user. The modeling of the part required good interpretive skills and good ability to realize what the program could and could not provide.

Most of these early programs have been replaced by much more sophisticated 2½-D and 3-D programs that can provide much more detailed information on flow through the cavities (see Fig. 5.8). Detailed information on cavity conditions can be provided in easy to interpret colorized contour plots. Though these new programs present the impression that they are easier to use, they are significantly more complicated and compute-intense. They still require a skilled analyst to assure that the geometry and mesh is representing the critical regions to be analyzed. If sizing a runner and evaluating a gate design are the issue, these programs can be an over-kill and a waste of engineering time. This is particularly the case when one considers that they generally use the same

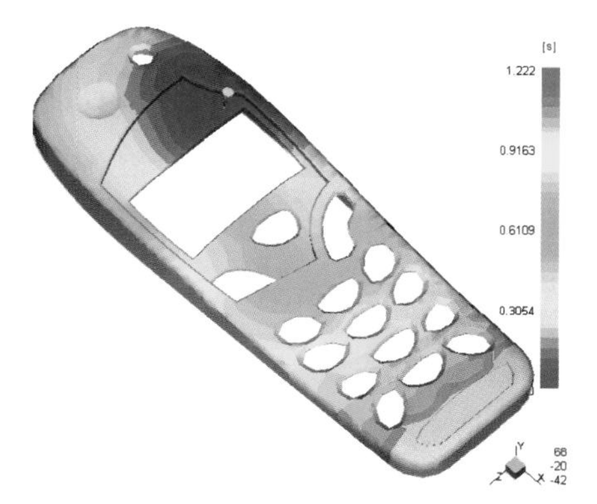

Figure 5.8 2½ D mold filling analysis output of fill pattern

1-D beams to represent their runners as the older 2-D programs. The primary advantages of the newer programs is studying the filling patterns and melt conditions throughout a cavity and for the further analysis of mold cooling, part shrinkage, warpage, and structural performance.

Some cautionary remarks regarding the use of any of the standard 1-D, 2-D, 2½-D and 3-D injection molding programs:

1. Mold filling analysis can provide good information on how small a runner can be while still allowing the mold to fill. With a cold runner, be careful that the size provided from a mold filling analysis is not too small to allow for the cavity to be properly packed out during compensation /packing phase. It is generally expected that the cold runner diameters should be no less than 1.5 times larger than the thickness of the part. (Part requirements and design must be considered)
2. One should be careful when trying to analyze an insulated or internally heated hot runner system. Most programs do not calculate the development of a frozen layer in these applications. Check with the software provider on how these conditions are handled.
3. The 1-D beams used in the 2-D and 2 ½-D filling analysis programs cannot pick up the shear-induced filling and melt imbalance in multi-cavity molds.
4. At this time, none of the newer 3-D filling analysis program predict the magnitude of the shear-induced filling and melt imbalances in multi-cavity molds. With careful meshing, some of these programs have been able to predict a slight variation in the melt and the filling. Filling imbalances of less than 1% are being predicted where the actual imbalance may be over 30%.
5. Mold filling analysis is commonly used to artificially balance the filling of a fishbone type runner layout. These programs can significantly reduce the effort required to manually balance these molds. However, a molder should realize that an artificial filling balance will not balance melt condition or packing.

5.5 Runner Cross Sectional Size and Shape

5.5.1 The Efficient Flow Channel

The most efficient flow channel cross section is a full round. This will result in the lowest pressure drop per material used. The full round and annular flow shape is almost exclusively used in hot runners. The annular shape is much less efficient but is a result when internal heaters are used.

Parabolic or trapezoid shapes are sometimes found in insulated hot runner molds.

A wider variety of shapes are used with cold runners, including full round, parabolic, trapezoid, half round, rectangular, or some careless modification of one of these (see Fig. 5.9). However, cold runners should almost always be either full round or parabolic. The parabolic provides the closest performance to a full round runner without the need to machine and match up the two halves of a full round runner. Other shapes such as half round, rectangular, and disproportioned trapezoid or parabolic should be avoided.

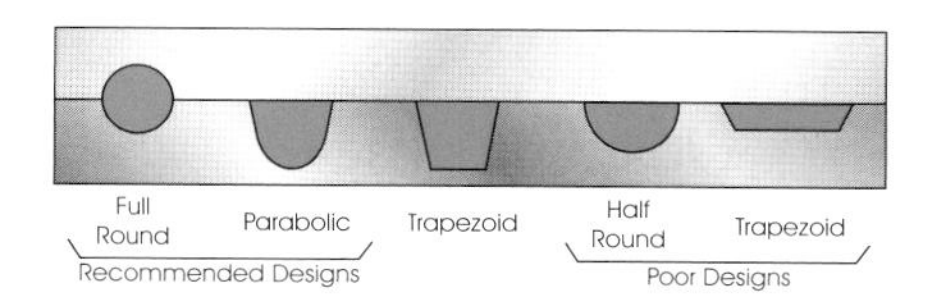

Figure 5.9 Runner cross-sections used in injection molds

The full round runner provides the largest cross section to surface area. This reduces surface drag and heat exchange between mold and material. The full round runner also minimizes low flow regions within a runner resulting in the most consistent melt conditions. This is particularly a concern in hot runners where low flow regions increase residence time and potential material degradation.

5.5.2 Pressure Development in the Runner

Pressure development in a full round runner can be calculated with

$$\Delta P = \frac{8Q\eta L}{\pi R^4} \tag{5.2}$$

Where P is pressure; Q is volumetric flow rate, L is channel length, η is the material viscosity, and R is the channel radius.

This apparently simple relationship is actually quite complex because the viscosity is almost continuously changing along the flow channel as it is affected by temperature and shear rate. In addition, in a cold runner the channel diameter is variable as it depends on the thickness of the developing frozen layer.

When calculating the pressure drop of a non-round cross section, such as a parabolic or trapezoid cross section, the same equations are generally used except that R is replaced by an equivalent hydraulic radius.

$$R = (2A)/\text{Perimeter} \tag{5.3}$$

where A is the cross sectional area of the flow channel and the perimeter is that of the same flow channel.

Annular-shaped cross sectional flow channels result in hot runners with internal heaters or valve pins. The flow channels are essentially thin rectangular/shells channels wrapped into an annular shape. This is much less efficient than a full round.

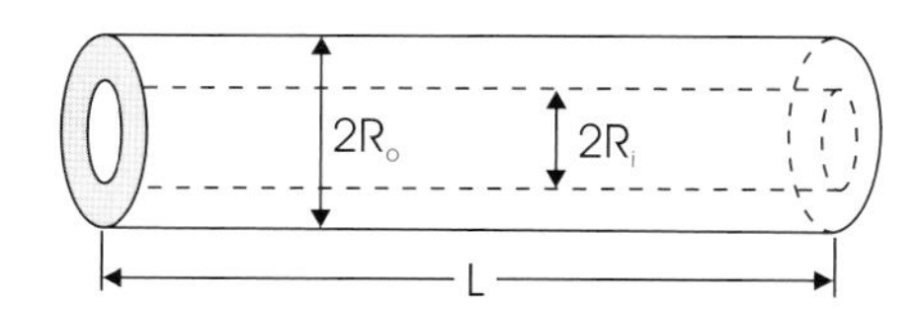

Figure 5.10 Annular flow channel

Pressure drop in an annular flow channel (see Fig. 5.10) can be calculated with:

$$\Delta P = \frac{8Q\eta L}{\pi R_o^{\ 4}\left[1-(\frac{R_i}{R_o})^4 - \frac{(1-(\frac{R_i}{R_o})^2)^2}{\ln(1/\frac{R_i}{R_o})}\right]} \tag{5.4}$$

The annular flow channel is essentially a slit wrapped around into a tubular shape. This shape is generally used with internally heated hot runner systems. In this case, similar to the cold runner, R_o is somewhat unknown, as it will be affected by the formation of a frozen skin.

Example

The following contrasts the shapes and pressure drop through a 0.5 in. diameter full round hot runner to the pressure drop through an internally heated hot runner system, which contains an equivalent amount of plastic materials (i.e., same residence time).

A. Find the cross sectional area of the full round channel.

$$A = \frac{\pi d^2}{4} = \frac{\pi (0.5)^2}{4} = 0.196\, in^2 \tag{5.5}$$

B. If the internally heated channel is heated with a 0.625 in. diameter heater (d_1), find what bore diameter (d_2 = runner channel diameter) would yield the same 0.196 in.^{2} cross sectional area for flow as the full round runner (see Fig. 5.11).

$$A = \frac{\pi d_2^2}{4} - \frac{\pi (0.625)^2}{4} = 0.196\, in^2$$

$$d_2 = 0.8\ in$$

$$if\ d_1 = 0.625\ in,\ then\ r_1 = 0.3125\ in$$

$$if\ d_2 = 0.800\ in,\ then\ r_1 = 0.400\ in$$

C. Knowing the channel thickness, you can now find the channel width. The channel width is the circumference of the mid-plane of the cylinder defined by the r_1 and r_2 radii (see Fig. 5.11).

$$\text{Channel Thickness} = r_2 - r_l = 0.400 - 0.3125 = 0.875\ \text{in.} \tag{5.6}$$

$$\text{Mid-Plane Radius } (r_3) = \frac{0.400 + 0.3125}{2} 0.356\ \text{in.} \tag{5.7}$$

$$\text{Channel Widths} = \pi(2r_3) = \pi(0.713\ \text{in.}) = 2.237\ \text{in.} \tag{5.8}$$

Therefore, the annular channel, containing an equivalent amount of material as a 0.625 in. diameter full-round is essentially a flat rectangular strip that is 0.0875 in. thick by 2.237 in. wide.

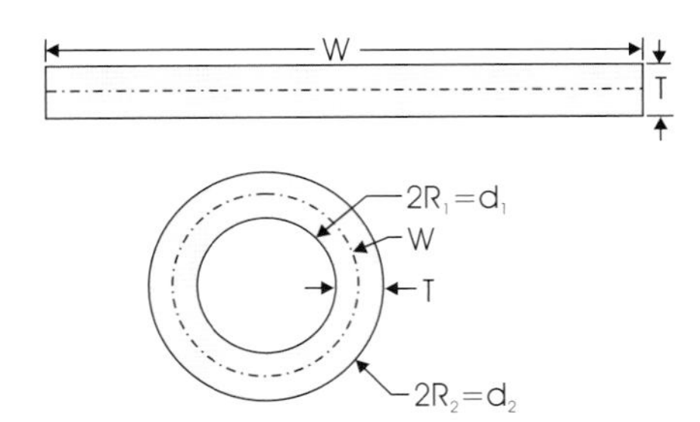

Figure 5.11 Cross-section of annular flow channel and the equivalent rectangular channel

D. Compare pressure drop through the two different channel shapes that contain the equivalent amount of plastic material (same cross sectional area).

Given:

Flow Rate (Q) = 2 in.^{3}/s

Viscosity (η) = 0.0304 (lbs.)/in.^{2}

Length of Flow Channel (l) = 10 in.

Pressure drop through a full round channel (with Eq. 5.2):

$$\Delta P_r = \frac{8Q\eta L}{\pi r^4} = \frac{8 \times 2 \times 0.0304 \times 10}{\pi (0.25)^4} = 397 \text{ psi}$$

Pressure drop through the equivalent annular channel:

Method #1 (with Eq. 5.4):

$$\Delta P_f = \frac{8 \times 2 \times 0.0304 \times 10}{\pi (0.4)^4 \left[1 - \left(\frac{0.3125}{0.4}\right)^4 - \frac{\left(1 - \left(\frac{0.3125}{0.4}\right)^2\right)^2}{\ln\left(1 / \frac{0.3125}{0.4}\right)} \right]} = 4.861 \; psi$$

Method #2: Pressure drop through the annular flow can also be determined by solving the pressure drop through a rectangular slit equal to the annular gap as established in Section 5.5.2.C and shown in Fig. 5.11.

$$\Delta P_f = \frac{12 Q \eta L}{w h^3} = \frac{12 \times 2 \times 0.0304 \times 10}{2.237 \times 0.0876^3} = 4.851 \text{ psi} \qquad (5.9)$$

Where ΔP_r = the pressure drop through the full round channel;

ΔP_f = the pressure drop through the annular flow channel.

Note that in the above example, pressure drop through the annular flow channel is nearly 12 times that found in an equivalent full-round flow channel. In actual applications, this will vary as the frozen layer development in the annular channel is not considered and the viscosity is considered Newtonian.

5.5.2.1 Flow through a Hot Runner vs. a Cold Runner

For the most part, the pressure development in the runner system is the same for hot and cold runners. Both types of systems experience laminar flow and fountain flow, which means there is no flow at the mold wall. In other words, there is no slip of the melt at the wall of the mold as the plastic is being injected.

Hot runner molds typically have slightly larger diameter runners because there is no concern with runner regrind or concern with its cooling time. These larger diameters allow for reduced pressure drops through the runner. Despite the surrounding cold mold in a cold runner, the bulk temperature of the melt is very similar in both hot and cold runner systems due to the significant shear heating developed in a runner. This shear heating also minimizes the development of a frozen layer during mold filling in a cold runner.

5.5.3 Runner Effect on Cycle Time

5.5.3.1 Cold Runner and Sprue Cooling Time

The cooling time of the sprue and runner has the ability to affect the overall cycle time. Although the sprue and runner do not have to be frozen completely, they must cool long enough that they may be easily ejected. This rarely becomes an issue unless when molding thin walled parts. If the sprue puller region, which is normally the thickest area in the melt delivery system, is forcing the cycle time to be extended, a hot sprue may be a good replacement.

5.5.3.2 Hot Runner

Hot runners have a clear advantage over cold runners in most high speed thin walled molding applications. Time is saved as less material must be plasticated and injected to fill the runner, clamp stroke is reduced, runner ejection time and handling is eliminated as well as eliminating additional cooling time that might be required for the cold runner. However, the hot runner can potentially extend cycle time in some cases, as it not only adds heat to the mold but restricts the location of cooling channels. This is particularly true in the gate region. Here the hot drop reaches directly to the part. The addition of cooling to this area is physically obstructed by the hot drop itself. Though cooling can be designed and machined in special channels around the drop tip, this is commonly left out by the designer due to cost and complexity. In addition, direct gate cooling can potentially cause premature gate freeze.

5.5.4 Constant Diameter vs. Varying Diameter Runners

It is common practice, with geometrically balanced runners, to decrease the runner diameter at each branch as it progresses from the sprue. This is a practice that is often blindly performed without understanding its purpose, or the potential negative effects

Based on common practice, a varying diameter runner will require less pressure to fill than a constant diameter runner. At a given pressure, the varying diameter runner will require approximately 10% less material than a constant diameter runner. Therefore, the varying diameter runner may be seen as a more efficient design, which will either reduce filling pressure or reduce the amount of runner material required.

When varying runners are to be used, the runner sizing should progress from the gate back to the sprue. The smallest diameter runner section should be directly feeding the gate and must provide for both filling and packing. The diameter of each successive branch back towards the sprue would be increased.

One common method for sizing runner branches is:

$$d_{feed} = d_{branch} \times N^{1/3} \tag{5.10}$$

where d_{branch} is the diameter of a downstream branch runner, d_{feed} is the diameter of the runner section feeding the branch runner and N is the number of branches. In an eight-cavity mold, d_{branch} would initially be the tertiary runner and d_{feed} the secondary runner. When sizing the primary runner, d_{branch} becomes secondary runner and d_{feed} the primary runner.

Example

Given an eight-cavity geometrically balanced runner feeding 2 mm thick parts. The first runner sections attached to the gates (tertiary runner branches) are sized to be 3.0 mm (part thickness + 1 mm). The secondary runner branches are then found by:

$$d_{feed} = d_{branch} \times N^{1/3}$$

$$d_{feed} = 3\text{ mm} \times 2^{1/3}$$

$$d_{feed} = 3.78\text{ mm}$$

Table 5.2 shows the results of a study, which contrasts the performance of three different runner designs – a constant 0.125 in. diameter; a varying diameter runner designed using Moldflow's MFG (which can automatically design a runner to have a constant pressure gradient), and a varying diameter runner using the equations above. The length of the primary, secondary, and tertiary branches for each of the three design variations were 1.5 in. (38.1 mm), 0.35 in. (8.89 mm), 0.83 in. (21.08 mm) respectively. Each of the runner designs was designed to result in the same pressure drop according to MFG. The temperature rise of the melt as it flowed through the runners and the cooling times of the runners were also determined with MFG.

As can be seen, the computer generated runner and the calculated runner appear to be more efficient than the runner with the constant diameter, as they achieve the same pressure drop with 10% less material. Note that the varying diameter runner designed with MFG results in nearly the same volume as the one designed using the above hand calculations. Also note that the melt temperature rise in all three runner designs is similar.

When sizing a cold runner, its minimum diameter must allow for proper packing of the part. Therefore, if a runner is to have progressive runner branches with varying diameters, it must be designed from the gate back to the sprue. The smallest diameter runner would be attached to the gate and each successive branch back toward the sprue would be increased.

Table 5.2 Contrasting Constant vs. Diminishing Diameter Runner Designs

Runner type	Runner volume (in.³)	Temperature rise (°F)	Max. cooling time in the runner (s)	Min. cooling time in the runner (s)	Min. runner diameter (in.)	Max. runner diameter (in.)
Constant diameter	0.29	6.36	16.86	16.43	0.1250	0.1250
Diminishing diameter runner (Method 1)	0.26	6.64	21.14	11.74	0.1015	0.1435
Diminishing diameter runner (Method 2)	0.26	6.64	19.95	12.50	0.1050	0.1387

Now let's take a more critical look at the above designs. First note that as runner diameters increase as they recede back from the gate to the sprue, the diameters of the varying diameter runners (Method 1 and 2) range from 0.1015 in. to 0.1435 in. (identified as Min. and Max. runner diameters in Table 5.2). Note that the 0.1435 in. diameter runner section will take nearly 25% longer to cool than the smaller 0.125 in. diameter used in the constant diameter runner design. If this does not limit the cycle time of the part, and the 0.1015 in. diameter does not limit packing of the part, then the varying sized runners look to be the preferred designs. However, lets further consider if the above runners feed a part requiring a 0.100 in. diameter runner for adequate packing yet there was no problem filling the part if the entire runner were 0.100 in. diameter. In this case, the varying diameter runner would require about 12% more material and the thicker primary runner would take nearly twice as long to cool as the constant diameter runner.

When closely reviewed, the use of diminishing diameter runners is not always advantageous. When considering filling, packing and effect on cycle time, a constant diameter runner can often be more desireable.

Therefore, when designing a runner:

- If packing controls the size of the runner (the part can fill at the minimum diameter runner required to pack the part), then the runner should have a constant diameter.
- If fill pressure controls the size of the runner (the smallest diameter runner section of a tapered runner is larger than the minimum size required for packing), then a tapered runner should be considered. However, if the largest diameter primary runner sections controls cycle time, a constant diameter runner should again be considered.

5.6 Designing Runners for Shear- and Thermally-Sensitive Materials

When molding highly shear-sensitive materials, using larger diameter runner systems and radiusing the runner at branches is recommended to reduce potential for degradation. This is particularly the case when molding rigid PVC (RPVC). Runner diameter should be such that shear rates do not exceed 10,000 s^{-1}. An inside radius of 6 mm should be used to reduce potential for any negative corner effects.

Radiused intersections have also been used in powdered metal injection molding applications. Some molders had theorized that inertial effects resulted in a separation of binder and powdered metal as the melt took the corner at a runner branch. It was expected that radiusing the runner would reduce the inertial effect and the resultant imbalance found in multi-cavity molds. However, several studies have indicated that the imbalance actually results from the shear and temperature difference developed across the melt stream along the length of the runner (discussed in Chapter 6) and is independent of the corner. There was no evidence of binder separation in these studies. [6] [7]

Corners, commonly created at branches in a runner, are more of a concern in hot runners than in cold runners, where it is standard to have abrupt 90 degree corners. The only case where radiused corners are commonly found in cold runners are with RPVCs (as mentioned earlier) and with long glass-fiber filled materials. The sharp corners are expected to break long glass fibers as they attempt to bend around them. Because PVCs are sensitive to shear, it is expected that radiused corners will reduce shear that could cause degradation. However, the corners have a minimum shear effect when compared to the effect of the continuous shear experienced while flowing the length of the entire runner.

With hot runners, great effort is put into breaking the corners, particularly when shear- and thermally-sensitive materials are to be molded. At a minimum the inside corners should be smoothed. This is generally done after cross drilling by manually reaching in and working he inside corner. The increased demand on hot runners, compared to cold runners, is caused by the fact that the material in the runner is not ejected after being sheared. This increases the potential for degradation. In addition, the streamline flow around a corner results in low flow areas that will increase material residence time and potentially contribute to degradation of the material (see Fig. 5.12).

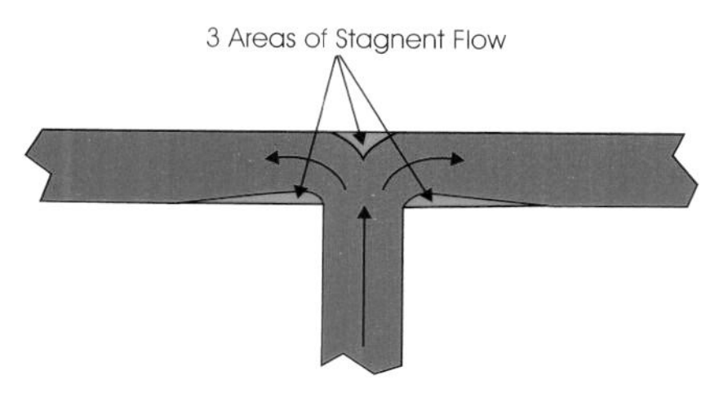

Figure 5.12 Slow flow regions in a branching hot runner due to streamlining

5.7 Runner Layouts

The runner layout in a mold is generally based on providing the shortest path from the sprue to the gate, minimizing runner volume, and considering simplicity of machining operations required to create the runner, as well as providing a geometrically balanced flow path. The following sections provide a general overview of different runner layout styles. These layouts are discussed in more detail in Chapter 6 and 7.

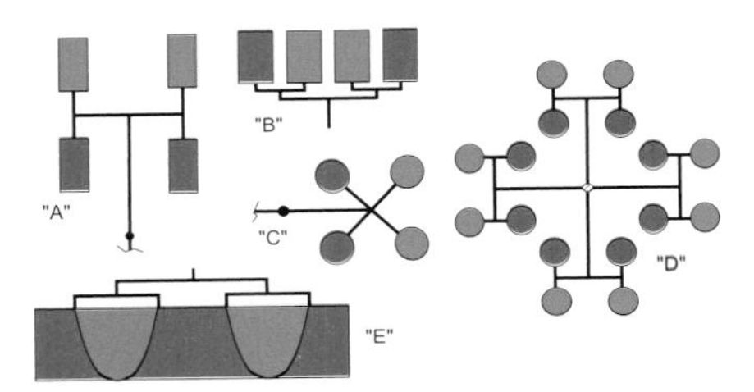

Figure 5.13 Sample of geometrically balanced runner layouts resulting in unbalanced filling and melt conditions; red regions will receive high sheared melt

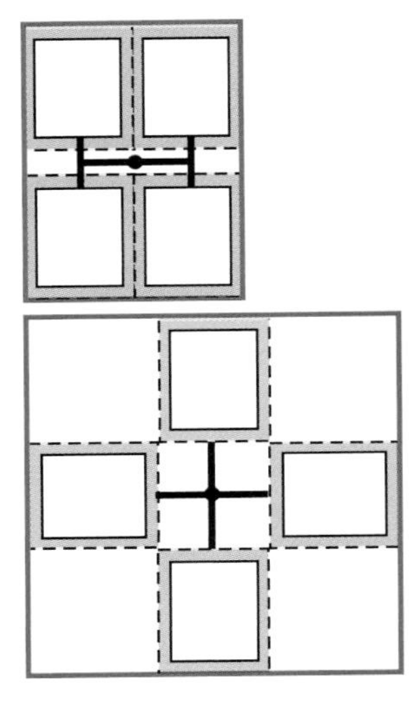

Figure 5.14 Contrast in mold size of "H" vs spoke runner

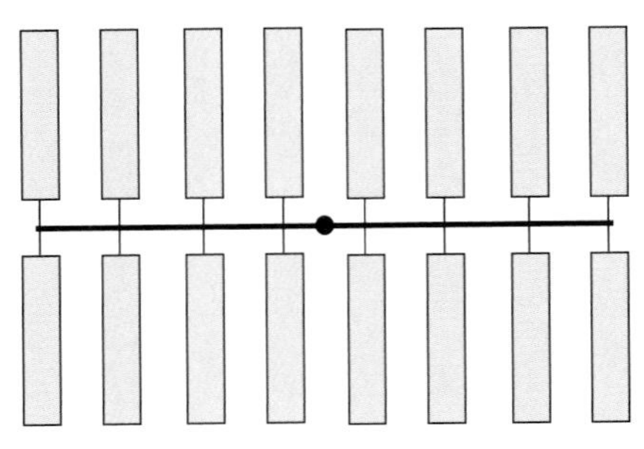

Figure 5.15 Non-geometrically balanced 16-cavity runner layout (commonly referred to as fishbone, ladder, or tree layout). Not recommended for most applications

5.7.1 Geometrical Balanced Runners

Historically, the common "H" or "X" geometrically balanced runner were expected to provide an ideal balance of flow, pressure and melt conditions to each cavity (all runner branches being the same length with symmetrical cross-sectional shapes, see Fig. 5.13 A–D). In fact, this type of runner has been dubbed "naturally balanced". The same concept of a naturally balanced runner system is also often applied to multiple runner branches, which may be feeding a single part (Fig. 5.13E). However, these geometrically balanced runners will normally result in parts being formed with significantly different melt properties and under significantly different processes. The significance of the imbalance is often not understood, but will affect part size, weight, and mechanical properties as well as the molding process. This is true in most cold and hot runner molds.

Despite the geometrical balance there is almost always some variation between cavities nearer to the center of the mold (red cavities in Fig. 5.13) compared to cavities further away from the center. In most cases, this imbalance does not occur until there are more than four cavities in the mold. However, the imbalance is actually dependent on the number of branches in the runner and can occur in molds with as little as a single cavity, dependent on the layout of the runner.

This imbalance is caused by the branching of the melt laminates that have varying shear and thermal histories. The development of this phenomenon is discussed in detail in Chapter 6. The spoke type runner eliminates branching, but at the cost of much larger molds and higher mold build costs (see Fig. 5.14).

5.7.2 Non-Geometrically Balanced Runners

Non-geometrically balanced runners are not recommended for most applications. Though conventional geometrically balanced runners will result in material and filling imbalances, the imbalances created within non-geometrically balanced runners are even worse and cannot be corrected.

Non-geometrically balanced runners are often laid out in a "tree", "fishbone," or "ladder" pattern as seen in Fig. 5.15. These layouts are the most sensitive to process variations and will normally result in the largest variation in filling of and melt conditions in the cavities. However, these layouts are still used today, particularly in high-cavitation, low-tolerance parts. Figures 5.16 and 5.17 contrast a geometrically and non-geometrically balanced runner layout used in a 64-cavity mold. The logic behind using the non-geometrically balanced runner is that it will result in less regrind of the runner and is simpler to machine. In hot runner molds it simplifies machining and may eliminate the need to use stacked manifolds. In addition, some molders have recognized that the geometrically balanced layout was not providing the desired balance it was supposed to deliver.

One strategy in designing the fishbone type runner is to design the primary runner as large as possible. This causes the primary runner to act as a manifold, minimizing the variation in pressure drop between cavities. The problem with this approach in a cold runner mold is that the melt reaching the first gates and cavities will hesitate. If it hesitates too long, it may freeze off. In addition, the long cooling time required for the larger primary runner may increase cycle time. An alternative smaller diameter primary runner will develop more flow resistance to the cavities further downstream on the fishbone. This can help force the melt into the cavities closest to the sprue during mold filling. Although there may be less risk of melt hesitation as these cavities nearest the sprue fill out, as soon as they fill, the melt pressure in them will become quite high and approach a hydrostatic condition. Meanwhile, as each progressive cavity fills, the flow rate in the remaining flow front will increase, thereby further increasing fill pressure. This will compound the high pressure hydrostatic condition in the early filling cavities. This high pressure can result in flash during mold filling. In either of these cases, the size of the cavity and gate will affect how these runners fill. In a hot runner mold, hesitation is less likely to occur, as the runner is already full and melt is already at each of the gates. Otherwise, the imbalanced conditions will be the same as with the cold runner.

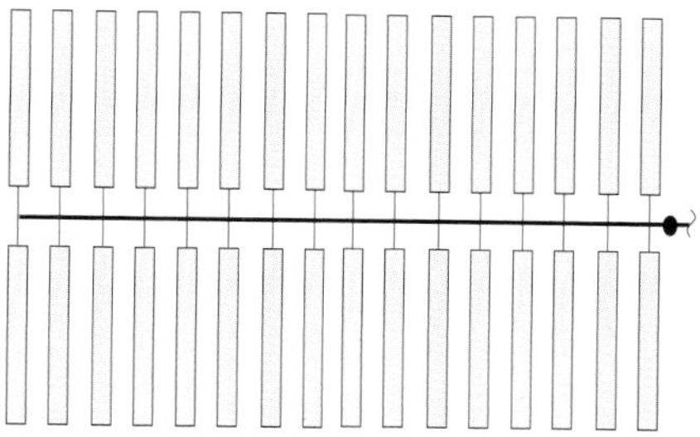

Figure 5.16 Non-geometrically balanced 64-cavity mold. *Note: half of layout shown

Artificial balancing of fishbone type runners has regularly been used in industry. Here, the progressive branching runners along the primary runner are sized in an attempt to balance the filling. This is best done through the use of mold filling analysis software. Artificial balancing is discussed in detail in Section 7.2. (Also see Section 4.2.7.2 for more discussion.)

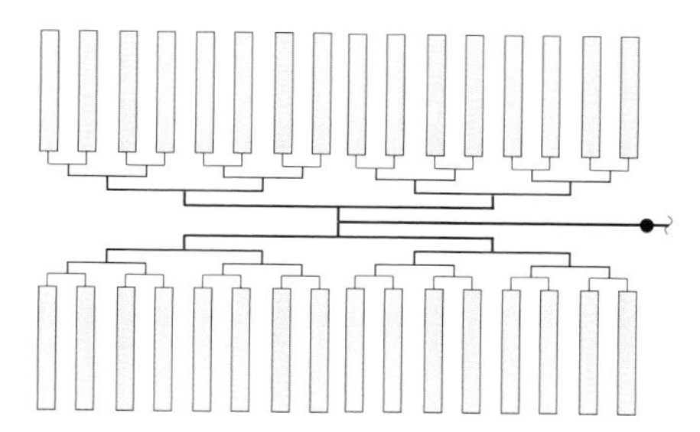

Figure 5.17 Geometrically balanced 64-cavity mold *Note: half of layout shown

5.7.3 Family Molds

Many molders use family molds to make molding more economical when producing low volume parts. The requirements for the runners are different for a family mold than for a multi-cavity mold producing the same parts. It is not as important that the melt entering each of the cavities have the same properties as the parts are no longer expected to be identical. However, it is still important that the parts fill at the same time. Otherwise, first filling cavities will become highly pressurized, over packed, and potentially flash as the flow front accelerates to fill the unfilled cavities.

Traditional family molds are balanced by adjusting the runner diameter. This is best done through the use of mold filling analysis software. With hot runner molds, some molders will attempt to achieve a fill balance by adjusting the temperature at various locations in the manifold and gates. Not only is this not a very effective method, but it is not recommended as varying drop temperatures corrupt the design of the hot runner system resulting in possible leaking and degradation of material from the hotter drops. Sequential valve gating can also be used to fill family molds by providing control and flexibility to fill all parts in the mold equally.

Murray Feick developed a patented fill balancing technique called the Adjustable Thickness Overflow Molding (ATOM) system to help balance the filling of a family mold [8]. The objective of this technique is to make sure that the cavities finish filling at precisely the same time. In order to achieve this, an adjustable auxiliary overflow cavity is utilized. This extra cavity is placed as an appendage at the end-of-fill position of the first-filling cavity of the family mold. If there are more than two main cavities in the family mold, multiple overflow cavities of different dimensions can be used on the cavities that would naturally be the first cavities to fill. An advantage of this extra cavity is that it can be adjusted while the tool is in the machine. It is "tuned" to the proper size by running a series of short-shot trials. This way, if process conditions or molding materials change, the extra cavity can be readjusted and molding can resume.

A particularly effective method for filling family molds with a hot runner system was developed by Synventive and is marketed as the "Dynamic Feed System". Due to its high cost this method would be impractical for most all lower tolerance, low volume production.

References

[1] J. P. Beaumont, J. H. Young, M. J. Jaworski, "Mold Filling Imbalances in Geometrically Balanced Runner Systems," SPE Annual Technical Conference, Atlanta, p. 599 (1998)

[2] B. H. Maddock, *SPE J.*, July 23–29 (1067)

[3] R. S. Golab, "Energy Loss in Screw Transport as Related to Injection Molding Flow Analysis Software," SPE Annual Technical Conference, New Orleans (1993), p. 2746

[4] J. P. Beaumont, R. Nagel, R. Sherman, Successful Injection Molding – Process, Design, and Simulation, Hanser Publishers, Munich, 2000, p. 241

[5] C. L. Tucker, III, Computer Modeling for Polymer Processing Fundamentals. Hanser Publishers, Munich (1989), p. 96

[6] M. Baily, A. Schenck, "Evaluation of Runner Design for Powdered Metal Injection Molding," SPE Annual Technical Conference, San Francisco (2002)

[7] B. Martonik, A. Schenk, "Powder Metal Injection Molding – The Effect of Runner Design on Material Properties and Filling Imbalances," Annual Technical Conference, Nashville (2003), p. 3358

[8] M. Feick, "Balancing Cavities in Family Molds," *Injection Molding Magazine*, August 1997, p. 71–72

6 Filling and Melt Imbalances Developed in Multi-Cavity Molds

Chapter 6 presents not only the negative impact of mold filling imbalances, but also reveals the *mold gremlin* that has haunted the plastics industry for decades. Mold filling imbalances may be the most costly and misunderstood source of problems in injection molding today. The primary source of this problem is the uncontrolled distribution of non-symmetrical melt conditions developed in the melt as it flows through, what has been called, the "naturally balanced runner." Its deceptive geometrical balance has masked this runner as the source of some of the most mystifying and disabling problems that molders encounter every day.

The Mold Gremlin has been revealed within the industry standard "naturally balanced" runner. It results in the development of both melt and process variations/imbalances within a mold. This melt variation may be the biggest variable in injection molding today.

It is a simple matter of economics that by increasing the number of quality parts produced during a given molding cycle in a given mold, that these parts will cost less to produce. This assumes that each part produced in each cavity of the multi-cavity mold will be almost identical. However, despite almost identical cavities, runner flow paths, and cooling significant variations often exist between each of the molded parts. These variations significantly limit the benefit of a multi-cavity mold.

Right now, many molders are just realizing that the industry standard geometrically balanced runner layouts, commonly referred to as "naturally balanced", are the dominant source of the variation found in multi-cavity molds. Non-uniform melt conditions developed by the runner not only create variations from cavity to cavity, but they also create melt variations within a given cavity. These variations will affect shrinkage, warpage, and mechanical properties of a molded part. Though the effects of the melt variation are most recognized in molds with eight or more cavities, the phenomenon actually creates variations in almost all four-cavity molds as well as in molds with as few as one cavity. The non-symmetrical melt conditions universally effect cold runners, hot runners, conventional single material molding, gas assist injection molding, co-injection, microcellular, structural foam, and two-shot molding. Understanding and recognizing the development and distribution of these non-symmetrical melt conditions is fundamental to the evolution of the molding industry.

6.1 Source of Mold Filling Imbalances

6.1.1 Imbalances Developed from the Runner

Mold filling imbalances have always been recognized with fishbone, or ladder, type runner layouts (see Fig. 6.1). The variation in flow lengths between the two cavity groups shown here will result in a readily anticipated filling variation. Often, the parts formed off the branches nearest the sprue will initially fill with a faster fill rate, and with less pressure than the branches downstream. During packing stage, these cavities nearest the sprue will see the highest pressures. However, under certain circumstances (slow fill, restrictive flow channel), the opposite can occur due to filling hesitation at the early cavities. In either case, this runner design will result in a very narrow process window and the production of distinctly different families of parts in the same mold. With the introduction of injection molding simulation in the mid 1970s, the analyst was provided with even more insight into the resultant variations in pressures, melt temperature, shear stresses, and shear rates that developed between the various flow groups in the fishbone type runner.

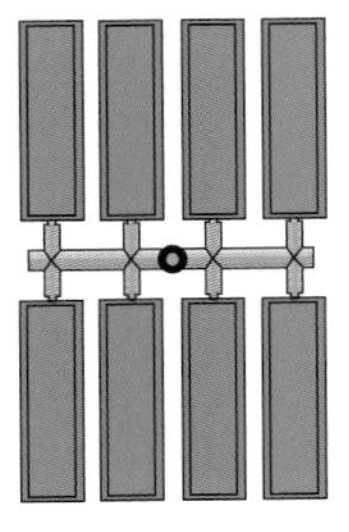

Figure 6.1 Typical fishbone/ladder runner layout

In Fig. 6.1, two flow groups can be distinguished: the four cavities closest to the sprue represent one flow group and the four cavities downstream form a second flow group. A twelve-cavity fishbone type runner would have four flow groups, and a sixteen cavity runner would have eight flow groups. Each of the flow groups would produce variations in the molded product and narrow the process window.

In response to the obvious imbalances resulting from these designs, over the past several decades most of the industry adopted geometrically balanced runner layouts as a near-standard (see Fig. 6.2B). These were perceived to provide improved consistency between cavities in most multi-cavity molds. The processing difficulties and product variations, which are still found with these, were assumed to be related to issues such as variations in cooling, steel variation, and mold deflection. The industry's acceptance of these layouts resulted in them being referred to as "naturally balanced" runners. The popular mold filling simulation programs, which use 1-D (one dimensional) modeling techniques for runners, confirmed the perception of a naturally balanced runner. These programs predicted a geometrically balanced runner would provide uniform fill and melt conditions to each cavity. However, these programs were not, and still are not, capable of recognizing the shear-induced imbalance developed in "naturally balanced" runners nor its contribution to imbalances in fishbone layouts. The industry pretty much saw what they wanted to see, which was that a geometrically balanced runner system provided an ideal runner. It was not until 1997 that the first publications were released, which pointed to major flaws in these geometrically balanced runners and the simulation programs [1].

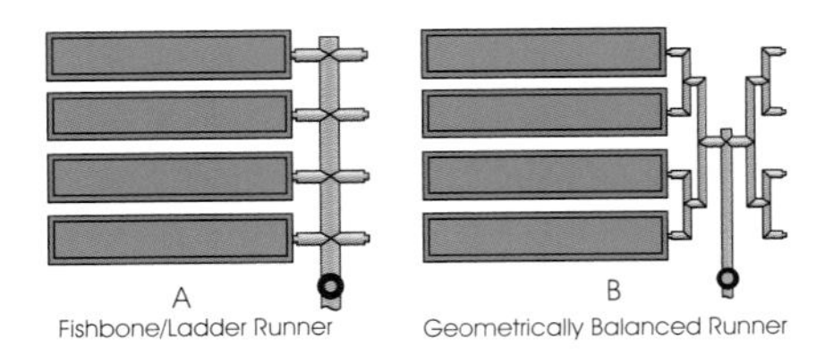

Figure 6.2 Typical fishbone (A) and geometrical balanced (B) runner layouts

It is now well documented in numerous publications that the conventional geometrically balanced runners create very distinct flow groups, similar to those in the old style fishbone layouts. The existence of these flow groups result from shear and thermal variations created across the laminates of the flowing material in a runner. As the runner branches in the geometrically balanced runners, the nature of the laminar flow causes melt property variations across the flow laminates. These laminates with varied melt conditions are then separated into the downstream branches of the runner. The deceptive geometrical balance of these runners had hidden this flaw, which is probably the single most costly and damaging phenomenon effecting the plastics injection molding industry today.

All the developments during the past several decades in new plastic materials, injection molding machines, machine control, process monitoring, processing techniques, and computer simulation cannot overcome the shear-induced melt variations developed in the runner as it flows through the mold. These issues must be addressed within the mold. Due to the significance of shear-induced and flow-induced melt imbalances, the majority of this as well as the next chapter are dedicated to their discussion.

6.1.2 Imbalance Caused by Non-Runner Layout Issues

There are a number of factors other than the runner that can cause variations in parts produced within a multi-cavity mold. As most of these are more obvious, they are also commonly falsely blamed for imbalances created from the shear-induced variations in the runner. Though in some cases their contributions to imbalances can exist, their effects are generally small compared to those caused by the runner.

The three most common contributions to mold filling variations not caused by the runner are variations in mold steel, variations in mold cooling, and mold deflection. The most dominant of these are the mold steel variations, in particular variations in runner and gate cross sectional areas. Not only is the actual gate opening diameter critical, but just as important is it that the gate land geometry be identical from cavity to cavity. Though cavity dimensions are often toleranced to as little as 0.0002 in. (0.005 mm), runners and gates are often built with no specified tolerance. As a result, mismatched runners, sloppy alignment, and design can often cause significant variations. Detailed discussion of these issues is presented in Chapter 8.2. Variation in mold temperature is most commonly blamed for the classic imbalance patterns developed by shear-induced melt variations in the runner. Common theories are that the cavities near the center of a mold run hotter than cavities further away owing to the thermal load from the high concentration of cavities, the sprue, and the runner (see Fig. 6.3). Because the center cavities are warmer, it is expected that they will fill more easily than the outer cavities. Though this temperature imbalance may sometimes exist, its effect on filling and product are generally minimal. Mold temperature variation must be extreme to have any effect on filling. Interestingly, the industry had widely accepted this theory as the primary cause of mold filling imbalances, despite its weak theoretical base and lack of any research to support it. However, the acceptance of the theory was probably the single most significant contributor to deflecting detection and understanding of the less obvious shear-induced imbalances developed in runners.

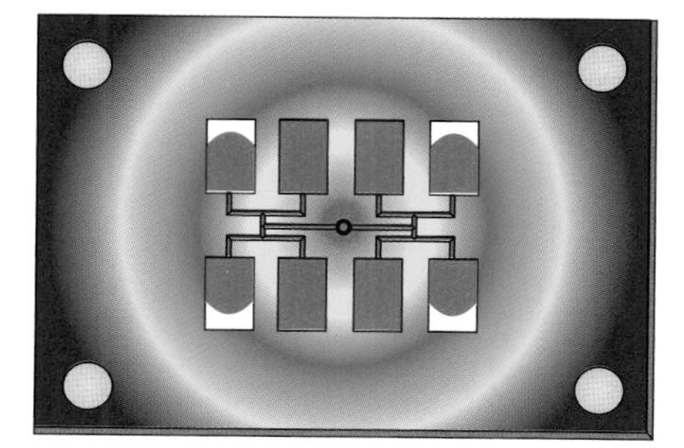

Figure 6.3 Theoretical high center thermal load in an injection mold

Mold deflection can also contribute to a patterned product variation in multi-cavity molds. The high injection and packing pressures within a mold can cause deflection of the mold and even the surrounding mold platens. Maximum deflection, causing opening of the mold's parting lines, is expected to happen in the center of the mold. This could cause cavities in this region to fill first and for the melt to flash in the center region of the mold. In addition, as the melt in the cavities freezes, pressure will dissipate and the deflected mold will return to its normal position. This elastic mold closing would cause the parts near the center to be packed out better. The closing parting line would act to *coin* the parts. The resulting variation in mold filling and product, from the center to the outer regions of the mold, follows a similar pattern as the one that might occur from the cooling variations discussed earlier. Despite the possibility of mold deflection creating product variations, as with the cooling, these effects are generally negligible in well-designed and constructed molds. The theoretical pattern of product variations, which could result from mold deflection and cooling variation are very similar. The fact that they both mimic the pattern resulting from runner-induced shear imbalances shrouded the major source of product variation in multi-cavity molds.

The following table summarizes the cause, identification, and solution of the above three non runner-induced filling variation. The table also presents four additional factors that can contribute to mold filling imbalances.

Table 6.1 Non-Runner Induced Filling Variations, Their Cause, and Prevention

Cause of Variation	Characteristics of Variation	Solutions to Variation
Variation in mold steel	Results in a continuous patterned product variation other than that created by the runner or cooling layouts. Cavities, thin regions, or restrictive runners or gates, will consistently fill differently than other cavities.	Often, this type of imbalance results from inconsistencies within the runner and the gates. Tolerances are generally specified for cavity components and provide good control. However, this is not the case with cold runners and gates. Variations in gate sizes, alignment along the length of the runners, as well as alignment of the two halves of a full round runner should be checked. Tolerances of +/– 0.002 in. should be specified for all alignment issues related to the runner. Gates should be toleranced to within 0.001 in.
Mold cooling	Mold cooling variations will have very little impact on mold filling imbalances. This is particularly true at normal fast fill rates. However, mold cooling variations can affect the packing stage and the shrinkage and warpage of the parts. Product variations from poor cooling will follow a continuous pattern if the original cooling design was flawed, or develop when a circuit becomes blocked. Cooling variations can affect part size, weight, and shape and will follow the relative variations in mold cooling.	All cavity inserts should have direct cooling (coolant should be in direct contact with the inserts). Coolant delivery to each cavity should be at the same flow rate and should be within 3 °C. Coolant flow rate should be turbulent with a Reynolds number of 10,000 or more. Flow rate through individual circuits in the mold should be monitored. Variations in flow rate would indicate blocked branches within a parallel circuit or partially blocked circuits.

Cause of Variation	Characteristics of Variation	Solutions to Variation
Mold deflection	Problems resulting from mold deflection are generally similar in pattern to those resulting from fishbone- and most shear-induced runner imbalances. The center cavities will normally fill first. Unlike the runner-induced imbalances, these center cavities will almost always be heavier and larger.	During injection and packing, the high resultant forces can drive the two mold halves apart. The center region of the mold is generally the least supported and may tend to deflect. The deflection opens these cavities, and the runners feeding them, more than those around the better supported perimeter. This may even occur when center support columns are used in the ejector housing. The deflection can be minimized by designing the support columns so that they are slightly proud, resulting in a preload, which can better resist the deflection. This deflection is generally minimal with structurally sound mold designs.
Non-homogeneous melt conditions entering the mold from the plasticating unit	The non-homogenous melt conditions will be randomly dispersed between cavities. This will result in a non-patterned, random variation in filling and product.	Most injection molding processes use a generalized screw design. This is in contrast to the extrusion industry, which has recognized that each different material has different plasticating characteristics and therefore, prefers a special screw design for each. Though the nature of the injection molding process is more forgiving of the screw design, improvements to the design can be a benefit. Use of a plasticating screw with a mixing head, or use of static mixers downstream from the screw can improve the melt.
Clogging of vents	The additional air trapped in a cavity resulting from a clogged vent will increase the resistance to the melt filling a cavity. This can affect flow rate and create an imbalance. This product variation develops with the vent becoming damaged from wear or from clogging from material residue or flash.	Provide regular maintenance to the mold to clear clogged vents. Design vents to be on the parting line of the mold for easy service. If non-parting line vents are required, attempt to vent along moving parts, like an ejector pin or slide, which will have a chance to wipe any residue away with each ejection stroke.
Cold slugs	Cold slugs normally result in a random pattern of imbalance. However, in cold runner molds, the cold slug will often affect entire cavity groups fed by a runner branch where the cold slug has restricted, or blocked, flow through a runner. The restriction will commonly result in an obvious no-fill of blocked cavities. In hot runners, the effect of a cold slug from the gate tip is often unnoticed during molding. It can initially block, or restrict, flow through one or more gates, and then open part way through the shot. All cavities may fill, but they were formed under different conditions.	For cold runner molds, the sprue puller should always be recessed to catch the cold slug developed at the nozzle/mold interface. A recessed sprue puller has been found to catch most cold slugs in a cold runner [2] Additional cold slug wells may be positioned at progressive branches of the runner. However, it is likely that if the cold slug got past the sprue puller, it will create some variation in product. Cold slugs developed at the gate tips of hot runners are heavily dependent on the gate tip design. They are most commonly found with semi-crystalline materials. The smaller the gate tip orifice, the more likely the melt in the tip will freeze, forming a small cold slug, which can cause the various nozzles to open erratically.

Table 6.1 *Continuation*

Cause of Variation	Characteristics of Variation	Solutions to Variation
Thermal variations in hot runners	Thermal variations will result from poorly designed systems, failure of a heater or thermocouple, or by operator interference. Most commonly, the cause stems from component failure or operator input. The imbalance in a poorly designed system will be seen from the beginning and follow a regular pattern. Variations in component temperatures will create variations in mold filling and product size, weight, and mechanical properties.	The hot runner system should be designed and built by a credible company with a history of designing successful systems. However, the buyer of the system should critically review the proposed design. High/low temperature alarm control systems can quickly alert the molder to component failures. Operator variations normally result from an operator changing nozzle or manifold temperatures in reaction to an observed filling imbalance. The filling imbalance, which is most likely caused by a non-geometrically balanced runner layout or shear-induced imbalances, can sometime be addressed by varying component temperatures. However, these methods generally create their own variations in material shrinkage, warpage, and mechanical properties.

6.2 Imbalance Effects on Process, Product, and Productivity

The imbalance effect on process, product, and productivity is significant and must be understood by any conscientious molder, particularly those seeking 6 Sigma. The most dramatic impact comes from the little understood effects of imbalanced mold filling and melt conditions that result from shear-induced imbalances created in the runners and the melt and packing imbalance that results in common artificial fill balancing methods.

Given the simple four-cavity mold shown in runner layout "A" in Fig. 6.4. To achieve the specified size, weight and/or shape of the part, the molder will normally adjust the following process conditions;

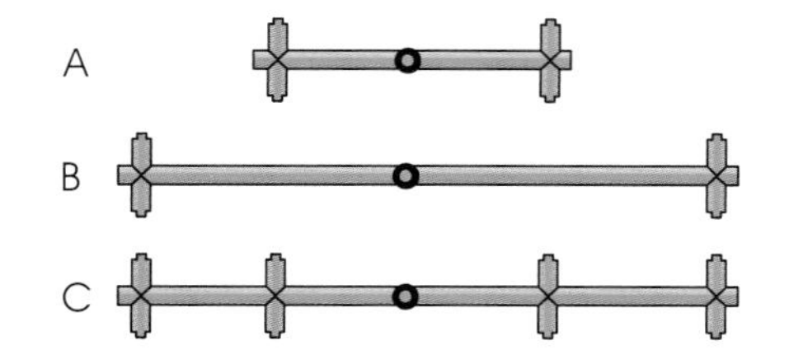

Figure 6.4 An 8-cavity ladder runner (C) is comprised of two different geometrically balanced 4-cavity runner layouts (A and B), each requiring their own unique process to mold the same parts. The ladder runner does not allow for the two different processes to be run at the same time

- Fill rate
- Pack time
- Pack pressure
- Melt temperature

Variations in any one of these parameters will create a variation in the molded part. The effects of these process variables were discussed in detail in Chapter 3. Once the molder has achieved the desired part, he records his process. Depending on how critical the part's tolerances are, the difficulty of molding the part, and the competitiveness of the market for producing the part, some variation in the process may be tolerable. The more demanding the specifications, complexity of molding the part, and competitiveness of the market, the less the process can vary.

Now consider the same molder, trying to produce the same parts with the same material and molding machine but with the cavities positioned as shown in runner layout "B" in Fig. 6.4. In order to produce the parts to specification, the operator will again adjust fill rate, melt temperature, pack time, and pack pressure. Most certainly, the resulting process will be different from that required for the runner layout "A." Now, if the same molder is asked to mold parts in an eight-cavity mold using the layout shown in runner layout "C", you might imagine the increased challenge faced to identify a process, which will produce quality parts from each of the eight cavities. Even if it is possible, you can expect that the process window is extremely small, providing for little drift in any of the process variables or material characteristics. In addition, it should be realized that there will be two sets of products, at least somewhat different from each other. Each group will have been formed at different fill rates, melt temperatures, pack pressures, and pack times.

Let's look more critically at the above case to understand the less obvious variations that are developed. The melt traveling through the fishbone runner layout will first reach the four cavities closest to the sprue. This will be referred to as Flow Group #1. The parts formed in flow group #1 should be very similar. The four cavities further from the sprue, Flow #2, will produce very similar parts, but somewhat different from the parts formed by Flow #1. Unless there is a hesitation effect created by a slow fill rate, or a particularly restrictive gate or cavity, Flow #1 parts will fill first. Given a constant flow rate delivered to the mold by the molding machine, the Flow #1 cavities will initially be filling at a faster flow rate than the Flow #2 cavities. The instant that the Flow #1 cavities are filled, the flow front velocity in Flow #2 will approximately double. The variations in flow rate alone will create variations in the shrinkage of the parts.

Further, consider the shear and thermal history of the melt forming the parts in each flow group. Again, these will be quite different and this alone will create variations in the products produced. Figure 6.5 show the predicted melt temperature variation from a mold filling analysis (though only two cavities are shown, the results are representative of the two flow groups in an eight-cavity fishbone runner). Figure 6.5A shows that the melt temperatures are initially fairly similar when both parts are still filling. Figures 6.5B and C show the increasing difference in melt temperatures developed once the first cavity is filled. At the instant the first-filling cavity is filled, flow rate, shear rate, and melt temperature in the unfilled cavity significantly increase resulting in processing problems and product variations.

In addition to flow rate, shear, and thermal variations in the parts, pressures will also vary considerably. During mold filling, pressure distribution within a mold is a maximum at the entrance of the sprue and decreases to zero at the flow front. Within the first-filling cavity, the

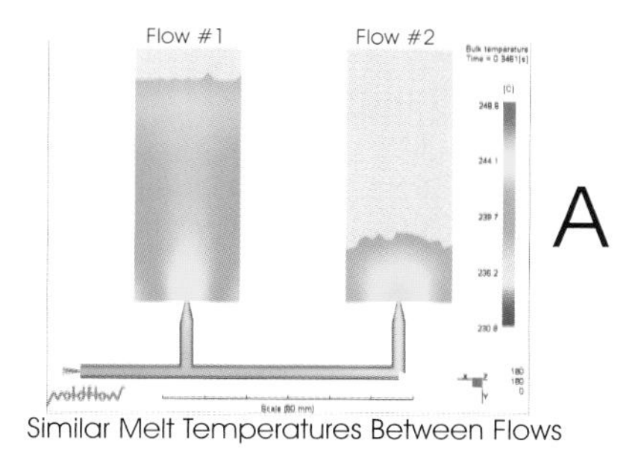

Flow #1
Flow #2
B
moldflow
Flow #2 Melt Temperature Increases as #1 Fills

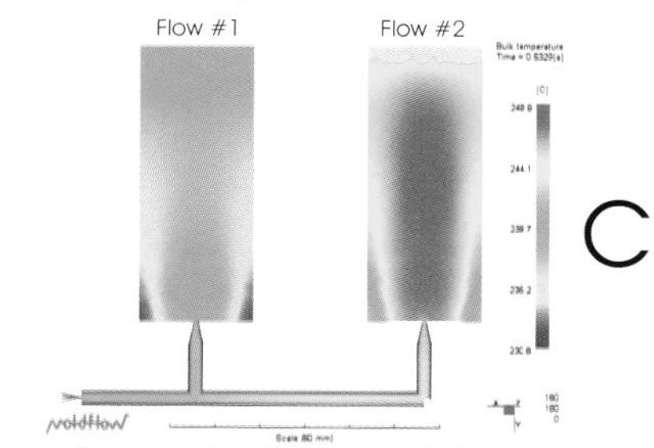

Figure 6.5 Melt temperature imbalances developed from unbalanced mold filling. Note melt temperature spikes in last-filling cavities after first-filling cavities are full

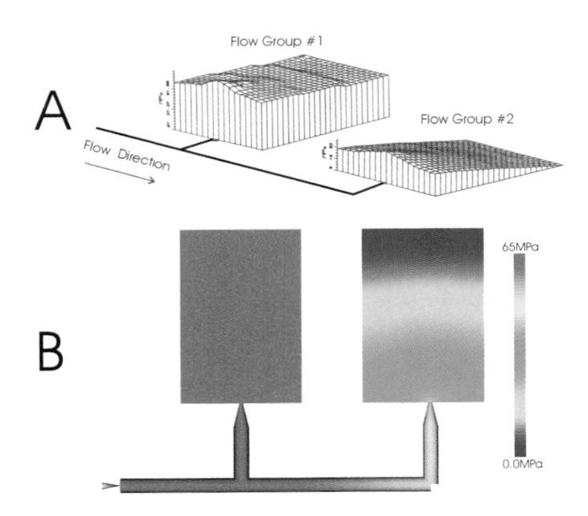

Figure 6.6 Pressure graph (top) derived from mold filling analysis (bottom) showing development of hydrostatic pressure in first-filling cavities resulting from a mold filling imbalance

pressure is a maximum at the gate and zero at the flow front. The instant the cavity fills, the pressure at the end of fill immediately rises to a level much closer to the gate pressure. Essentially, a somewhat hydrostatic condition develops where you have a fluid in a closed vessel with a pressure applied at the gate. The result is a sudden increase in the reactive force trying to open the mold. At the same instant the Flow #1 cavities filled and the graduated pressure disappeared, the flow front velocity in Flow #2 doubled. As fill pressure is directly related to flow rate, the fill pressure will also nearly double. As the runners between these cavities are the same, this sudden spike in pressure will also act on the Flow #1 group further intensifying the pressure within these cavities. The synergy of these effects creates a dramatic effect on the pressure build-up within Flow #1 cavities, while the pressure distribution in the Flow #2 cavities will still be graduated to zero at the still advancing flow front. Without careful control of the flow rate, the first-filling cavities will likely flash while zero pressure conditions still exist at the flow front in the later-filling cavities. Figure 6.6 shows the results of a mold filling analysis of a fishbone runner. Figure 6.6A is a graphical representation of the more traditional output shown in Fig. 6.6B. The output shows the pressure distribution within both cavities at the instant of mold fill. In the cavity nearest the sprue, the pressure has become nearly hydrostatic at something close to 66 MPa. In contrast, the pressure in the downstream, last-filling, cavity ranges from zero at the flow front to a maximum of only about 36 MPa at the gate end of the part.

Regardless of when the operator switches over from fill to pack control, the pressure and flow conditions within these cavities will be quite different. Now, in addition to parts being formed at different flow rates, melt temperatures, and shear conditions, the pressures within the cavities will also be different. Further compounding the resulting variations is the fact that the difference in compression and temperature of the melt in the two flow groups will continue into the packing stages of the molding cycle. Again, melt temperature and flow difference will result in different pressures transferring to the cavities. It should also be realized that these variations will result in variations in the flow rate of the compensating flow that occurs during the packing stages. This in turn will cause the gates to freeze at a different time, thereby creating a variation in our final key process variable – pack time.

Finally, consider the effect of the above variables on cycle time. As parts are formed with two different processes, you can expect that the minimum cycle time for each will be different. This difference will be compounded by the inability to achieve an ideal process for both parts. Both will be molded under compromised conditions. Therefore, the resulting residual stresses in the parts may require them to be held in the mold longer, possibly at a higher temperature, in an attempt to stabilize their size and shape. The cycle time for this eight-cavity mold will now be dictated by the part requiring the longest time to stabilize.

The situation is further complicated by the sensitivity of the imbalance to flow rate and geometry. As mentioned in the beginning of this discussion, cavities closest to the sprue will fill first if they are not affected by hesitation effects. If the flow leading into the Flow #1 cavities meets significant resistance from the gate or cavity, the flow front will hesitate, causing the flow in the runner to the Flow #2 cavities to accelerate. The accelerated flow will spike the melt temperature. The combined high flow rate and higher melt temperature will decrease the viscosity of the non-Newtonian material and allow it to pass more easily through the restrictive geometry of the Flow #2 cavities and gates. Meanwhile, the viscosity of the hesitated material in Flow #1 cavities has increased significantly due to cooling and the low flow rate. The result is that the Flow #2 cavities may fill first. If the injection flow rate is increased, the hesitation may decrease and an improved balanced may be achieved. At a very fast fill rate, hesitation may be virtually eliminated and the balance reversed. This appears as though adjusting the injection flow rate may be able to achieve balance. However, be aware that if this process condition is found, the melt entering each of the cavities will still be quite different, and packing will still be different to each group. In addition, the molder's focus will be on seeking a filling balance rather than identifying the process that creates quality parts in the most economical way.

All in all, this is not a pretty picture. Despite this, many molders are able to successfully produce product with this imbalance. However, the days of producing product under these conditions are disappearing with the ever demanding and competitive nature of the injection molding industry. This is particularly true with the increasing competition from third world countries.

So far, we discussed the conditions developed in a fishbone, or non-geometrically balanced runner. Almost the same conditions exist in most of today's industry standard geometrically balanced runners, once referred to as the "naturally balanced" runners. Interestingly, the imbalance can often be just as bad as in a fishbone runner and, under some conditions, even worse. Early studies found that imbalances between Flow #1 and Flow #2 in geometrically balanced runners could be as high as 1:19. [1] However, the imbalance is normally closer to 1:1.3. This 30% imbalance is still very significant with most products and is in excess of the 5% that many molders are seeking.

The development of an imbalance in a geometrically balanced runner layout is quite different from that in a non-geometrically balanced system and is discussed later in this chapter; however, the effects are the same. Flow groups are developed based on a logical pattern, which will also be presented later. The Flow #1 cavities will fill at different flow rates, melt temperatures, and pressures and will be packed at different pressures and pack times. In addition, the same negative effect on increased clamp tonnage and cycle time will occur.

6.3 Shear-Induced Melt/Molding Imbalances in Geometrically Balanced Runners

The dominant cause of product variations in multi-cavity molds is a result of melt and filling variations developed within the current industry standard geometrically balanced runners. The problem is compounded by the fact that many molders are still not aware of the runners' contribution to product variations or of the significance of this negative contribution. The source of the runner-induced problem is the combination of the stratification of shear and thermal conditions developed across a runner, owing to the laminar flow conditions, and the sorting of these conditions into various regions of a mold.

6.3.1 Development and Stratification of Melt Variations Across a Runner Channel

"Design for Six Sigma (DFSS) compliments the Six Sigma improvement methodology, but DFSS takes one step back, ferreting out the flaws of the product and process during the upstream design stage, not the quality control stage or even the production stage" [3]. Shear-induced melt imbalances are a fundamental flaw in today's injection molds, molding and process design that prevent molders from taking full advantage of this manufacturing method.

The flow of plastic in a runner is quite complex, as the shear rate and temperature, and therefore, the viscosity, vary both along and across the runner channel. At all flow rates, the shear rate is highest along the outermost regions of the flow channel and is zero at the center (see Fig. 3.12). It can also be expected that the shear rate will be highest in the runner and gates of a mold, where the velocity of the melt is normally at its highest.

The high shear rate region near the outer wall has a combined effect on the viscosity. Viscosity in this region is reduced because of

- Its non-Newtonian characteristics and
- The resulting frictional heating.

The resulting frictional heating will cause the melt in these outer laminates to be well above the temperature of the melt in the center of the runner channel. Injection molding simulation has shown that it is common for these outer laminates to be more than 100 °C hotter than the center laminates. This higher temperature in the outer laminates will almost always exist, even in a cold runner in which there is some heat loss to the cold mold wall through conduction. In thermoset injection, or transfer, molding, the frictional heating in the outer laminates is compounded by the heat gained from the heated mold wall. As flow throughout a mold during injection molding is laminar, the variations in shear, temperature, and viscosity across a runner's cross section remain in their relative position as they proceed down the length of a runner, gate, and cavity.

6.3.2 Laminate Separation in Branching Runners Causing Cavity-to-Cavity Filling Imbalances

Though runner-induced melt variations can affect even single-cavity molds, their effects are most commonly recognized when there are more than 8 cavities in a mold. As the polymer melt flows down the primary runner, it develops a highly sheared region around its outer perimeter (shown as cross-section C-C in Fig. 6.7), which represents the conditions in a hot runner. In a cold runner, the conditions would be quite similar, except for a very thin frozen layer developed along the channel wall. When the melt stream is split at a runner branch, the highly sheared (hotter) outer laminates on one side of the primary runner (region "A" in Fig. 6.7) will flow along the left wall of the left branching secondary runner. The low sheared (cooler) center laminate from the primary runner (region "B" in Fig. 6.7) will go to the opposite, right, side of the left turning secondary runner. Similarly, the highly sheared (hotter) outer laminates of the primary runner and the low sheared (cooler) center laminate will create side to side melt variations in the right turning runner branch. The result is that the velocity distribution along the branching runner is corrupted and one side of the secondary runners will consist of material with different properties than the other (cross-section D-D).

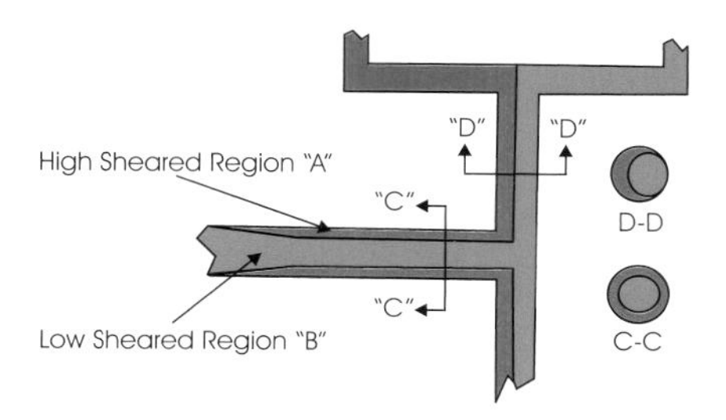

Figure 6.7 Development and distribution of high sheared material in a branching runner

If the melt experiences a second flow branch by flowing into a tertiary runner, a material property imbalance will be created between the cavities being filled by these branches. The highly sheared (hotter) material from the secondary runner will follow the tertiary branch on the inner side of the mold. The low sheared (cooler) material will follow the tertiary branch on the outer side (see Figure 6.8). The result is that the material traveling down each of these branches will be of a different viscosity, temperature, and flow rate. Figure 6.9 shows the results of a 3-D finite element flow simulation of a branching runner. Note the formation of the melt in the shape of a crescent moon as illustrated in Figure 6.7.

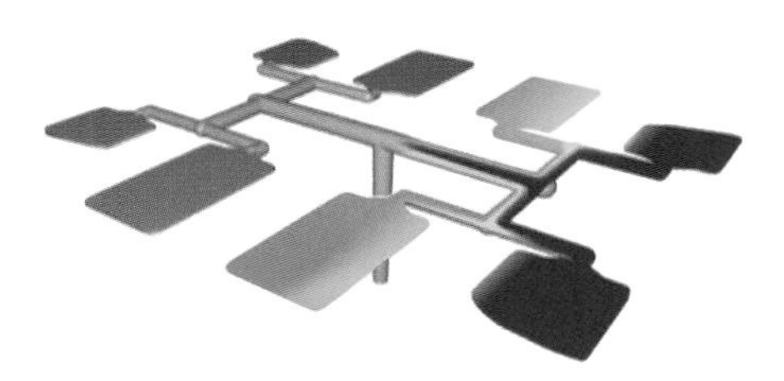

Figure 6.8 Development and distribution of high and low sheared material in an 8-cavity mold and the resultant melt and filling imbalance

This imbalance can be seen by molding a series of short shots. Figure 6.10 illustrates the development of two different flows and the short shot pattern expected in a conventional geometrically balanced eight-cavity mold. Flow #1 cavities (the four closest to the sprue) will normally fill first. If short shot samples are compared, as shown in the figure, the Flow #1 parts will be larger and heavier. If the parts are compared *after* all cavities are filled and packed out, it is sometimes found that the Flow #1 parts are lighter and smaller owing to variations in packing.

The fill imbalance can sometimes be reversed if the Flow #1 melt experiences excessive hesitation when it hits a restrictive gate, cavity, or feature within the cavity. Comparative short shots are best taken at normal processing conditions and when the best filling cavities are approximately 80% full [4]. This method of evaluation is referred to as the *5 Step Process*™ and is detailed in Chapter 15.1.

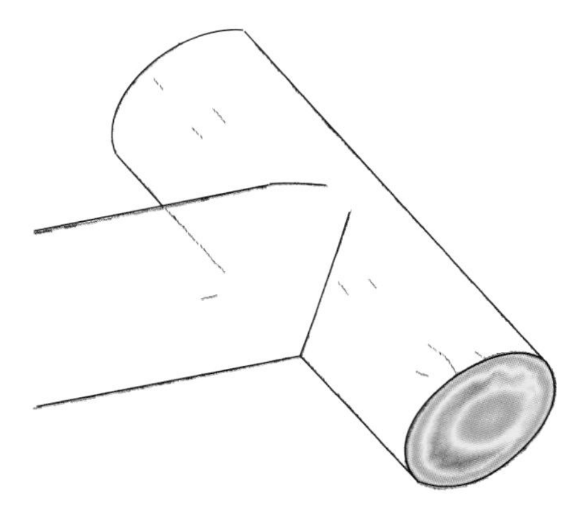

Figure 6.9 Results of a 3D fluid dynamics analysis showing the melt temperature distribution in a branching runner

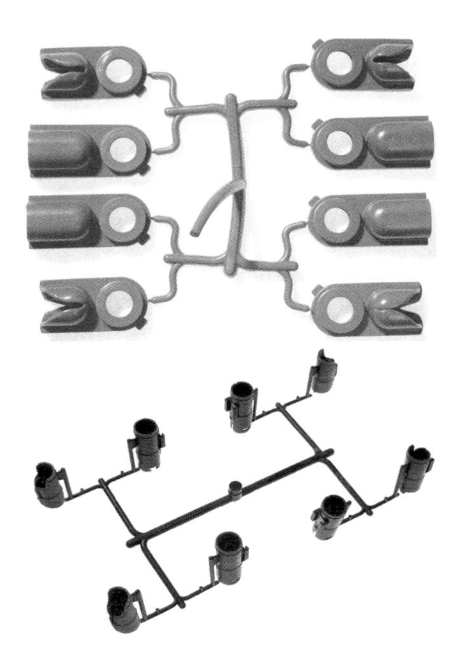

Figure 6.10 Mold filling imbalance developed in two different geometrically balanced runners

The imbalance created in the runner system is affected by runner geometry, material, and process. In general, a larger diameter runner decreases the variations in shear that develop in the runner. The result is that the larger the runner diameter the less the imbalance. However, this method of reducing the shear imbalance creates its own problems in both cold and hot runner molds. In a cold runner mold, the larger diameter increases the amount of material that must be recycled and potentially increases the required cooling time. Most plastic parts are relatively thin and take only a few seconds to cool. Although the runner does not need to be cooled to the same degree as the molded part, it still needs to be sufficiently frozen so that is does not stick in the mold or stick to the molded parts when the runner is ejected with the parts. In addition, the ejected runner must then be dealt with. This can include regrinding, controlled metering with virgin material to be fed back into the process, storage, and reselling. In many cases, a certain amount of the material can be fed back into the process from which it originates. Although the ability to reuse the material may seem attractive, many hidden problems may arise from it, such as controlling its reuse and its related variation in the molding process, the possibility of contamination during handling, increased need of granulators and maintenance, storage of unused material, increased labor, particulate fines, proper mixing into the virgin material, cleanliness, and so forth.

Hot runner molds commonly have larger diameter runners than cold runner molds because there is no concern for runner regrind or effect on cooling time. Despite the larger runner cross-section, these systems are still known to regularly cause shear-induced runner imbalances of over 20% [5]. Contributing to their imbalance may be the fact that, unlike in the cold runners, there is no cold channel wall that might help to offset the frictional heating developed in the melt's perimeter. In addition, there is a limit to the acceptable increase in the diameter. Increasing channel diameter increases the melt's residence time and decreases its flush rate.

Fill speed is the most significant process variable affecting the imbalance. The imbalance is most dramatic at very slow and very fast injection rates. At low fill rates, the shear difference is decreased but the filling is affected by other factors such as thermal hesitation and the material's elastic characteristics that increase the imbalance. Increasing fill rates may reduce the imbalance initially and then increase it again as the fill rate continues to increase. Despite this sensitivity to fill rate, there is very rarely a fill rate that achieves a balance. Even if one is found, it is even rarer that the fill rate is one that will produce dimensionally and cosmetically acceptable parts. Also, it should be realized, that as discussed with the non-geometrically balanced fishbone runner, a fill balance achieved by this means does not eliminate all of the material and pressure variables that exist, nor does it allow the operator to tune the process for an optimum product and productivity. Rather, the efforts are focused on trying to achieve an elusive and deceptive fill balance.

Figure 6.11 shows the results of a study contrasting the sensitivity of various materials to shear-induced imbalances. The materials included a nylon, PBT, PPS, and acetal. Each of the materials was molded in an eight-cavity mold according to the *5 Step Process*. The mold cavities were a small rectangular plaque, measuring 25 mm wide by 50 mm long by 2.5 mm thick. A representative short shot of this study is shown in Figure 6.13. An injection rate sweep was conducted for each material to see how sensitive the imbalance for each material was to injection rate. Note that the imbalance varies with materials and process. The minimum imbalance of 10% is found at a very slow fill rate. Under the normally higher fill rates used for producing plastic parts, the imbalance ranges from a minimum of 20% to over 50%.

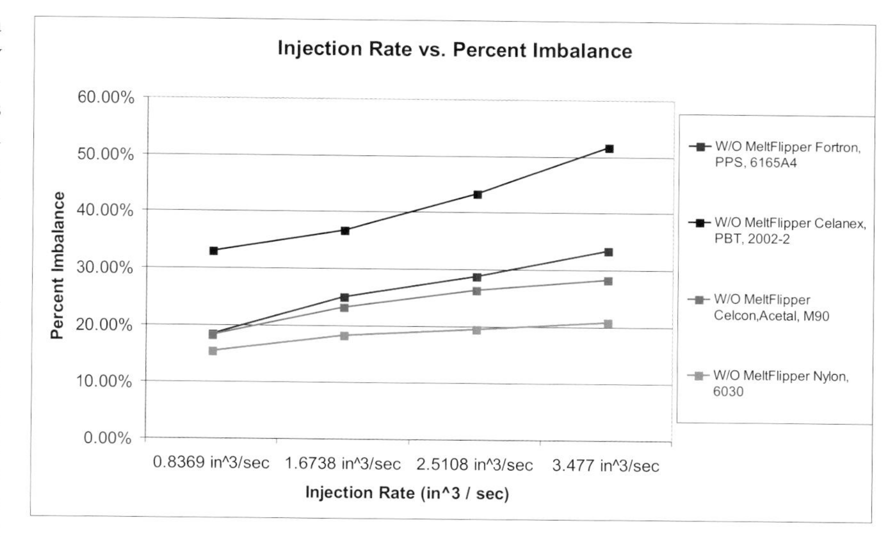

Figure 6.11 Imbalances developed in a conventional 8-cavity geometrically balanced mold

Studies on different materials have found that there are two main material properties that dictate the sensitivity of a material to shear imbalances. These include how viscous a material is and how sensitive the material's viscosity is to temperature [6].

6.3.3 Shear Induced Melt Imbalances in Stack Molds

Stack molds also exhibit a classic imbalance as a result of shear and thermal melt variations developed during flow through its hot runner system. The problem is primarily caused by the extended sprue bar used for these molds. As the melt travels through the extended hot sprue bar, it develops the same melt variation presented earlier in this chapter. As the melt branches into the primary runners, the high-sheared material from the perimeter of the sprue bar follows along the sprue side of the branching runner sections (see Fig. 6.12). This highly sheared material will then be fed into the drops directed back toward the molding machine's injection unit. The result is that the cavities along the first parting line of the stack mold, those nearest the injection unit, will fill under different melt conditions as those along the second parting line. This classic imbalance can easily be seen using the *5 Step Process*™ presented in Chapter 15.1.

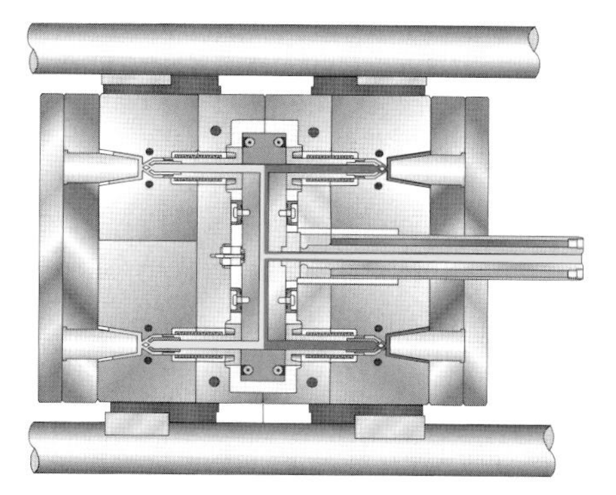

Figure 6.12 Common imbalance developed in stack molds. Cavities nearest the injection unit fill first

6.3.4 Development of Intra-Cavity Variations

Shear induced melt variations developed in a runner not only create variation in products formed by different flow groups within a multi-cavity mold, but will affect the flow and material properties within a given cavity. These "intra-cavity" influences can create product variations even in two-cavity molds. They also can have a significant influence on the formation of a part molded in a single-cavity mold. This shear-induced intra-cavity flow influence is a major contributor to core shift and non-concentricity of cylindrical parts formed in center-gated cavities.

Intra-cavity shear-induced filling imbalances present serious manufacturing difficulties for injection molded products. These imbalances create dimensional and weight variations across the part, which are difficult to diagnose. As with mold filling imbalances, these intra-cavity imbalances are also unaccounted for by today's most popular injection molding simulation software.

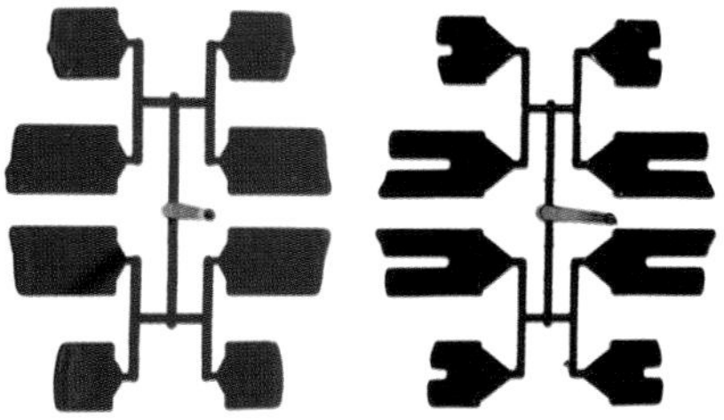

The runner shear-induced intra-cavity imbalances developed in virtually any molded part may be present in single-cavity molds with two or more gates, and two-cavity molds with one or more gates. Although the runner layouts in these molds may appear simple, the melt variations developed in the runner will continue into the part-forming cavities, having similar influences as in the balance of a multi-cavity mold. The high and low sheared laminates will spread to different regions of a part, corrupting the expected filling pattern and distribution of material properties.

Figure 6.13 Two different 8-cavity molds illustrating cavity filling imbalance and side-to-side filling imbalance within the cavities which are flat and have a constant wall thickness

Figure 6.13 presents short shots taken in two different test molds. In each case, the expected cavity-to-cavity filling imbalance can be seen. Flow #1 cavities are filling well ahead of the Flow #2 cavities. Closer inspection reveals the intra-cavity filling imbalances, most evident in the U-shaped cavity. In Flow #1 cavities, the highly sheared material developed from the sprue and primary runner can be seen filling the inner leg of the U well before the outer leg of the same U-shaped cavity. Even the inner extreme edge of the inner leg can be seen to be leading. The same effect is taking place in the rectangular plaque. The Flow #1 cavities are not only filling before the Flow #2 cavities, but the fill pattern across the parts is corrupted by the shear developed from the runners. The Flow #1 flow front is dominated by the shear developed in the sprue and the primary runner, whereas the flow front in the Flow #2 cavities are led by the shear developed in the secondary runners.

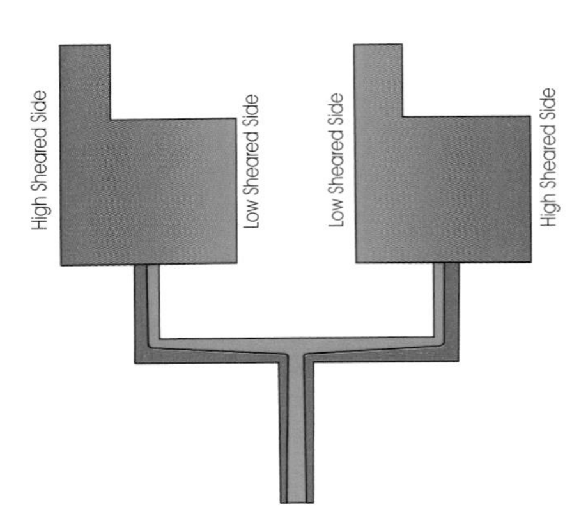

Figure 6.14 Development of asymmetric melt conditions creating product variations in a 2-cavity mold

The development of intra-cavity imbalances, and its potential negative impact, can be evaluated by considering the simple two-cavity mold shown in Figure 6.14. This could represent a parting line injection mold or the conditions in a two-cavity hot runner. The highly sheared laminates that develop as the polymer flows through the primary runner split at the intersection of the secondary runner. Laminar flow dictates that these highly sheared laminates will follow the primary runner side of the secondary runner, while the less sheared laminates from the center of the primary runner will follow the opposite side of the secondary runner. When the melt reaches the cavities, the high shear material will fill one side of the cavity and the low shear material will fill the other. The resultant melt variations from side to side in this part will cause shrinkage variation from side to side. It will also result in the extended feature on the left side of both parts to form differently (also see the cost of melt imbalances discussed in Section 6.6).

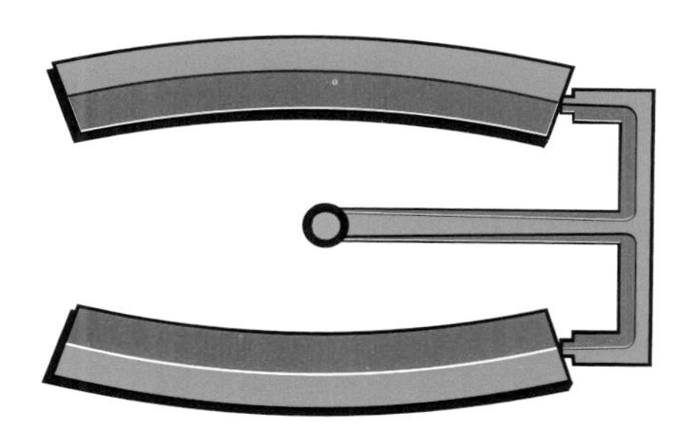

Figure 6.15 Warpage as a result of asymmetric melt conditions entering a cavity

6.3.4.1 Warpage

Warpage is a result of variation in shrinkage occurring in different regions of the same part, which will create a stress between these regions. If the stress can overcome the rigidity of the plastic, the part will warp. Since intra-cavity filling imbalances are a result of variations in a material's shear and thermal history, it can be expected that regions filled by low and high sheared material will shrink differently and thereby potentially cause the part to warp (see Fig. 6.15)

Even the filling pattern in a simple single-cavity edge-gated disk is affected by the runner. Figure 6.16 shows a short shot of a 100 mm diameter, 2 mm thick disk. The filling does not radiate out from the gate as expected. This filling pattern is a result of the high shear developed in the sprue and runner being concentrated around the perimeter of the cavity. The resulting perimeter flow causes the part to warp in a bowl shape.

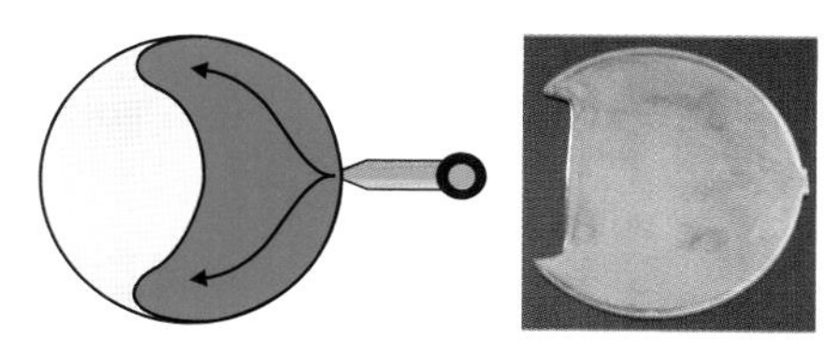

Figure 6.16 Effect of high sheared material on filling pattern in a single cavity, 4 in. diameter impact disk. Figure on left illustrates concentration of high sheared material causing the filling pattern shown in the actual short shot photo on the right

It was previously demonstrated that the high and low sheared material entering a cavity will affect the filling pattern (see Fig. 6.13) These non-symmetrical filling patterns in center-gated symmetrical parts such as test tubes, preforms for blow molding, gears, and fans can result in non-concentricity of these parts.

6.3.4.2 Core Deflection

Core deflection is caused by unbalanced pressures developed from the melt on a core. The location of the gate has a significant impact on core deflection. Figure 6.17 shows two cores with three different gating locations. Gate locations 1 and 2 will both result in high pressure developing on the side of the core near the gate. This will cause the core to bend away from the gate. Gate location 3 is preferred when gating concentric parts. Not only will gate location 3 reduce the potential for core bending, it should also help prevent air traps, weld lines, and non-concentricity. However, despite this apparently ideal center gating location, filling patterns in center-gated parts in multi-cavity molds are almost always unbalanced. Shear-induced melt variations again will create side-to-side filling and packing variations, which can deflect the mold core forming the part.

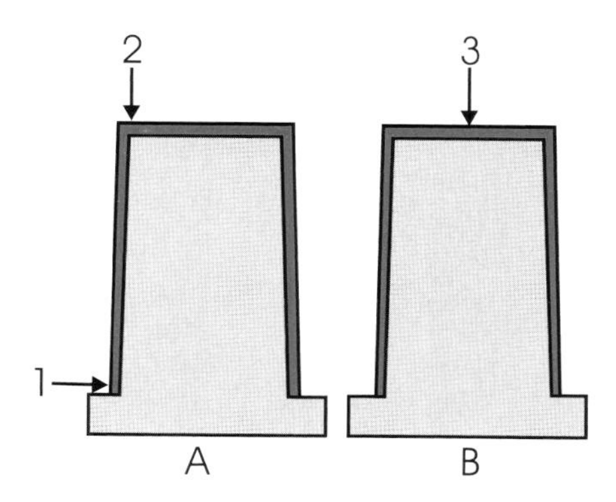

Figure 6.17 Gating locations 1 and 2 will cause core deflection. Gating location 3 should not contribute to core deflection as long as the melt entering the cavity has symmetrical temperature and shear conditions

Figure 6.18 illustrates the development of a side-to-side filling variation that can develop in a simple four-cavity three-plate cold runner or hot

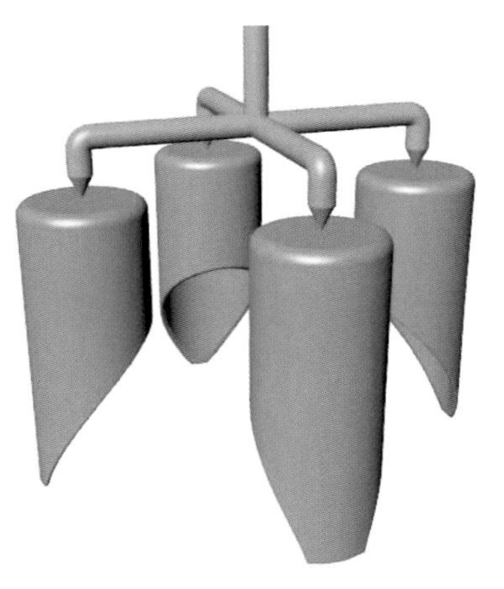

Figure 6.18 Potential intra-cavity filling imbalance, resulting in core deflection, developed in a simple 4-cavity runner layout

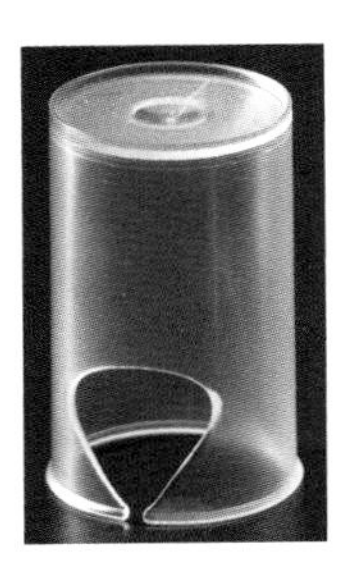

Figure 6.19 Center-gated canister molded in a 16-cavity hot runner. Asymmetric melt conditions resulted in the intra-cavity filling imbalance shown

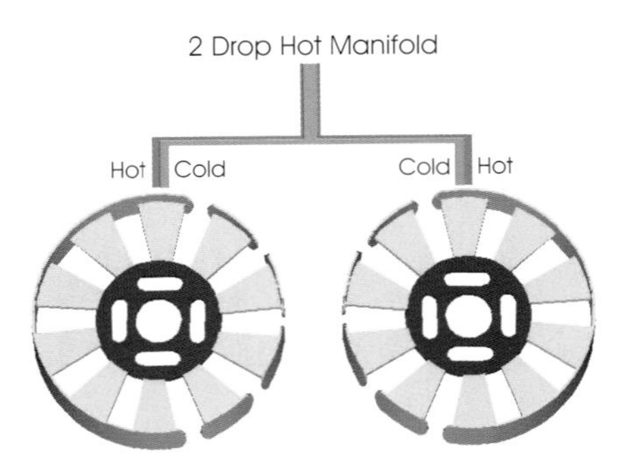

Figure 6.20 Unbalanced filling of automotive fan found in a 2-cavity hot runner mold. The hot manifold layout (top) shows the development asymetric melt conditions resulting in the asymetric filling of the corresponding two cavities (bottom)

runner mold. The highly sheared laminates, developed from the machine's nozzle and sprue, are split at the primary runner. This creates a bottom to top (sprue side to core side) melt variation in the primary runner, which continues into the part forming cavity. This can potentially deflect the core during both the filling and the packing stages and result in variations in wall thickness within the part. This wall thickness variation can then cause the part to warp. The resulting wall thickness variations and warpage can often be traced to be directly related to the expected position of the high and low sheared materials. Interestingly, it is often found that the actual core deflection is away from the low sheared material side of the core. This is analogous to the condition where the last filling cavities in an unbalanced mold can sometimes end up producing the largest and heaviest parts.

Even if the core does not deflect, significant problems can develop from the melt variations entering a cavity. Figure 6.19 shows a small, center-gated canister molded in a 16-cavity hot runner mold. Despite the ideal center gating location, a significant filling imbalance can be seen. The lead flow on the side of the part is fed from the high sheared regions of the runner. The flow in this case actually races down one side, around the flange at the open end, and creates a gas trap along the side of the part.

6.3.4.3 Effect on Concentric Parts (Gears, Fans, and Others)

The continuation of unmanaged shear-induced melt variations into any centrally-gated part can create significant challenges that are commonly misunderstood. This is particularly the case with high precision parts such as gears and fans. Both of these require excellent concentricity. Figure 6.20 is based on an industrial case of a large fan produced in a two-cavity, hot-to-cold runner system. Each drop of the two-drop hot runner is feeding a wagon wheel cold runner with 10 spokes, each directly feeding the fan. Despite the perfectly geometrically balanced runner system, each cavity was filling eccentrically. The half of each cavity toward the edges of the mold was filling before the half in the center of the mold. The resulting eccentric filling and packing caused a disabling imbalance in the finished molded part. The part weight imbalance was severe enough that the part had to be hand balanced using weights following molding. Initially, it was thought that the mold's cores were deflecting outward from the mold, thus opening the flow channel and reducing the pressure drop in those areas. However, it was found that when one cavity was shut off, the parts filled without a side to side imbalance. The layout of the entire hot-to-cold runner system was then analyzed. It was found that the shear from the nozzle and sprue portion of the hot runner was favoring the outside of each drop. This imbalance then followed to the outside of the wagon wheel cold runner and into the parts.

6.3.5 Alternative Theories of the Cause of Mold Filling Imbalances

There are a number of theories as to the causes of the imbalances seen in geometrically balanced runners. Mold cooling and plate deflection are two older classic theories that are probably the biggest contributors to delaying the true discovery of the cause of the imbalance. Though they may sometimes contribute to a molding imbalance, their effects are generally minimal in a well designed mold.

6.3.5.1 Cooling Variations

Variation in mold temperature used to be the dominant explanation for observation that the cavities nearest the sprue were filling ahead of the outboard cavities. The logic being that the center of the mold has a larger concentration of heat from the sprue, runners, and cavities. The heat is more easily dissipated from the outboard cavities along the extremities of the mold. The result is that the center of the mold should be naturally warmer than the outer regions. As a result, the melt would more easily fill the warmer inside cavities.

This theory was first shown to be unfounded in a study published in 1996 [1]. In this study, a purge mold was used with thick cavities at the end of a geometrically balanced branching runner. The 19 mm thick cavities provided virtually no resistance to flow. Under the best process conditions possible, this mold showed an unbalanced condition in which 63% of the material went to the Flow #1 cavities (cavities filled by high sheared melt developed in the runner). With variations in material and process, Flow #1 cavities received up to 95% of the flow. This could be duplicated when the mold was allowed to thermally stabilize for an extended time to assure that all portions of the mold were at virtually the same temperature. The first shots taken after the thermal stabilization would show the same imbalances. In addition, mold temperature changes of 64 °C were shown to have a minimal effect on flow.

This theory can easily be tested by either thermally stabilizing a mold as described above, or by re-plumbing the cooling circuits so that the center region of the mold can be cooled separately from the outer region. When chilling the center cavity group and removing cooling from the outer group it will be found that the filling pattern is virtually unaffected at normal fill rates. *Note* that non-uniform cooling as described here may create other variations during packing and cooling phases that can affect part consistency; however, these variations are independent of the shear-induced filling and packing variations.

6.3.5.2 Plate Deflection

Plate deflection has also provided a common explanation as to why the center cavities fill first. The center of the mold is often the least supported

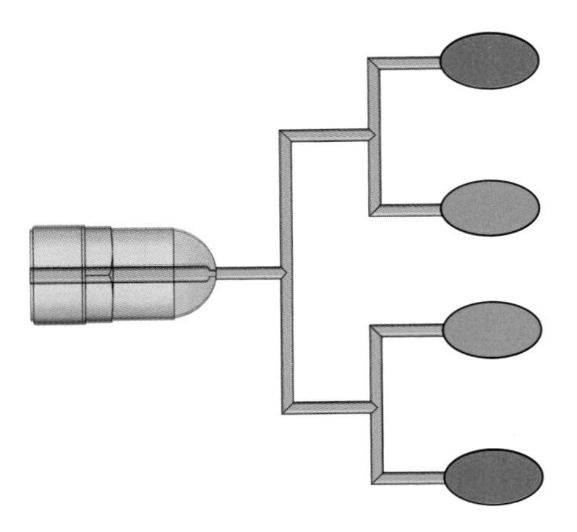

Figure 6.21 Development of two flow groups in a parting-line inject 4-cavity mold

and it is anticipated that the high injection pressure can cause the center of the mold to deflect. This would open up flow channels, creating variations in flow between inner and outer cavities in the mold. Again, this was proven incorrect by the same study mentioned in Section 6.3.5.1. As described, the thick-walled cavities provided virtually no resistance to flow. In addition, the cavities were only partially filled in order to determine the amount of imbalance by weighing the short shot parts. It was determined that there was virtually no pressure buildup in the cavities to create a reactive force. The only force developed to deflect the mold was in the runner. In addition, the cavities were positioned such that the Flow #2 cavities were located toward the open end of a U-shaped support frame. Therefore, it was the Flow #2 cavities that were most likely to open from any plate deflection if it were to occur. Despite this, it was the Flow #1 (closest to the sprue) cavities that were filling first.

In addition, it will be found that when molding with parting line injection molds, like the one in Figure 6.21, the two outer cavities will be the first to fill. The least supported inner cavities will fill second.

6.3.5.3 Corner Effect

Attributing filling variations to a corner effect is a more recent theory, which has developed in response to the realization of shear development in the runner. The corner effect theorizes that most of the shear is developed from the inside corners of branching runners, rather than from the runner perimeter. Therefore, by radiusing the runner intersection the imbalance is supposed to disappear; however, both mold runner designs shown in Fig. 6.10 discredit this theory. Even though both of these runners are radiused at their branches, theoretically to decrease shear at a sharp corner, distinct filling imbalances are still occurring. Anyone who has radiused the inside corners will find that it has virtually no effect because shear is a continuous development that starts back in the injection barrel and nozzle.

6.3.5.4 Melt Pressure as the Cause of Filling Imbalance

It has also been theorized that the mold filling imbalance in geometrically balanced runners is a result of heat generated by pressurization of the melt rather than from shear. This is based on the gas law, which has a temperature rise with compression. This alternative theory reasons that frictional heating is negligible. A recent study by Magenau and Cleveland has shown that the temperature rise in an ABS at 5,000 psi static pressure was only 2 °C, and 4 °C at 10,000 psi [7]. However, when the same material was injected through a runner at an average pressure of 5,000 psi, the melt temperature raised by 15 °C vs. the 2 °C rise under the static pressure conditions at the same pressure. The difference in melt temperature rise at the same pressure is concluded to be from shear and is clearly the more dominate factor.

6.4 Runner Layouts

It is important to recognize the development of imbalances in a runner system. The imbalance of fishbone runners are obvious and are not further discussed here. Rather, this section addresses the less obvious layouts that result in imbalances. The first group of layouts is the most common geometrically balanced runner layouts, illustrating the development of shear-induced melt variations. The second group of layouts is the apparent geometrically balanced systems that, despite their appearance, would be imbalanced even if shear-induced melt variations were eliminated.

It is particularly important to recognize the development of shear-induced imbalances in the common geometrically balanced runner designs used in industry today. By recognizing its development, and the location of the various flow groups within a mold, a molder will be able to more quickly isolate the source of product variations occurring within their mold. The "*5 Step Process*"™ presented in Chapter 15.1 helps to isolate and quantify shear-induced imbalances vs. steel variations within a mold, thereby enhancing a companys quality methods.

6.4.1 Identification of Various Flow Groups in Common Geometrically Balanced Runners

The runners illustrated in Fig. 6.22 are representative of some of the common cold runner layouts used in industry today. They were all once thought to be "naturally balanced". The illustrations are intended to help recognize the development of shear-induced imbalances and the resultant location of the flow groups, or product groups, within the mold. It should be expected that the various flow groups will result in various product groups. Each product group will be formed from material with somewhat different melt conditions. Flow #1 groups will generally be most distinctly different from all the others and are colored in red.

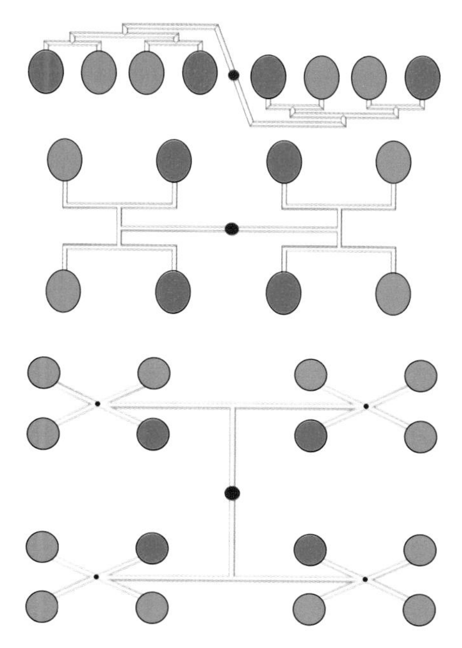

Figure 6.22 Development of multiple flow groups in common geometrically balanced runners. Highest sheared material is shown in red

Figures 6.23, 6.24, and 6.25 illustrate three variations of simple eight-cavity hot runner layouts that might be seen in industry. Each of the three variations shows the distribution of highly sheared material into the branching tertiary runners. The green illustrates the distribution of the highly sheared material developed from the primary runner. The red illustrates the distribution of the highly sheared material that would be developed from the sprue and the injection molding machines nozzle. This can also be quite significant as the flow rate in these locations is twice as fast as in the branching runners. Figure 6.23 is the most common eight-cavity runner layout. Note that the highly sheared material developed in both the primary runner and the sprue end up feeding the inner four cavities closest to the sprue. An alternate design, which is sometimes seen, has the secondary runners branching into an "X" at the end of the primary runner. The effects in this case would be essentially the same as shown here.

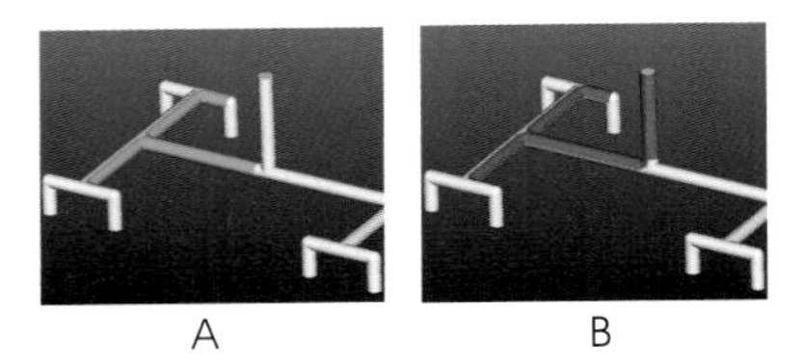

Figure 6.23 Distribution of high sheared material from the primary runner (A) and sprue (B) in a conventional 8-cavity hot runner

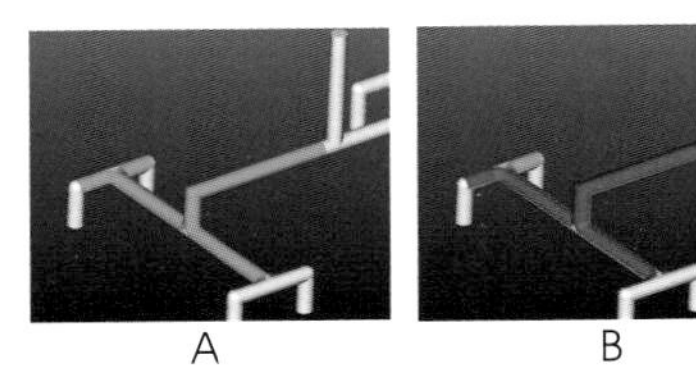

Figure 6.24 Distribution of high sheared material from the primary runner (A) and sprue (B) in an 8-cavity hot runner mold using a level change

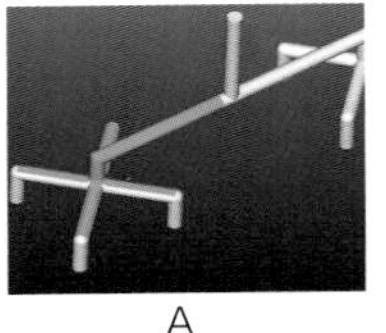

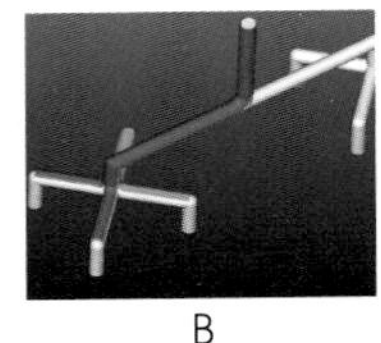

Figure 6.25 Distribution of high sheared material from the primary runner (A) and sprue (B) in an 8-cavity hot runner mold

Figure 6.24 depicts a variation of Fig. 6.23 where the primary and secondary runners are at different levels and are connected by a bridging runner section positioned at 90° to the main runners. In Fig. 6.24A, the highly sheared material from the primary runner becomes uniformly distributed between the two branching secondary runner. However, Fig. 6.24B illustrates how the highly sheared material from the injection nozzle and hot sprue are fed to the outboard cavities (those furthest from the sprue). Here Flow #1 is opposite from the layout shown in Fig. 6.23.

Figure 6.25 illustrates the distribution of highly sheared material in an "X" patterned branching runner at a separate level with the primary runner. The results here are very similar to those with the previous "H" pattern illustrated in Fig. 6.24. The highly sheared material from the primary runner is distributed uniformly to the branching secondary runners. However, again the highly sheared material from the sprue and machine nozzle is fed to the outboard cavities.

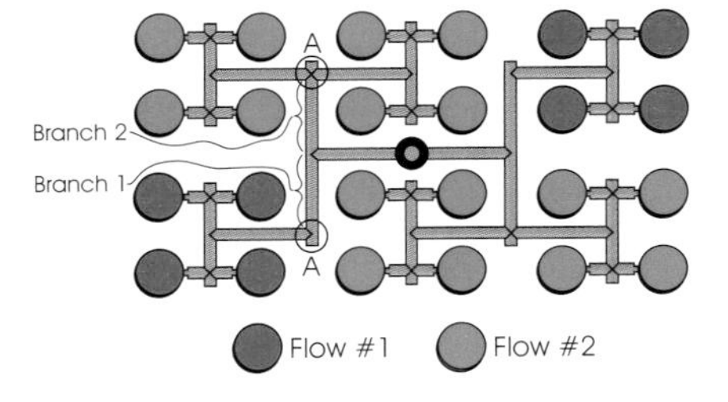

Figure 6.26 Common 24-cavity runner layout illustrating development of two distinct flow groups despite lack of shear-induced melt variations being predicted

Figure 6.27 24-cavity mold filling analysis displaying pressure differences within the two flow groups

6.4.2 Apparent Geometrically Balanced Runner Layouts

Figure 6.26 illustrates a geometrically balanced runner that, despite its appearance, would be imbalanced even if shear-induced melt variations were eliminated. Each cavity is fed by a runner whose lengths and diameters are the same. However, the runner branches into various cavity groups, which would see different flow rates even if the melt were perfectly homogeneous at each of the branches. In Fig. 6.26 the Flow #1 cavities are fed by a runner that experiences three branches, whereas Flow #2 is fed by a runner with four branches. At locations "A" in the figure, the melt feeding the Flow #2 cavities is split, which is not the case for Flow #1. This causes the flow rate to each of the two cavity groups to be different after it passes location "A". The resultant difference in flow and melt conditions will be continued into the cavities resulting in two families of parts. Figure 6.27 shows the pressure distribution from a mold filling analysis after the first filling (Flow #1) cavities have filled (note that the mold filling analysis does not pick up the effects of shear-induced imbalances).

6.5 Effect of Shear-Induced Melt Variations on Two-Stage Injection Processes

Each of the following processes utilizes at least two stages to fill a mold cavity. During the first stage, the mold cavity is partially filled by a controlled amount of plastic material. During the second stage, the remainder of the mold is filled either by the injection of a second plastic material, gas, or fluid, or by the expansion of a gas dispersed in the first stage material. In each case, the amount of material injected during the first stage is critical to maintaining a consistent product. These two-stage processes are even less tolerant to filling variations developed in multi-cavity molds than conventional single-stage processes. As a result, these processes are not commonly used in molds with more than four cavities without the use of additional technologies to create a filling and melt balance. In addition, the effects of intra-cavity imbalances are exaggerated in these processes as discussed in the following.

6.5.1 Gas Assist Injection Molding

There are four primary stages that take place during gas-assisted injection molding. In the first stage, a predetermined undersized volume of plastic is injected into the mold cavity creating a short shot. Gas is then injected either immediately or after a short delay into the hot core of the thickest sections where the material remains molten and fluid. The gas displaces the molten fluid plastic, in mid-channel, driving it forward to fill the unfilled regions of the cavity. In the third stage, the set injection pressure of the gas holds the resin to the cavity surface during part cooling to prevent surface defects such as sinks. Finally, the gas is vented out and the solidified part is ejected from the mold.

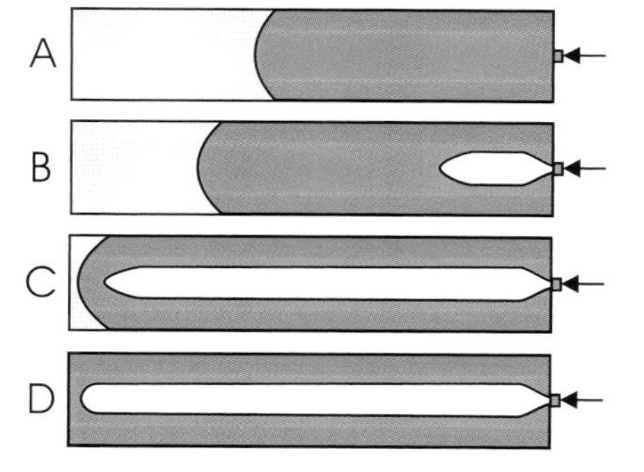

Figure 6.28 Progression of melt and gas filling during the gas assist injection molding process. Cavity is first partially filled with plastic (A), then gas is fed in, pushing plastic into remaining portions of the cavity (B, C, and D)

Figure 6.28 illustrates the two stages of this process in a single cavity. In this example, during the first stage approx. 60% of the cavity is filled with plastic (Fig. 6.28A). During the second stage (Fig. 6.28B–D), a nitrogen gas is injected into the melt, progressively driving the plastic forward. Depending on the method, the gas may be injected from the injection molding machine or from a gas pin placed in the cavity or runner. Regardless of the method, the gas drives the still molten plastic in the center of the flow channel walls forward, filling the remaining portions of the cavity. The amount of plastic first injected into the cavity would be adjusted so that following gas injection the gas would be evenly distributed across the cavity. If less than 60% of the cavity were filled with plastic during the first injection stage, the gas may blow through the leading flow front (see Fig. 6.29A). If more than 60% of the cavity were filled with plastic, the gas could not penetrate as far (Fig. 6.29B). This would result in a thick uncored section at the end of the part. This uncored section would result in excess material being used, a heavier part, longer cooling time, and variations in shrinkage from the thicker part. Therefore, the amount of

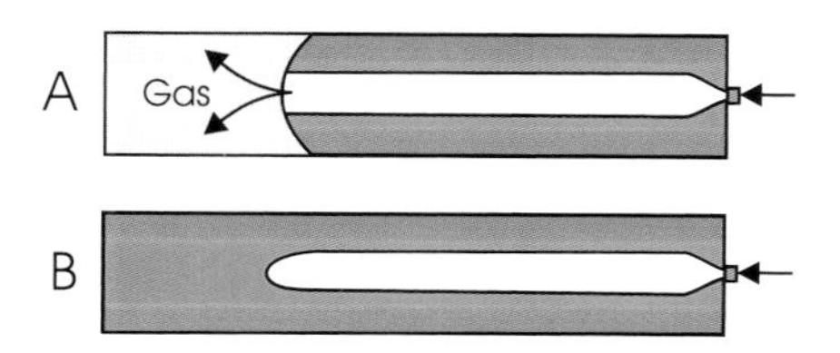

Figure 6.29 Improper amount of plastic fed into the cavity during gas assist injection molding can lead to gas blowout (A) or incomplete gas penetration (B)

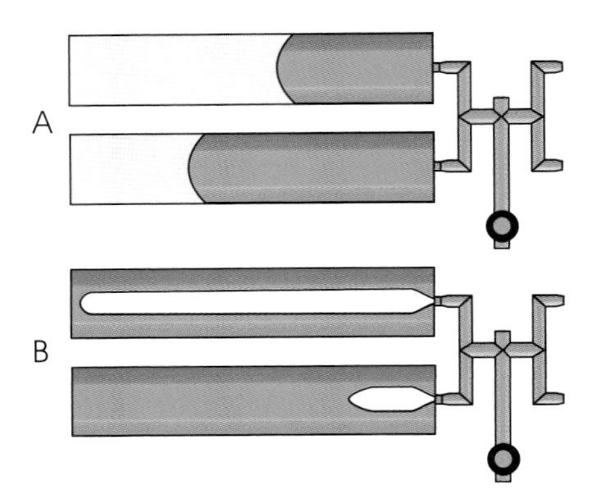

Figure 6.30 The classic filling imbalance developed from shear-induced melt imbalances in an 8-cavity mold (A) will result in the gas distribution shown during the gas assist injection molding process

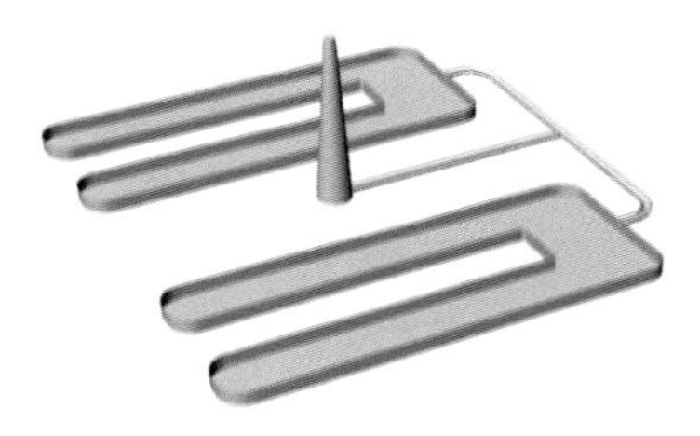

Figure 6.31 Illustration of 2-cavity U-shaped gas assist test mold

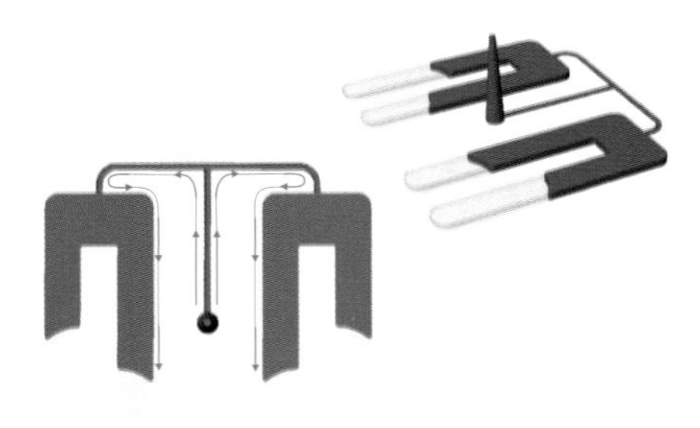

Figure 6.32 2-cavity gas assist mold showing filling imbalances developed from asymmetrical melt conditions developed in the runner (high shear laminates indicated by red arrows). This will result in a similar gas distribution across the two sides of the part as illustrate in Fig. 6.31

material injected into the mold during the first stage is critical to the part quality.

Figure 6.30 shows the characteristic filling imbalance found in a multi-cavity mold and the problem that would result when a gas is injected. Note that due to the filling imbalance, the first stage of the gas assist injection process will deliver different amounts of material to the various cavity groups. This will result in significant variations in gas penetration during the second stage gas injection as previously presented and shown in Fig. 6.29.

The effects of intra-cavity shear-induced variations on gas assist injection molding were presented in a study by Pollack and Miller [8]. This study used a two-cavity test mold with the runner layout shown in Fig. 6.31. The U-shaped parts were approximately 75 mm long, 10 mm thick and each leg of the U shape was about 20 mm wide. Gas was introduced directly into the cavities through a gas pin located just inside the gates. Figure 6.32 illustrates the short shot pattern created during the first-stage injection of the parts. The figure also illustrates how the highly sheared laminates, developed in the melt in the sprue and primary runner, are split at the runner branch and follow into the inside half of the U-shaped cavity. When the gas is introduced through the pins, it can only travel a short distance into the inside leg of the U-shaped part before that half is full. However, the gas is able to proceed much further into the outside legs of the U-shaped parts. The result is a significant variation in the gas distribution between the inside and outside legs of the same part, even though the gates and gas pins are located in the center of these two symmetrical regions.

The intra-cavity material variations, developed from the non-uniform distribution of the highly sheared material in the runner, will influence the gas distribution in most gas assist processes. This explains much of the previously unpredictable distribution of the gas in existing industrial molds. It can be expected that by controlling the position of the high and low sheared laminates, one can also provide control of the gas distribution within a gas assist molded part (distribution control is discussed further in Chapter 7).

6.5.2 Co-Injection Molding

Co-injection molding, or "sandwich" molding, is the sequential injection of two different but compatible plastic resins into a part cavity. It is often a cost effective technique that places a low cost material such as recycled, un-pigmented, off-spec, or foamed resin inside a higher quality outer layer. As with gas assist molding, laminar flow ensures that the primary material at the front of the shot forms the skin of the part while the secondary material makes up the core. It is usually necessary to follow up the secondary material with a small amount of the primary resin so that the surface near the gate area is consistent with the outer surface of the rest of the part.

Figure 6.33 illustrates the progressive steps used in one method of co-injection molding

Just as in gas assist injection molding, a first material is injected to partially fill a cavity. The second injection stage uses a second plastic material rather than gas. Like the gas, the second plastic material travels through the outer frozen skin of the first-stage material, driving the still molten stage-one material in the center, forward. If there is not enough first stage material in the cavity, the second material will eventually blow out of the flow front. If there is too much first stage material in the cavity, less of the second stage material will be introduced. Cavity-to-cavity, or intra-cavity shear-induced, imbalances create very similar limitations for the co-injection process as those presented above for gas assist. First stage cavity-to-cavity filling variations of only 20% can cause over 40% variation in the penetration of the second material.

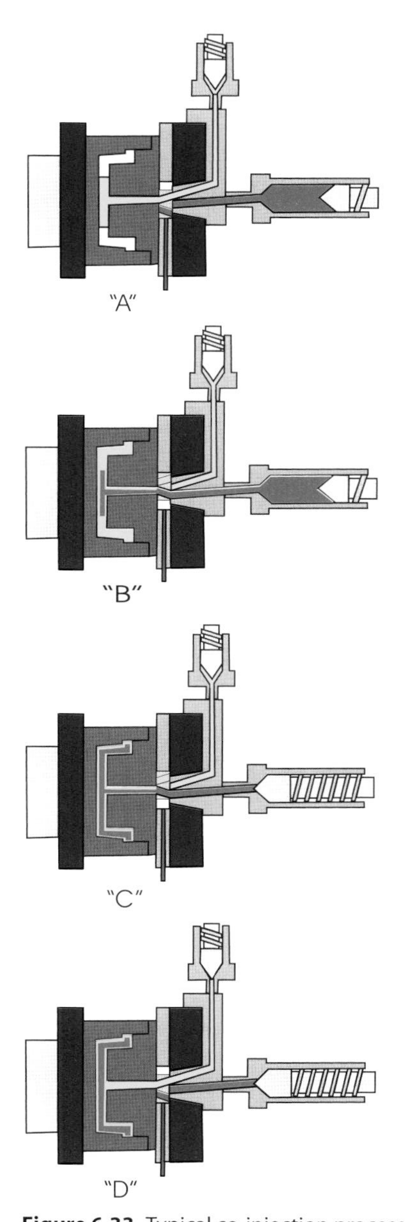

Figure 6.33 Typical co-injection process. A) Material 1 partially fills the cavity; B) Material 2 begins to fill the cavity flowing between the outer laminates of material 1; C) Material 2 finishes filling the cavity; D) Material 1 is re-introduced to finish packing and cover gate region

6.5.3 Structural and Miocrocellular Foam Molding

Some products call for the use of foamed plastic materials. The foam reduces the density of the part and provides such benefits as reduced part weight, decreased material usage, packing out thick regions, and reducing warpage and clamp tonnage. The foaming can result from either a chemical additive blended into the plastic pellets, or a gas, or supercritical fluid, injected directly into the molten plastic in the injection barrel. The chemical additive will go through a chemical reaction as it is heated and sheared with the plastic in the injection barrel. A result of the chemical reaction is a gas, which is dissolved in the molten plastic. The size of the gas bubbles is kept to a minimum under the high pressures in the injection barrel and while the melt is being injected.

Microcellular molding is a process that blends gas (usually nitrogen or carbon dioxide) with plastic resin creating a "supercritical fluid" in the machine barrel. During the molding process, the gas forms highly uniform micro-scale cells and the internal pressure developed from the foaming eliminates the need for packing pressure. The micro-scale cells continue to grow, their size being controlled by the processing conditions such as melt pressure and temperature. Similar to the foaming created by the chemical additives, the expansion of the gas is controlled by the high pressures in the barrel during mold filling.

Regardless of which method is used to introduce the gas into the melt, these processes keep the gas compressed at the high pressures found in the injection barrel, and during injection into the mold. The first stage of the molding process injects the plastic, containing the dissolved gas, into the mold. The mold is either partially filled or entirely filled, but after filling all flow and the pressure from the molding machine is removed. At the now low pressures in the cavity, the gas is allowed to expand. This is the second stage in the fill process and depending on the desired effect, the expanding gas either provides the pressure required to finish filling a partially filled mold and/or provides a pressure throughout the melt as it tries to shrink as it cools. The sustained pressure during shrinkage keeps the frozen plastic in contact with the wall of the cavity while the gas voids grow in the center, non frozen regions. The amount of material in the cavity during the first stage injection controls the final density of the part.

Injection molding with structural foam and microcellular molding processes fall victim to the same shear induced imbalance problems found in gas-assist or co-injection molding. Any inconsistencies in the first stage injection will result in variations in foaming between cavities.

6.6 The Cost of Melt Imbalances

As mentioned earlier, many molders are able to produce parts or meet production requirements even when imbalances exist in a given mold. Part tolerances may have been adjusted, or cycle time may have been sacrificed in order to meet specifications, or the molder puts up with the high scrap rates and the sorting or trimming of parts. However, in today's global market place, this practice is becoming less of an option if a molder wants to remain competitive.

Melt imbalances fly in the face of companies trying to optimize their operations and trying to exercises quality methods such as Six Sigma. Melt imbalances in multi-cavity molds increase spoilage; product variation; cycle time is controlled by the worst molded part; production start-up time is increased; mold commissioning time and time to market is increased; and molders are driven toward producing with a larger number of low cavitation molds – each creating its own variables. All of this leads to increased cost and customer dissatisfaction.

Increased setup time is also a problem caused by filling imbalances. The processor must take great care to find the narrow process window, if one exists at all, that will produce acceptable parts from all cavities of a mold.

Perhaps one of the most underestimated costs of mold filling imbalances is incurred during the mold commissioning, or mold approval, process. All too often, a great deal of time and money is spent sampling and debugging the mold and the molding process to produce acceptable product during initial mold commissioning. Some parts may flash while others have sinks or non-fills within the same shot. The end result is typically a part-to-production, or time-to-market, lead-time increase of weeks to months due to the problems experienced during the mold commissioning stage of the production process. This stage in the product life cycle quickly becomes an expensive and time-consuming issue. The industry is hard pressed to provide molds that start up flawlessly and acceptable parts come out of the first trial. The mold commissioning time obviously increases with cavitation. It has been found that this increased effort is primarily a result of the shear-induced material variations developed in the runner.

Melt imbalances in multi-cavity molds are the main reason some companies limit themselves to using only 4-cavity molds, even when production volumes clearly indicate the use of a higher cavitation mold. This not only limits productively but creates whole new challenges when higher production volumes require use of multiple low cavitation molds. The multiple low-cavitation tools result in higher tooling and part cost, additional required floor space and man-power, additional capital equipment costs (machines, dryers, conveyors, thermolators, etc.), and an additional inherent variation created from running multiple molds in different presses operated by different process technicians.

A telecommunications company reported an annual savings of $ 104,247 (based on press-time alone) when molding with a single 16-cavity mold versus multiple low-cavitation molds [9]. These savings did not even account for additional savings experienced from lower auxiliary equipments costs, operator costs, sorting costs, floor space, or setup time. An additional $ 133,646 per year could be saved on the higher-cavitation piece price versus the lower cavitation tooling piece price. It is important to note that the 16-cavity mold will only be successful if the runner's shear-induced material property variations are eliminated as presented in Chapter 7. The savings documented do not even take into account the lower mold build costs and reduced demand on auxiliary equipment (conveyors, robots, mold temperature controllers, etc.).

Figure 6.34 Electronic connector showing troublesome region requiring tight tolerance for assembly

The electrical connector in Fig. 6.34 was produced by Osram Sylvania in a four-cavity parting line inject mold, requiring metal inserts and using the runner layout shown. It was found that there was a significant (0.2 mm) dimensional variation occurring at location "A," which had a critical specification that could not be met consistently [10]. By adjusting the process, either cavities #1 and #3 could meet specification *or* cavities #2 and #4. However, regardless of process, switching of cavity inserts, and adjustment to cooling, Osram could not produce all four cavities to specification nor could this product pattern be changed.

The cause of the problem can be diagnosed by tracing the shear variation created in the runner into the cavities. Even though the runner should result in a reasonable distribution of melt to each of the cavities, the flow within the cavities was quite different. Figure 6.35 shows a photograph of the runner and cavities. Here it can be seen that the highly sheared laminates in cavity #1 and #3 end up on the left side of the part, away from the critical feature. This shifts the weld line to the right side of the critical feature, opposite to the conditions in cavities #2 and #4.

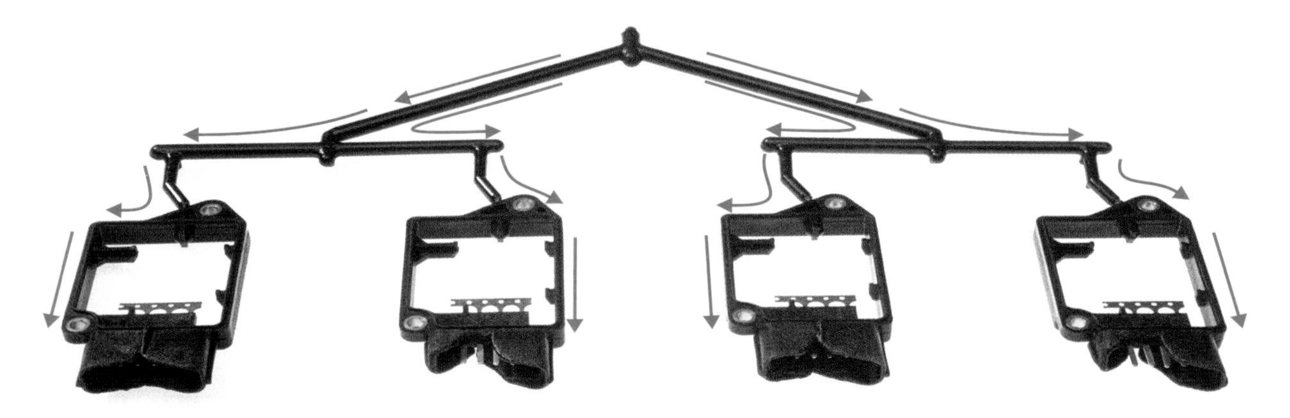

Figure 6.35 Short shot and runner layout showing development of two different molding conditions in the 4-cavity parting line inject mold. The result was that the 1st and 3rd cavities (left to right) were the same but different from the 2nd and 4th cavities

The result of the intra-cavity imbalance was 50% scrap because only one of the two cavity groups could produce parts to specifications. The cost went far beyond the direct costs of over $ 30,000/year. The molds were part of a multimillion dollar automated work cell, which was reduced to half of its capacity. Significant engineering resources were diverted for over a year, trying to diagnose and solve the problem. Also keep in mind that these costs did not cover the increase of required inspection, environmental issues of non-reusable metal-filled scrap, and increased cycle time used to try to address the parts' dimensional problems.

After nearly a year of chasing the problem, one of the engineers recognized the pattern illustrated in Fig. 6.35. Melt rotation technology (presented in Chapter 7) was installed and the product variation was virtually eliminated. Without any other alterations to tooling or process, the 0.2 mm dimensional variation was reduced to 0.05 mm, thereby bringing the parts within specification.

References

[1] J. P. Beaumont, J. H. Young, M. J. Jaworski, "Mold Filling Imbalances in Geometrically Balanced Runner Systems," SPE Annual Technical Conference, Atlanta (1998), p. 599

[2] P. Auell, B. Martonik, "Cold Slug Wells in Injection Molding," SPE Annual Technical Conference, San Francisco (2002)

[3] S. Chowdhury, The Power of Design for Six Sigma, Dearborn Trade Publishing, USA, 2003, p. 37

[4] Beaumont, J., Ralston, J., Shuttleworth, J., and Carnovale, M.: Troubleshooting cavity to cavity variations in multi-cavity injection molds, *J. Inj. Mold. Techn.*, 3(2) (1999), 1–11

[5] J.P. Beaumont, K. Boell "Controlling Balanced Molding through New Hot Runner Manifold Designs," SPE Annual Technical Conference, Dallas (2001)

[6] Beaumont, J. P. and Young, J. H.: Mold filling imbalances in geometrically balanced runner systems. *J. Inj. Mold. Techn.*, 1(3) (1997), 1–11

[7] S. R. Cleveland, A. J. D. Magenau, "Calculating and Quantifying the Development of Shear vs. Pressure Generated Heat in the Plastic Melt During Injection Molding," SPE Annual Technical Conference, Chicago (2004)

[8] Pollack, Thomas, Miller, Garrett, "Designing Runners to Control Gas Distribution in Gas-Assist Injection Molding", SPE Annual Technical Conference, San Francisco (2002), p. 3485–3489

[9] Beaumont, J. P, "New Runner Design Concepts Boost Quality & Productivity", Plastics Technology Magazine, April 2001, 64–69

[10] Sloan, J., "Mold Imbalances Goes With the Flow ... Literally", Injection Molding Magazine, April 1998

7 Managing Shear-Induced Melt Variations for Successful Molding

Chapter 6 revealed the *mold gremlin*, created by shear-induced melt variations, that has been hidden in nearly every mold since the beginning of the modern plastic industry. The chapter dispels the concept of the "naturally balanced" runner. Chapter 7 reveals the solution. The gremlin, reveled to be non-symmetrical melt conditions developed as the melt travels through a mold, creates product variations despite the use of perfectly symmetrical and balanced runners and cavity placements. The phenomenon is universal and its effects can leave as much of a melt variation as the old plunger machines, which were replaced by the reciprocating screw.

In the 1950s, the reciprocating screw injection molding machine revolutionized the molding industry through significant improvements in melt control over the older plunger machines. Today, we are finding that shear-induced melt variations developed in the runner *recreates* much of the melt variations once thought to be eliminated in the injection unit. Managing this mold gremlin may be the next evolutionary stage in molding.

The most broadly used and accepted means of addressing this problem is based on a patented process referred to as melt rotation technology and is licensed as the MeltFlipper®. This technology can be universally applied to both cold and hot runners and is now licensed worldwide. The method may appear to be quite simple but is commonly misunderstood and can easily be misapplied.

Sections 7.1 and 7.2 present static mixing and artificially balancing of runners. These methods have sometimes been used in an attempt to address some of the symptoms of shear-induced imbalances. Sections 7.3 and 7.4 present melt rotation technology.

7.1 Static Mixers

Static mixers provide a relatively simple means of mixing fluids and are commonly used in the plastics industry. The mixer itself is stationary while the fluid passing through it provides the dynamics necessary for mixing. Static mixers are most commonly seen in extrusion and sometimes in the injection nozzle of a molding machine. However, they can be used in hot runners to help improve the melt homogeneity. This can be beneficial in addressing the issues of shear-induced imbalances between cavities in the mold and intra-cavity imbalance issues. Static mixers are not normally offered by hot runner suppliers because past experiences have generally resulted in more problems than solutions.

Generally, static mixers are used in extrusion to reduce problems with color changes and sometimes to improve the homogeneity of the plastic material from the machine's screw. Some of the older and still popular static mixers are the Kenix mixer (by Chemineer, Inc.) and the Koch mixer (by Sulzer Chemtech USA, Inc.).

Figure 7.1 Kenics mixing elements (Courtesy: Chemineer, Inc.)

Figure 7.2 Melt blending from a Kenics mixer (Courtesy: Chemineer, Inc.)

The Kenix mixer consists of a series of spiraling blades, each twisted 180 degrees. Each of the spiraling blades alternate position so that the exit angle of each blade is at 90 degrees to the entrance of a downstream twisted blade (see Fig. 7.1) The melt flowing through each stationary blade is split and rotated by the spiraling blade. As it emerges from the first blade, it is split again by a second blade, which is at a 90 degree angle to the exit of the first blade. This is repeated through a progression of 6 to 8 blades. Increasing the number of blades improves the mixing. By the time the melt exits the last blade, it is significantly mixed. If there had been any streaking from a color change, it would be well blended into the new color and thus be undetectable. Fillers or other material additives would also be well distributed. Figure 7.2 shows the results of two colors being mixed by the Kenix mixer. Note that on exit from the flow channel, the two colors have been blended into one.

These types of mixers have proved to be highly effective in extrusion. They are normally positioned in a removable adaptor at the end of the extruder barrel just prior to entering the die. The removable adaptor provides for easy servicing. If any degraded material develops in one of the slow- or dead-flow areas of the mixer, the assembly of the adaptor and mixer can be removed and cleaned in a fluidized bed.

Many other varieties of static mixers exist today and are regularly used in the extrusion industry; however, most of them are not well suited for use in a nozzle of an injection molding machine. Here, they would cause major problems:

- First, flowing through the mixer would cause a high pressure drop. Even the limited resistance of the Kenix mixer is reported to increase the pressure drop by 600% through channels of equivalent channel length and diameter [1]. To help minimize the pressure drop issue, the mixer can be placed in a channel with an increased cross-section. This may be tolerable with some materials, but the decreased flow rate and increased residence time can contribute to degradation of the material. With time, degraded material can break free and clog the small restrictive gates.
- A second problem is simply the destructive power of the pressures developed in injection molding. The high pressure can be too much for some of the mixing element designs. The highest pressure of the injection molding process is experienced in the injection nozzle and can exceed 30,000 psi, thus causing parts of the static mixer to break off, clogging and potentially damaging the system.

Despite some of the limitations of static mixers in injection molding, they can still provide some benefits if properly designed and matched for a given plastic material. As stated, static mixers are normally applied prior to a mold to improve melt conditions leaving the injection barrel of the molding machine. However, they have also been used to help address the issue of shear-induced imbalances generated in the runner. In recent years,

Husky Injection Molding Systems introduced a static mixer for their hot runner nozzles. The company reports that their patented UltraFlow (see Fig. 7.3) improves the homogeneity of the melt entering a cavity, speeds up color changes, and reduces cavity-to-cavity imbalance. The mixer reduces variations in shear and temperature history, and thereby improves the uniformity of the melt delivered to each cavity. The technology uses a helical groove within an annular flow path where the melt stream is broken up and mixed.

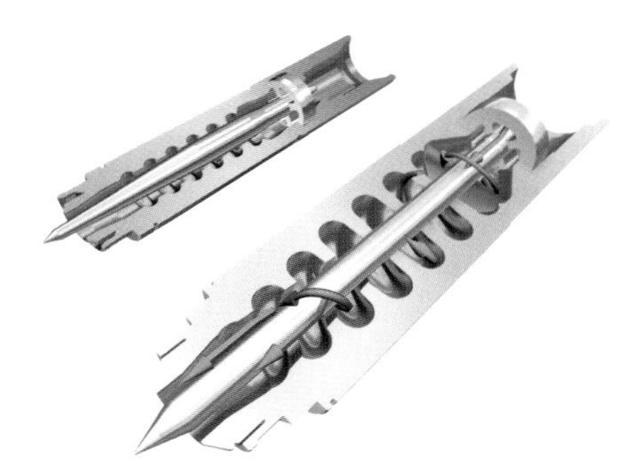

Figure 7.3 Ultra-Flow® mixing devise patented by Husky

7.2 Artificially Balancing

7.2.1 Varying Sizes of Branching Runners or Gates to Achieve a Filling Balance

Both geometrically balanced and non-geometrically balanced runners can be artificially balanced so that the cavities fill at the same time. In both cases this would require that the runner diameters or gates feeding the different cavities vary from one another. Typically, in a non-geometrically balanced mold, the runner branches feeding the cavities closest to the sprue would have a restrictive cross-section. The cross-sections of the continuing runner branches would then become progressively larger as they get further away from the sprue. Artificial balancing of a geometrically balanced runner is less intuitive because runner sizing is dependent on the distribution of the varied melt laminates as explained in Chapter 6.

In either of these cases, a mold filling balance may be achieved only for a very specific and sensitive process. Minor changes in process or material will jeopardize the balance.

It is important to realize that the results of an artificial filling balance are quite deceptive. Despite achieving a filling balance, the melt conditions entering each of the cavities will be quite different as they have been exposed to very different flow and shear conditions. In addition, the varying melt conditions will cause the part to be packed out under different conditions. The runner sizes in artificially balanced runners are developed for a very specific fill rate and melt condition. These conditions change dramatically during the critical compensation/packing phases of the mold filling cycle.

If artificial balancing is required, it is suggested that the ratio of the lengths of the shortest to the longest runner branch be kept as small as possible. In some cases, this will require that the shorter runner branches, closest to the sprue, be purposely lengthened.

In industry, artificial balancing is commonly attempted by modifying the gates, because the gates can be more easily modified with a small hand grinder, while still in the molding machine. However, this approach should be discouraged. Balance achieved through adjusting the cross-sectional size of a short gate is even more sensitive to process and material variations than the longer runner branches. In addition, even if a filling balance is achieved by modifying the gate, it can be expected that the different gate sizes will cause each of the parts to pack out differently as gate freeze characteristics will be altered. Gates should be left for their intended purpose.

7.2.2 Varying Temperatures to Control Filling Balance

A common practice to balance filling in multi-drop hot runner manifolds is to adjust the zone temperatures of individual nozzles or drops. In an eight-cavity hot runner system utilizing a single level manifold, the four cavities closest to the sprue would tend to fill first (Flow #1). Therefore, the processor may attempt to decrease the inner drop temperatures and/or increase the outer drop temperatures to try to counter the imbalance. (The opposite could occur if a two-level, 8-drop hot manifold is used). Though a filling balance may be achieved, this method of changing temperatures can create numerous other problems including extending cycle time. This approach will create variations in thermal expansion, which will compromise the design of the manifold system, possibly causing leakage in the manifold system. This method may also cause degradation of the material if the required temperature rise to achieve a balanced fill is too high. The design intent of any manifold system is to have each drop set at identical temperatures to ensure uniform thermal expansion. Further consider that this method varies melt temperature in order to attempt to balance mold filling. Variations in melt temperature will vary shrinkage, warpage, and mechanical properties of a part. A study presenting the problems associated with this approach is presented in Section 7.3.1.

7.3 Melt Rotation Technology

Melt Rotation is Patented Technology: It should be understood that melt rotation technology is patented and encompasses a broad range of methods and apparatus of which only a few are presented in this book. Presentation of this technology should not be misconstrued as granting rights to its use. Applying the technology without authorized approval, is unlawful, and in most societies, unethical. Despite its apparent simplicity, the reader is cautioned not to apply this technology without licensing or written approval, nor to attempt similar approaches without first having the various patents reviewed by a patent attorney [2, 3].

The simplest method of addressing shear-induced imbalances in multi-cavity molds is through the use of melt rotation technology (e.g., MeltFlipper® technologies, based on a patent by Beaumont and marketed by Beaumont Technologies Inc.) [2, 4]. MeltFlipper technology uses a free flowing non-invasive method that can be applied to virtually any type of hot or cold runner as a means of managing the asymmetric melt conditions across a runner. This universally applicable method is now even being evaluated to solve similar shear-induced melt problems in profile extrusion and blow molding.

The technology was developed to avoid the classic shortfalls of static mixers. It recognizes that a random melt arrangement was not necessary to achieve a homogeneous distribution of melt to each of the cavities in a multi-cavity mold. Rather than the random approach of mixing, this method takes advantage of the structure of the laminar flow conditions existing during injection molding. In its most basic form, it does not eliminate the asymmetric melt conditions in the melt; rather, it manages them so that they are balanced relative to a downstream branching runner. This technology has now been further developed, marketed as MeltFlipper MAX™ [3] to eliminate the asymmetric melt conditions for use in applications such as stack molds and for improving intra-cavity melt balancing.

By balancing the melt within multi-cavity molds, the technology not only improves product consistency but creates a more forgiving process with a larger process window, faster cycle times, lower clamp tonnages, and generally reduced mold filling pressures.

The technology is generally referred to as melt rotation technology as it achieves its objective by rotating the stratified melt variations of the melt stream in a circumferential direction to a desired position. In addressing shear-induced imbalances in a multi-cavity mold, the desired position would be arranging the stratified high and low sheared melt conditions so that they are significantly parallel to a downstream branching runner. By this means an equal ratio of the high and low sheared material would be distributed to each of the branching runners. This would eliminate the imbalance discussed in the previous chapter.

> The variations in melt conditions across a melt channel can be established by a number of things, though the dominate contributors are expected to be from shear developed during flow and/or from thermal exchange between the melt and the surrounding flow channel. For simplification of presentation, this book generally refers to the development of melt variations as shear induced.

This apparently simple method is based on very specific science. Its successful application depends on a clear understanding of the development of the stratified shear and thermal laminates and their weighted distribution across the melt stream. The correct rotation of the melt stream can then be determined and the appropriate method to achieve the rotation can then be applied.

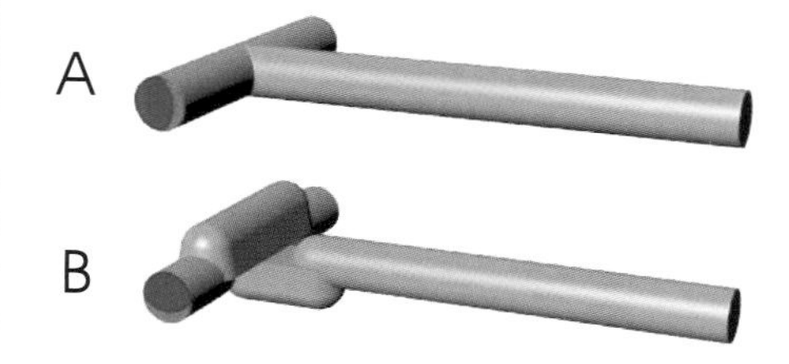

Figure 7.4 Typical runner branch (A) and common MeltFlipper® design (B) at intersection of primary and secondary runner of a cold runner mold

Figure 7.4 shows the effect of one means of melt rotation on the positioning of the asymmetric melt conditions in the secondary runner.

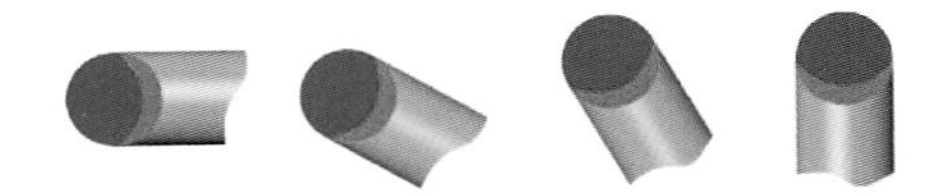

Figure 7.5 Control of non-symmetrical melt conditions by adjusting flow angle (melt rotation technology method)

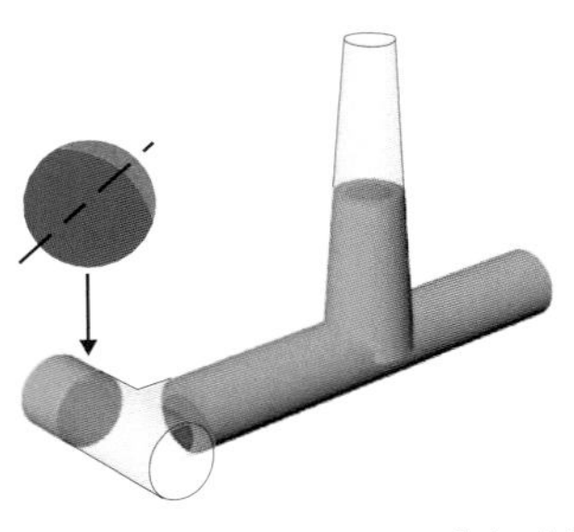

Figure 7.6 Positioning of the high sheared laminates in a runner as a result of shear developed only in the molding machine's nozzle and mold's sprue (does not include effect of the primary runner)

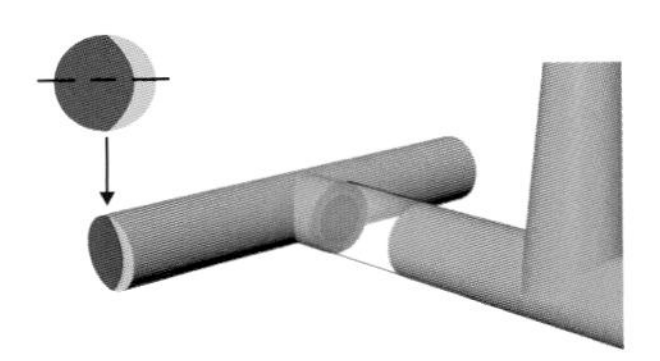

Figure 7.7 Positioning of high sheared laminates in a runner resulting from the primary runner only (does not include the effect of the molding machine's nozzle or the mold's sprue)

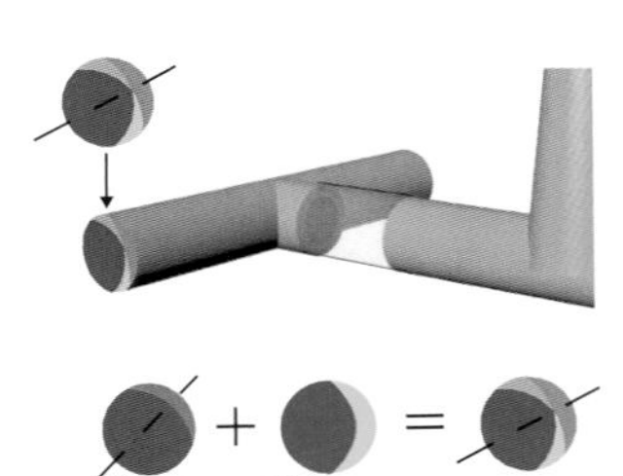

Figure 7.8 Combination of nozzle, sprue, and runner effects on the creation and positioning of high shear laminates in a runner

Figure 7.4A shows the normal distribution of the highly sheared material in a branching runner. Note that there is a side-to-side variation in the melt conditions. If this continues into a further branching runner (tertiary runner branch), the highly sheared material will flow into one branch and the less sheared materials into a second branch. Figure 7.4B shows the repositioning of the melt stream resulting from the MeltFlipper®. Here the high and low sheared materials are now positioned top to bottom; however, they are symmetrical side-to-side relative to a down stream side-to-side branching runner. Melt rotation technology realizes that the asymmetric melt conditions do not need to be eliminated to achieve the desired effect of balancing melt conditions. Rather it manages them to achieve a desirable position of the high and low sheared materials.

Figure 7.5 shows how the position of the high and low sheared material in a runner can be controlled by specifying the angle of flow between intersecting runner sections. This simplified illustration assumes that the distribution of melt conditions in the primary runner is fully symmetrical. However, this symmetric condition generally exists only at the intersection of the sprue to the primary runner, where there has been no prior branching in the runner. The flow angle, direction of flow into the intersected runner, and the prior melt history must be considered and the design adjusted for each case, including each progressive runner branch in a multi-cavity mold.

In order to optimally apply melt rotation technology the position of the asymmetric conditions in a runner should be known. This can be quite complex as it must consider the prior effects of the entire runner including the runner's sprue.

Figure 7.6 shows the effect of the mold's sprue only (shear developed in the primary runner is ignored). Here, the high sheared material is distributed onto the sprue side (top) of the primary runner and then onto the top-inner side of the branching secondary runner. Figure 7.7 shows the effect of the mold's primary runner only (shear developed in the sprue is ignored). Here, the high sheared material developed in the primary runner is axi-symmetric until it branches into the secondary runner. The high sheared material from the primary runner is positioned on the sprue side of the secondary runner. The combined effects of the sprue and primary runner are shown in Fig. 7.8. The result is that the melt feeding a downstream side-to-side branching runner will receive asymmetric melt conditions, which will cause the high sheared material to feed the cavities on the sprue side of the secondary runner.

Figure 7.9 shows two methods of branching runners, which have been used by the molding industry. Design A would be *the standard* in cold runners, whereas both designs might be found in hot runners. Note the resultant positioning of the high and low sheared material in the secondary runner as developed in both the runner's sprue and its primary runner sections. The design in A results in a majority of the high sheared material to be fed to the inside cavities (those closest to the sprue), which will normally cause them to fill first. Design B would have the opposite effect, causing a majority of the high sheared material to be fed into the outside cavities of the mold. In this particular example it appears that Design B will provide a better distribution of the asymmetric melt conditions between the two cavity groups than Design A.

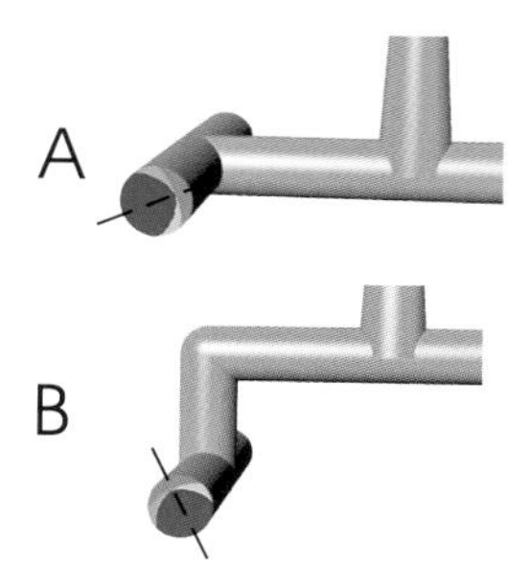

Figure 7.9 The non-controlled positioning of the high sheared laminates resulting from conventional runner designs, results in asymmetric melt conditions with will cause a melt and filling imbalance

Melt rotation technology would be able to control the position of the asymmetric melt conditions such that they are symmetrically positioned relative to a downstream runner branch or part-forming cavity. The effect of the non-managed shear distributions shown helps explain the imbalances illustrated previously in Figs. 6.23, 6.24, and 6.25.

If it is desirable to continue to reposition the melt in later runner branches, similar flow adjustments need to be made at each of the progressive branches. The flow angle required to achieve desirable melt positioning will change at each progressive runner branch as the asymmetrical melt conditions will change after each branch.

Melt rotation technology has been found not only to provide balanced filling in a mold but also to provide a significantly more stable balance. One of the first published studies on shear-induced imbalances used a 4-cavity geometrically balanced runner mold (see Fig. 7.10) [5]. The cavities were oversized purge pockets that provided virtually no flow resistance. The purpose of this design was to isolate the effect of shear-induced imbalances to the runners. Six different materials were evaluated as to their sensitivity to shear-induced imbalance and the effect of injection rate on the imbalance (see Fig. 7.11). Without melt rotation, the imbalance caused between – 8% to + 95% of the material to flow to the inside cavities. The imbalance not only varied between materials, but significantly varied with a change in injection rate. At the very slow fill rates, the effect of thermal hesitation further influenced the imbalance. The study was repeated with two of the more sensitive materials using MeltFlipper inserts at the

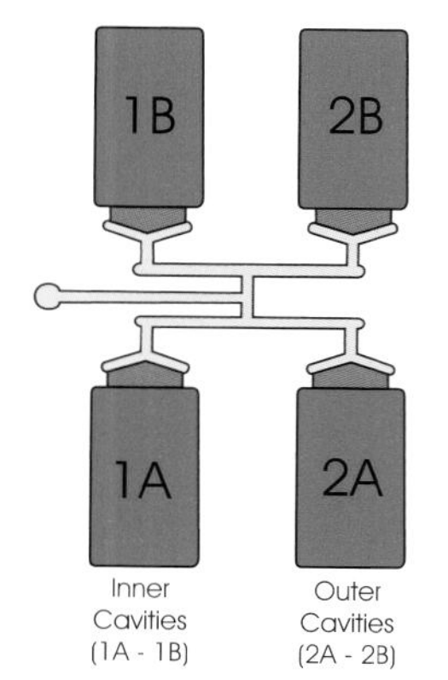

Figure 7.10 Layout of 4-cavity test mold used in runner study

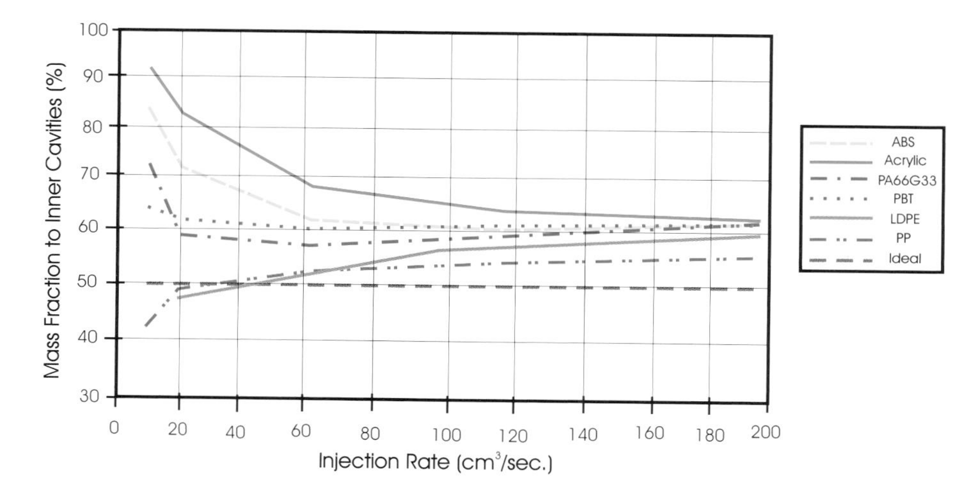

Figure 7.11 Filling imbalances as affected by fill rate and material from a 4-cavity mold study using a geometrically balanced runner and purge pockets

intersection of the primary and secondary runners. When the melt rotation technology was applied to the same runner, a nearly perfect balance was achieved, which was virtually independent of change in materials or fill rate (see Fig. 7.12).

Figure 7.13 provides the results of a similar study, except that an eight-cavity mold was used, which included more conventional 2.5 mm thick cavities. Again, with a conventional geometrically balanced runner a significant imbalance was found that varied with material and injection rate. Adding non-adjusted melt rotation technology resulted in a maximum imbalance of only 5%. [Note: the melt rotation apparatus was not adjusted to optimize the balance further]. The same melt rotation design was used for all materials. Here again, the melt rotation stabilized the imbalance such that it was virtually insensitive to material or process changes.

These two studies verify the robust nature of the melt rotation technology. The method achieves a balance that is extremely stable and tolerant of changes in both material and process. After application, no further instrumentation or adjustments are needed.

The means of achieving this solution to mold filling imbalances with melt rotation technology are numerous. The simplest method is to control the angle of flow at the intersection of a branching runner, similar to what was described earlier. This method can use a fully open free-flowing melt channel that has virtually no negative impact on molding. The method and

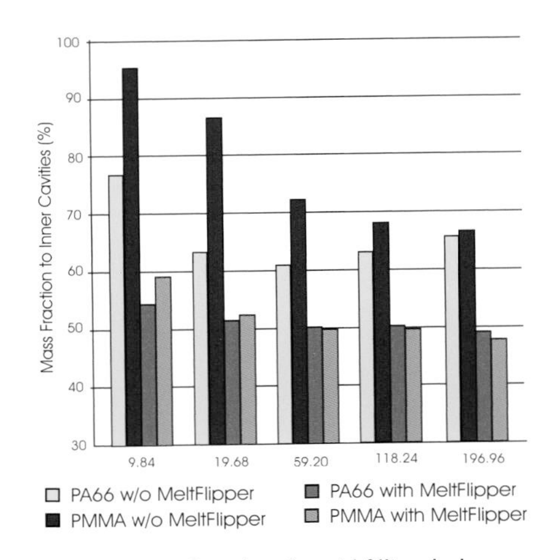

Figure 7.12 Results of mold filling balance study with and without MeltFlipper® technology

Figure 7.13 Results of mold filling study in an 8-cavity mold. Effect of fill rate on four different materials, with and without melt rotation technology

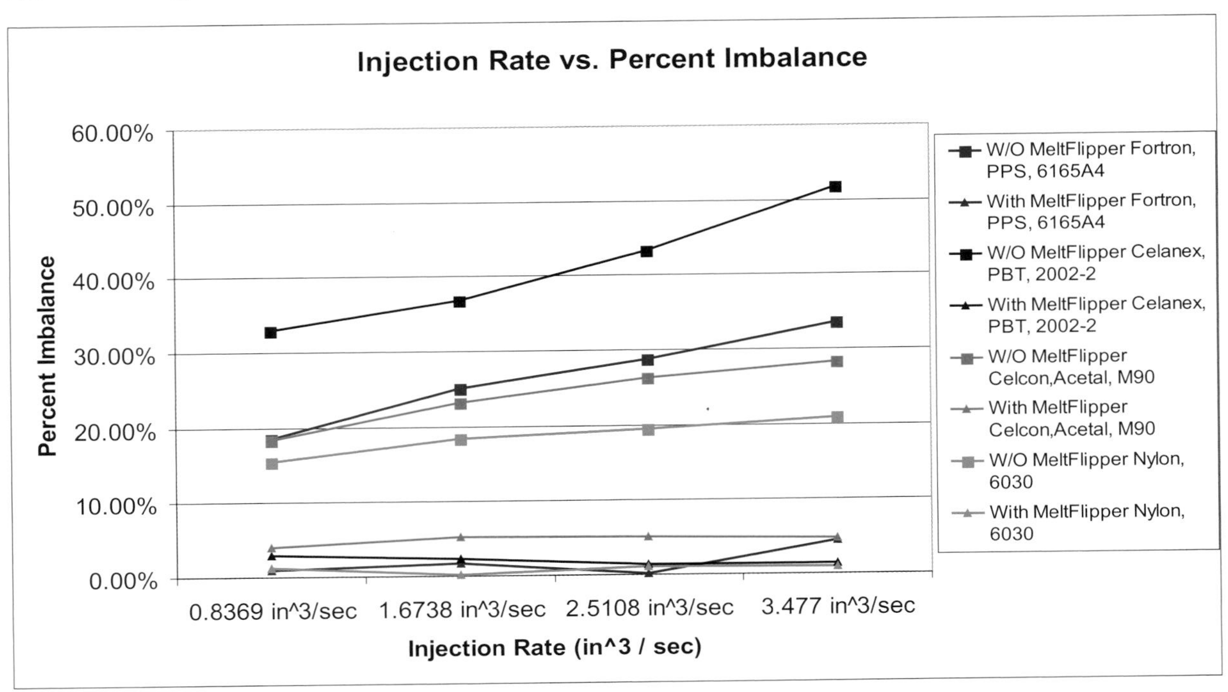

apparatus for achieving the desired positioning of the asymmetric melt conditions may vary depending on whether a cold or hot runner is being used, and whether the mold is a conventional 2- or 3-plate runner versus a parting line inject runner.

7.3.1 Melt Rotation Technology in Hot Runner Molds

Though MeltFlipper® technology has been readily available through licensing in cold runners since 1998, its application in hot runners is relatively new. In 2003, INCOE introduced their Opti-Flo™, the first hot runner to utilize this technology.

Numerous methods can be used to incorporate MeltFlipper® technology in hot runners. This could include controlling the angle of intersecting branches or by controlling the relative intersection of branching runners. The amount of intersection will control the angle of flow into the branch and thereby the rotation of the melt stream to a desirable position relative to a branch further downstream. Traditionally designed hot runners do not try to direct the asymmetric melt condition to a desirable position. They either branch along a single level or they branch at distinctly different levels bridged by a connecting runner section, which was not designed to control the circumferential positioning of the asymmetric melt conditions. Often, the branches at different levels are in distinctly different layered manifold blocks. The limitation of any of these methods is that they are not designed to address the shear-induced imbalance discussed in Chapter 6.

Melt rotation technology provides a means of balancing melt conditions without the need to create any flow restrictions, stagnant flow regions within the flow channel, or alter manifold and drop temperatures. It achieves the objective of balancing flow by simply managing asymmetric melt conditions rather than mixing them. Newer variations of melt rotation technology have been developed expanding its capability in both stack molds and conventional hot runners. These newer methods recreate melt symmetry when it is desirable (see Section 7.3.4).

A study presented by Blundy [6] contrasted the filling of an Opti-Flo manifold, utilizing melt rotation technology, with a conventional manifold. Blundy used INCOE's Unitized™ system that allowed easy exchange of the two manifold systems in the same mold. The sixteen cavity mold was instrumented with pressure transducers in each cavity and monitored with RJG's E-Dart System. The mold produced small 12-gram canisters from a nylon material. The study contrasted balancing the mold using the Opti-Flo manifold vs. using the traditional method of varying nozzle temperatures to achieve a balanced filling. Filling balance was evaluated based on the *5 Step Process* and review of the pressure trace data.

Following normal warm up, the Opti-Flo manifold had achieved a balance within the mold of 3% within only 20 minutes. This was achieved while keeping all manifold and nozzle temperatures at a constant 635 °F (see

Fig. 7.14A). Following this the manifold and nozzle temperatures were progressively lowered to evaluate potential for cooling time reduction and how robust the balance would be with process change. Temperatures were uniformly lowered to 560 °F with virtually no effect on the balance thereby providing a process window of 75 °F. At temperatures lower than 560 °F, there was evidence that the gates were freezing off thus creating the instability seen in Figure 7.14 B.

In contrast to the manifold with melt rotation technology, the conventional manifold took over 90 minutes to achieve the same 3% balance through altering nozzle temperatures. In order to achieve this degree of balance the nozzle temperatures had to be varied from 560 °F to 680 °F. This 120 °F variation in nozzle temperatures can cause numerous potential negative effects. These include molding parts with varied melt temperature and increased potential for leaking with common compression fit nozzles (the INCOE Unitized system used for this study uses a nozzle that is threaded to the nozzle and thereby does not rely on the thermal expansion assisted compression-fit commonly used with hot runners). In addition, the process window was extremely small providing virtually no opportunity to optimize indiviual part requirements.

Of particular interest in this study is the potential for lower cycle times as provided by the lower melt temperatures that could be used with the Opti-Flo manifold. While the manifold containing melt rotation technology could be run with a uniform melt temperature of 560 °F, the conventional manifold required nozzle temperatures that were over 100 °F higher. These high temperature nozzles will increase cooling time of the parts they are feeding, which in turn will control the cycle time of the mold.

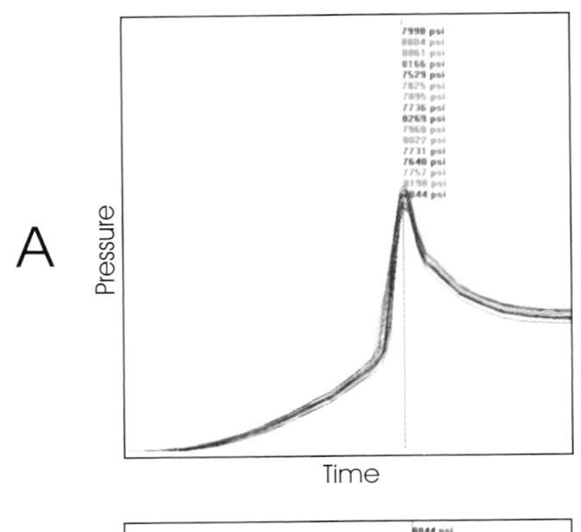

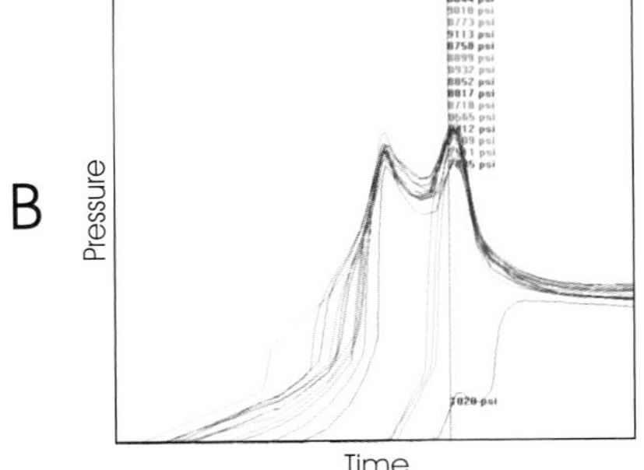

Figure 7.14 Pressure traces from 16-cavity Opti-Flo manifold found balanced filling despite a 75 °F temperature variation (A). Instability from gate freeze developed below 560 °F (B)

7.3.2 Melt Rotation Technology in Cold Runner Molds

Any application of melt rotation technology in cold runners must consider the ejection requirements of the runner. Therefore, the designs are generally different from those that can be used in hot runners. A common approach is the one shown in Fig. 7.15. The angle of flow at the intersection that controls the rotation is determined by the relative position of the intersecting runner sections. If possible, a rotation should always be applied at the first runner intersection. If rotations are not possible at runner intersections, in-line melt rotations may be used. Progressive rotations are added depending on the number of cavities and the geometry and requirements of the part being molded. As with hot runners, rotations on downstream branches will always vary from the first, because the position of the asymmetric conditions of the melt will vary. The correct rotation is determined through a proprietary knowledge-based system developed by Beaumont Technologies, Inc. The designs are very tolerant of design variants such as cold slug wells, recessed ejector pins at the branch, and diminishing runner diameters that are commonly used with cold runners.

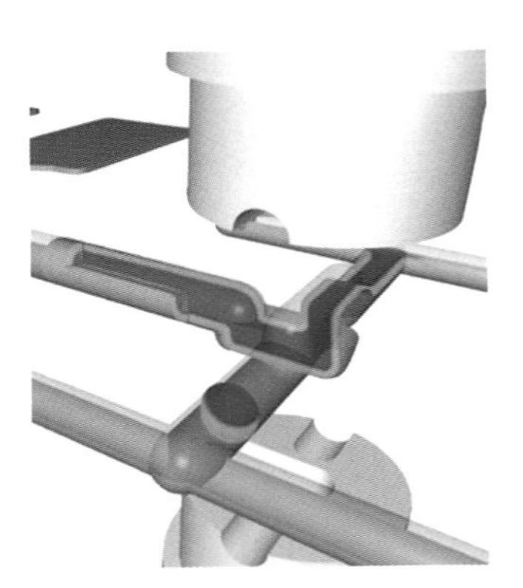

Figure 7.15 Repositioning of melt laminates in a circumferential direction through use of melt rotation technology (Courtesy: BTI)

7.3.3 Melt Rotation for Intra-Cavity Imbalances

The previously mentioned melt rotation methods can be used to address filling and melt imbalances between cavities in multi-cavity molds and to address filling and material variations within a given cavity. Figure 7.16 shows the results of the progressive use of melt rotation. Figure A shows both a cavity-to-cavity variation and a side-to-side (intra-cavity) variation that occur without melt rotation. Figure B shows a MeltFlipper® at the first runner intersection. This results in all cavities filling at the same time but still exhibiting intra-cavity imbalances. Figure C shows a modified melt rotation at the second runner intersection resulting in both a cavity-to-cavity and an intra-cavity balance. More consistent parts and a significantly improved process window are the result as all parts are filled with the same melt and process conditions.

Melt rotation was used in the troublesome four-cavity electrical connector mold discussed in Section 6.6 and shown again here in Fig. 7.17. After diagnosing the source of the product variations, melt rotation was introduced at the intersection of the primary and secondary runners (location "B" in Figure 7.17). The short-shot patterns before (Fig. 6.35) and after (Fig. 7.17) melt rotation are shown. As presented in Section 6.6, the 0.20 mm variation occurring in the critical feature between parts molded from the two-cavity groups was reduced to within the 0.05 mm specification. Scrap from the shear-induced product variation was virtually eliminated.

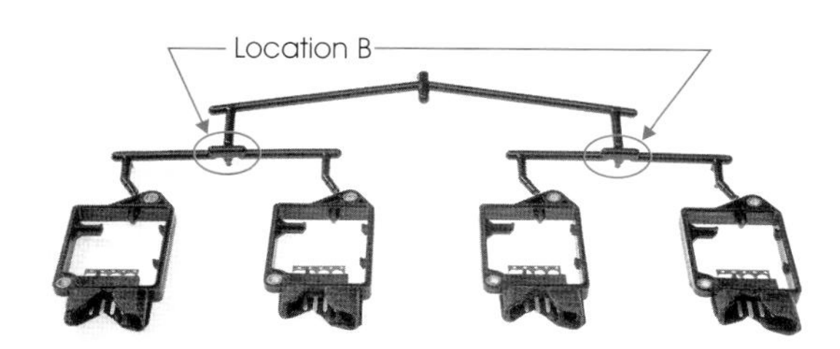

Figure 7.17 Parts molded in a 4-cavity parting-line injection mold that display consistent intra-cavity filling balance after introduction of melt rotation technology

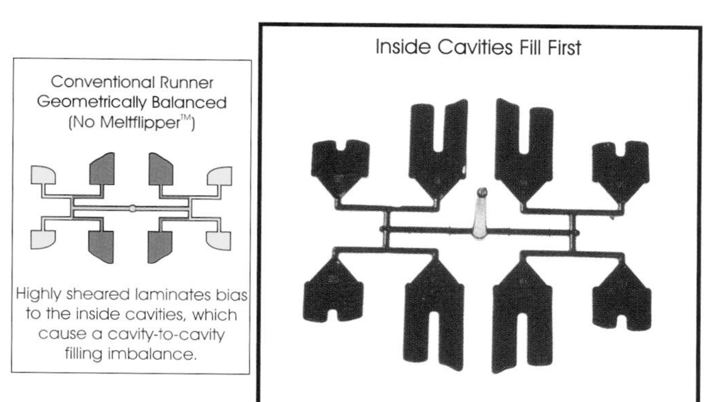

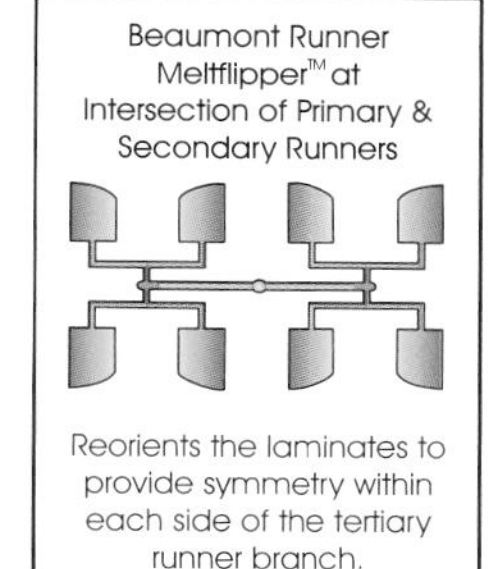

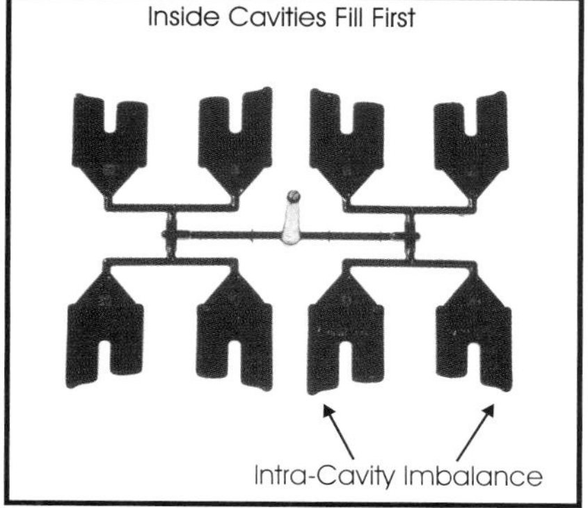

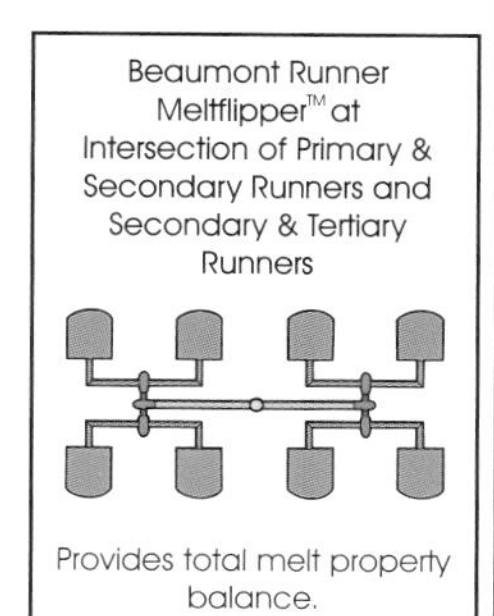

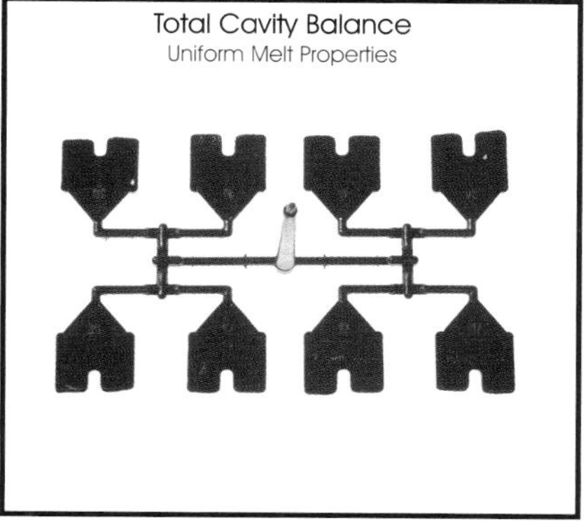

Figure 7.16 Effect of melt rotation on cavity-to-cavity filling balance and intra-cavity filling balance (BTI)

7.3.4 Multi-Axis Melt Symmetry

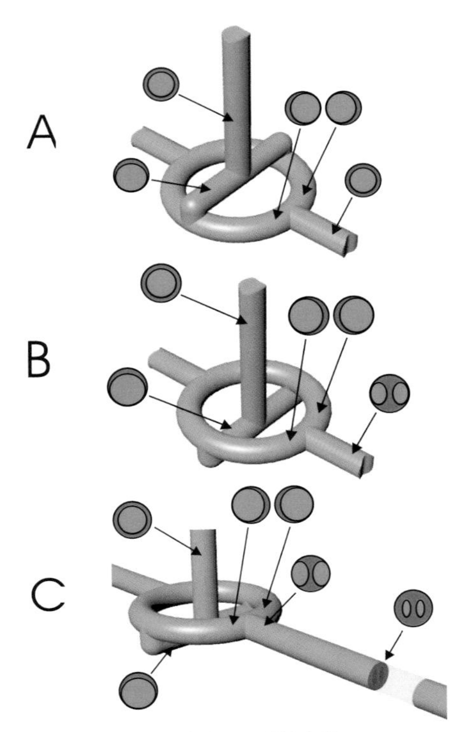

Figure 7.18 Variations of MeltFlipper MAX™ technology which recreates melt symmetry. Design A and B create the symmetrical melt conditions shown in the primary runner. Part C shows the melt conditions developed from design B, further downstream in the primary runner

Though conventional melt rotation technology can solve side-to-side variations in a melt stream, it cannot address the need for biaxial orientation that may be required in a part such as a center-gated tube, cap, or gear. It also has limitations when addressing the imbalance developed in stack molds. Stack molds generally require either a static mixer or MeltFlipper MAX™ technology. MeltFlipper MAX technology (MAX for Multi-Axis symmetry) is another new patented development that expands on conventional melt rotation technology [3, 4]. MeltFlipper MAX's melt management method uses melt rotation to strategically position the asymmetric conditions developed from a branching flow front and strategically recombine them to create melt symmetry on multiple axes. Again, numerous methods can be applied to achieve the desired melt condition. Variations of the design shown in Fig. 7.18 have been applied to cold runner molds, conventional hot runners, and stack molds. Depending on the desired effect, the asymmetric melt conditions developed in branching runners can be rotated such that when two melt streams are recombined, they are either on the inside or on the outside of a downstream runner. The design in Fig. 7.18A causes the highly sheared melt to be positioned to the perimeter of the runner. This recreates the multi-axis symmetrical melt conditions that existed prior to the melt exiting the sprue. Reversing the rotation of the melt causes the highly sheared melt from the sprue to be positioned as shown in Fig. 7.18B and C. Figure 7.18B shows how MeltFlipper MAX is used to position the highly sheared laminates into the center of the melt stream at the top of a sprue. As the melt continues downstream, the perimeter gains frictional heating resulting in the distribution shown in Fig. 7.17C.

Figure 7.19A shows the expected intra-cavity imbalance found in a 2-cavity hot runner mold. This could easily create a core shifting problem or jeopardize the desired concentricity of a part such as a gear or fan. By

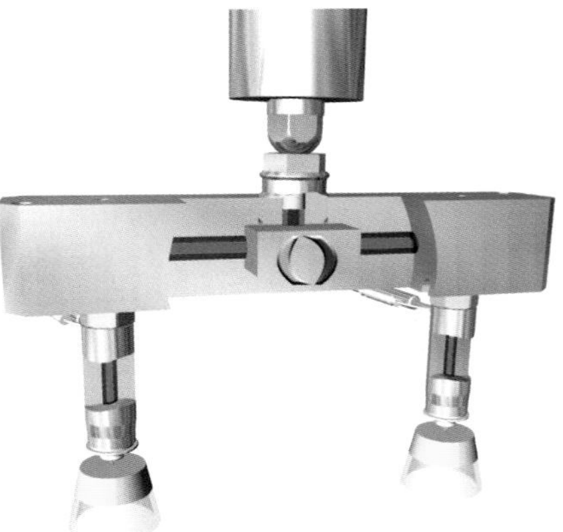

Figure 7.19 Expected melt distribution in 2-cavity hot runner mold without (left) and with (right) melt rotation technology at the intersection of the sprue and the primary runner (INCOE Opti-Flo manifold)

managing the melt as shown in Fig. 7.18A at the intersection of the sprue to the primary runner, the asymmetry in the melt is eliminated providing for uniform filling of the cavities (see Fig. 7.18B).

Figure 7.20A and B illustrate the expected melt conditions in an 8-cavity hot runner mold. Figure 7.20A shows the melt conditions with a hot runner using a conventional branching runner design, with an added full elevation change at the intersection of the primary and secondary runners. Note the differences in melt conditions that exist in the tertiary runner branches. This would result in two different flow groups and unbalanced filling. Figure 7.20B shows how melt rotation technology has been applied to achieve balanced melt conditions in each of the branching tertiary runners. This particular design has not only created balanced melt conditions in the each of the runner branches, but has also created a controlled multi-axis symmetry in the melt prior to it entering the cavities.

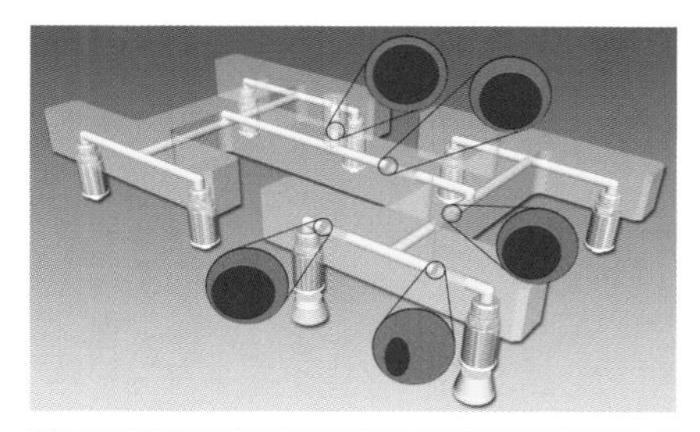

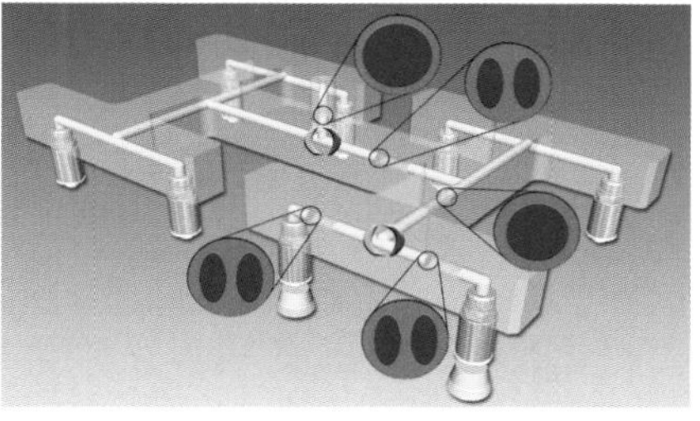

Figure 7.20 Expected melt distribution in an 8-cavity hot runner mold without (left) and with (right) melt rotation technology at the intersection of the sprue and the primary runner (INCOE Opti-Flo™ manifold)

Figure 7.21 shows the results of a study evaluating the effects of a MAX placed at the top of a sprue in an 8-cavity cold runner test mold. The intersection of the primary and secondary runner used a conventional runner branch (no melt rotation). Note that the filling of the eight cavities is significantly uniform relative to a 30%+ imbalance without the MAX (see photos in Fig. 6.13 and 7.16 for contrast). The design shown causes the melt to be positioned as illustrated on the outlet side of Fig. 7.18C prior to branching at the secondary runner. The repositioned high sheared laminates from the molding machine's nozzle and the mold's sprue would end up feeding the outer cavities and the high sheared laminates from the primary runner would feed the inner cavities. Though there remains a small imbalance in this early test shot, it demonstrates the ability of this method to manage the asymmetric melt conditions. Upon closer inspection of the photo, the further effects of the high sheared material from the sprue and primary runner can be traced into the cavities, where a side to side variation is seen at the flow front (similar to the center photo in Fig. 7.16).

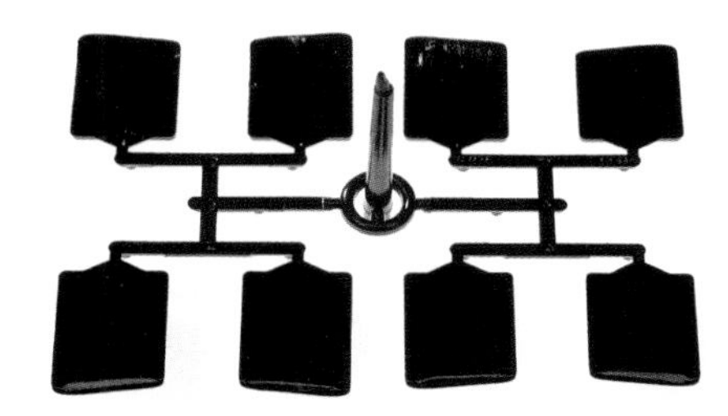

Figure 7.21 Filling balance resulting from MeltFlipper MAX in an 8-cavity cold runner mold (contrast to conventional runner in Fig. 6.13)

Another application for the MeltFlipper MAX™ technology is for hot runner stack molds. The resultant bi-axial symmetry would allow the melt to be balanced when branched along the same plane, as in a conventional hot or cold runner, or between the two parting planes of a stack mold (see Fig. 7.22).

The different variations of melt rotation can be combined to achieve many varied results. In addition, MeltFlipper MAX™, and variants of it, can be miniaturized to fit into the flow channel of a hot manifold or hot nozzle. Newer variations have now been miniaturized to fit into the nozzle tip of a hot drop.

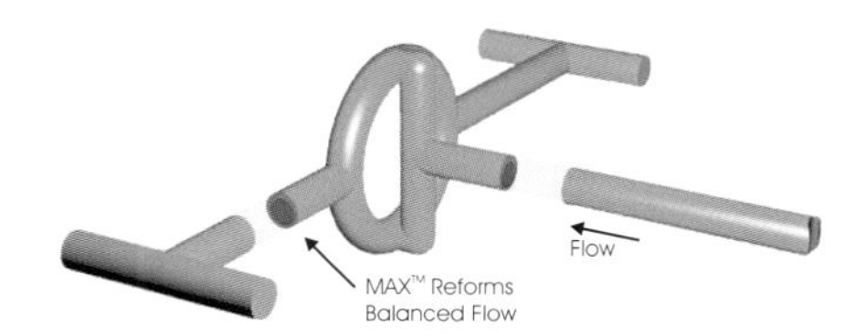

Figure 7.22 Intra-cavity distribution of gas in gas assist molded parts without (left) and with (right) melt rotation technology

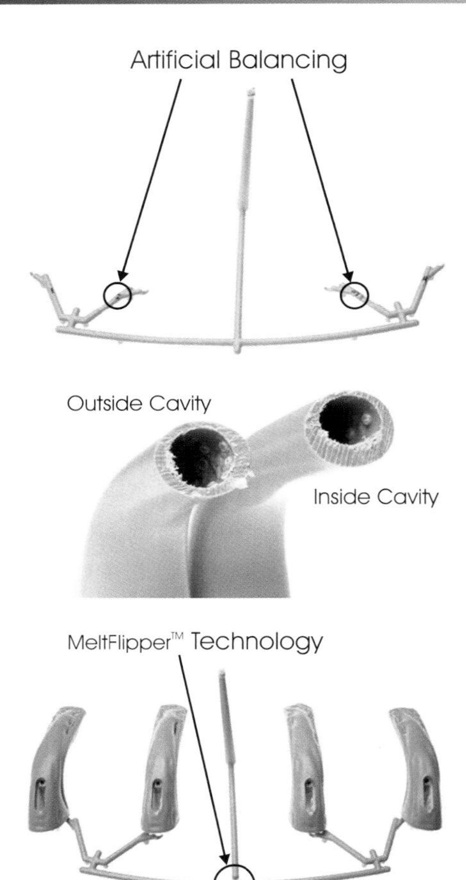

Figure 7.23 Results of study on shear induced imbalances on part molded using microcellular (MuCell) technology. The left three graphs show results before melt rotation and the three graphs on the right show the results after melt rotation technology was applied

7.4 Melt Rotation for Controlling Two Stage Injection Processes

As discussed in Chapter 6, shear-induced imbalances have a significant effect on two-stage injection processes such as gas assist, co-injection, structural foam, and the MuCell® process. As a result, these processes are normally limited to two- to four-cavity molds. Melt rotation not only provides the balance required to use these processes in higher-cavitation molds but it also provides a method to control the distribution of the second stage material within the formed part.

Figure 7.23 shows parts produced with the gas assist process. The parts were produced in a four-cavity mold and exhibited excess warpage. Inspection of the cross-section after molding found that the position of the gas void, created by the nitrogen, was different depending on cavity placement. By tracing the development of shear in the runners to their position in the part cavities, it was found that the inside two cavities filled before the outside cavities. Therefore, the inside cavities could not receive as much gas as the outside cavities, causing a thick section of plastic near the bottom of the parts. Even restricting runners to create an artificial balance could not solve this problem. Though the cavities were deceptively balanced for fill during one run, the process could not be repeated consistently due to the variations in the melt temperature and viscosity created by the shear imbalances. To remedy this situation, MeltFlipper® was used in the runner design and a balanced material fill to each cavity was obtained. Thus the flow of gas into each cavity was balanced and the thick section and its associated warp in the inner cavities were eliminated.

Figure 7.24 shows the results of a study conducted at Trexel in an eight-cavity test mold using the MuCell process. The three graphs on the left show the part weights of non-foamed parts as compared to parts molded with the MuCell process in which a conventional runner was used. The three graphs on the right show the effect of adding melt rotation to the same mold. With a conventionally designed runner, a filling imbalance of 24% was found. Following application of a non-adjusted cold runner melt rotation design, the imbalance was reduced to 2.6%.

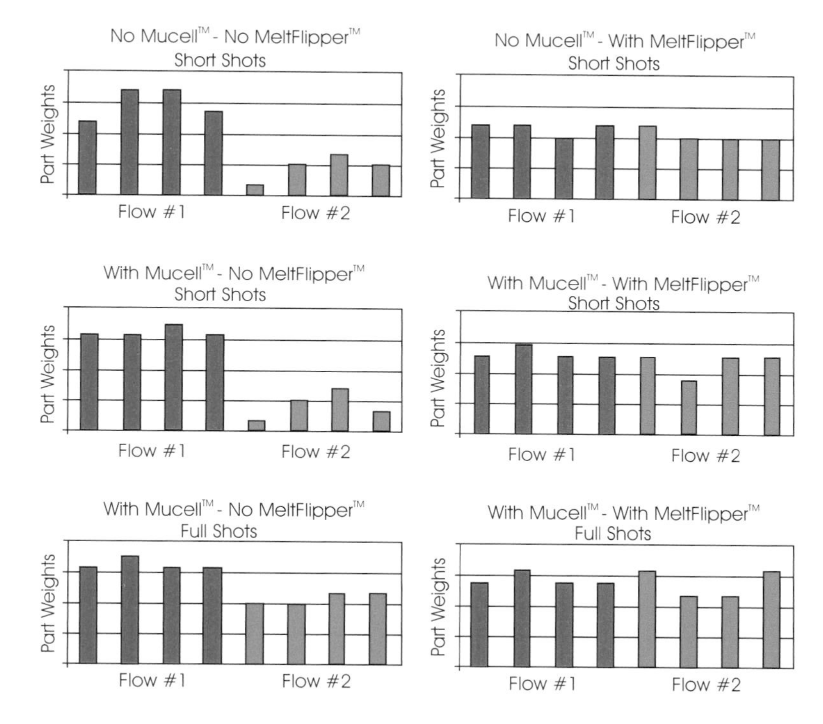

Figure 7.24 This figure contrasts the imbalances created in an 8-cavity test mold with and without MuCell and the effect of melt rotation technology on these imbalances. The left three graphs show the distinct development of two flow groups in the same mold (red vs. blue) with and w/o MuCell. The three graphs on the right show the effect of melt rotation technology in virtually eliminating the two flow groups. Effects of steel variations are still evident

References

[1] Kenics Static Mixers, KTEK Series, K-TEK-2 "Pressure Drop in the Kenics Mixer", Chemineer, Inc., May 1988

[2] Method and Apparatus for Balancing the Filling of Injection Molds, US Patent No. 6,077,470

[3] Method and Apparatus for Balancing Flowing Conditions of Laminar Flowing Material, US Patent No. 6,503,438

[4] Beaumont Technology Inc., Erie, PA, USA, www.meltflipper.com

[5] J. P. Beaumont, J. H. Young, M. J. Jaworski, "Mold Filling Imbalances in Geometrically Balanced Runner Systems," SPE Annual Technical Conference, Atlanta (1998), p. 599 (1998)

[6] J. Blundy, "Opti-Flo Hot Runners", technical conference presentation, SPI Structural Plastics Conference, Charlotte, N.C., March 22–23, 2004

8 Cold Runner Molds

For thermoplastic materials, a cold runner mold refers to a mold in which the runner is cooled, solidified, and ejected with the molded part(s) during each molding cycle. Approximately 70 % of molds today are cold runner molds.

The runner system in a cold runner mold normally consists of a sprue, runner and at least one gate. A single-cavity mold may only require a sprue, depending on part gating requirements. A hot sprue is sometimes used to feed a cold runner, yet the mold is still referred to as a cold runner mold. A three-plate cold runner mold will include a secondary sprue that bridges between the cold sub-runner and the part.

The cold runner mold is by far the most basic and most common type of mold. It is simpler, less expensive to construct, and easier to operate and maintain than a hot runner mold. Color changes are easily accommodated as the runner system and part are cooled and ejected each cycle, leaving no material in the mold.

The mold in Fig. 8.1 is a standard two-plate cold runner mold having a single parting line, or plane, which opens during each molding cycle to allow removal of both the runner and the molded part(s). The parting line is located between the core and the cavity halves of the mold. Figure 8.1 shows a sprue that provides a passage for the melt to travel from the injection unit, through the cavity half of the mold, to the parting line. At the parting line, the sprue connects to the primary runner, which either feeds the melt directly to one part-forming cavity or branches into additional runner sections that will feed the melt to multiple cavities.

In a single-cavity mold, the cavity is generally placed in the center of the mold, and the sprue delivers the melt directly to the center of the cavity. In a multi-cavity mold, the sprue delivers the melt to a runner, which in turn delivers the melt to the part-forming cavities. As the cavities and the runner are located along the same parting plane, gating is limited to the perimeter of the molded part. This limitation in gating location can sometimes lead to core deflection, gas traps, or undesirable weld lines. Some specialized gates, such as jump gates, provide some flexibility to gate away from the immediate perimeter.

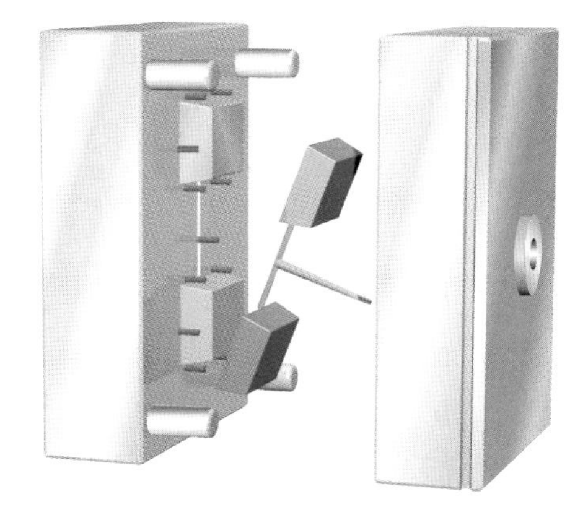

Figure 8.1 Conventional 2-cavity cold runner mold

Cold runner systems have several advantages over hot runner systems. Because of their simplicity, they are cheaper to build. In addition, the cost to maintain a cold runner system is less because there are no heaters, heater controllers, thermocouples, and other hot runner components to maintain. Operation is also much simpler as there is no need to tend to the various heater controllers or deal with the many potential problems such as gate drool, freeze, clogging, leaking, material degradation, or hang-ups in the runner manifold, and so forth.

A major disadvantage of the cold runner system is the fact that the unwanted frozen runner must be dealt with. This requires the need to separate the runner from the molded parts and then sell or grind the runner for reuse. This step of regrinding introduces additional potential for material contamination and the need for granulators and their maintenance. The reground material can often be fed back into the process, at a controlled ratio, with the virgin material and then be remolded. If the molded part requirements will not allow reground materials, the runner must be sold or thrown away. As the reprocessed material will differ somewhat from the virgin material, it can be expected that the method of reintroducing regrind can alter the molding process and the properties of the molded part. This may or may not be acceptable depending on the sensitivity of the plastic material, requirements of the molded part, and the ease of molding it.

8.1 Sprue

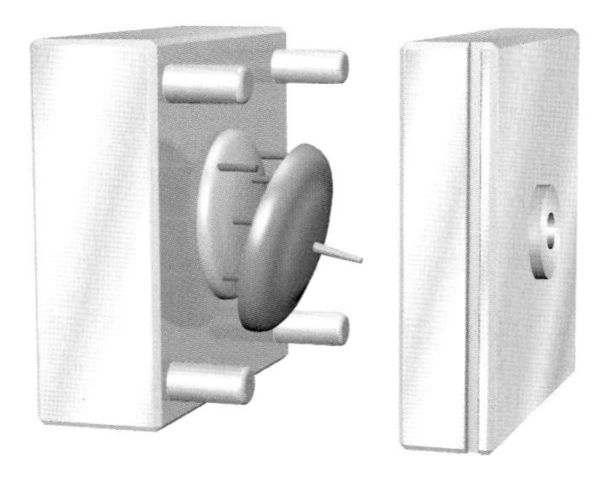

Figure 8.2 Single cavity mold with a direct sprue feed

The sprue delivers melt from the injection molding machine's nozzle to either a runner or directly to a part-forming cavity. With a single-cavity mold, the cavity is generally placed in the center of the mold where a sprue can deliver melt directly to the center of the cavity. This would be typical of a part like a bucket. In a single-cavity mold, one might be able to envision how the cavity could be gated nearly anywhere by offsetting either the sprue or the cavity. This is not generally recommended as it puts an offset load on the machine's tie bars. Figure 8.2 shows a single-cavity mold being fed directly by a cold sprue. The sprue provides a passage through the clamp plate and cavity plate to the cavity.

In a multi-cavity mold the sprue delivers the melt to a runner, which in turn delivers the melt to the part cavities. This approach can also be used in a single cavity mold where the part has a hole in the center (like a picture frame). The sprue would feed runners, which would gate into the inside edges of the part's hole.

8.1.1 Cold Sprue

A cold sprue is normally formed inside a sprue bushing. The sprue bushing is designed to be easily replaced and is normally purchased from a supplier of standardized mold components.

The replaceable sprue bushing provides a number of different attributes. Structurally it must withstand repeated impacts from the injection nozzle whenever it engages the sprue bushing. The interface between the sprue bushing and the nozzle must not be deformed as leaking or ejection problems might occur. The sprue also experiences some of the highest melt flow rates and melt pressures anywhere within a mold. To withstand these structural challenges, the sprue bushing is commonly made from hardened steel. The sprue must also provide for its ejection from the mold. For this purpose, the flow channel of the sprue is normally tapered and polished in the direction of draw.

The contact surface between the machine nozzle and the sprue bushing requires special attention. These surfaces are normally radiused rather than flat. The radiused surface helps with alignment of the flow channels between these components. If there were any misalignment, an undercut would be created and inhibit the sprue from being pulled from the bushing (see misalignment in Fig. 8.3A). In addition, the radius on the sprue bushing's interfacing surface should be slightly larger than that on the nozzle tip. This assures that there is a sufficient sealing force to provide a melt seal, helps with alignment and avoids the flash condition shown in Figure 8.3B.

To further avoid ejection problems the sprue "O" diameter (inlet diameter) must be slightly larger than the molding machines nozzle exit orifice. This helps avoid an undercut (Fig. 8.3C), which might be created at the junction between the machine nozzle and the sprue bushing if there were any slight misalignment between the two. Figure 8.3D is an example of all these concerns handled properly.

If there are multiple mold plates between the injection nozzle and the runner, the sprue must bridge these plates so that material does not flash between them. A flashed sprue will create an undercut and resist ejection.

During mold opening, the plastic sprue is pulled from the sprue bushing by either an undercut in the core retainer plate or an undercut in the sprue puller. The sprue puller is actually an ejector pin that, when moved forward, will either drive the sprue out of an undercut in the core retainer plate or, when pushed out of the retainer plate, will relieve an undercut allowing the sprue and runner to fall freely (see Section 8.2.4.1 for details on sprue ejection).

Both of these methods should provide a recess at the base of the sprue to act as a cold-slug well (see Fig. 8.4). The cold slug well provided by the sprue has been found to be the most effective placement for eliminating cold slugs. In a study performed by Auell and Martonik it was found that 100% of cold slugs were caught in the cold slug well at the sprue puller location [2]. Though the sample size in the study was relatively small, the study did provide good information on the effectiveness of the recessed sprue puller as a cold slug well.

A cold sprue is often the determining factor that controls the length of the molding cycle. The inlet diameter of a cold sprue is rarely less than 2.5 mm. Ejection requirements dictate that it be tapered. Tapering will increase the diameter and can result in a rather large intersection of the sprue and the runner. In thin-walled molded parts, this thick region could most likely extend the required cooling time. Given an entry diameter of 3.175 mm (0.125 in.) on a 76.2 mm (3 in.) long sprue, the draft on the sprue would increase the diameter of the sprue to about 9.525 mm (0.375 in.) at its

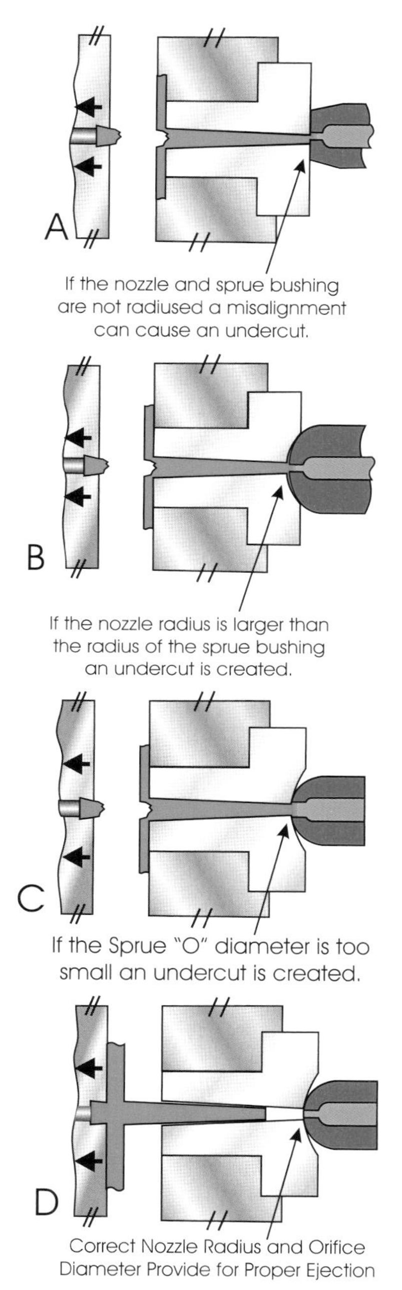

Figure 8.3 Effects of nozzle positioning, radius and orifice size on ejection of the sprue in a cold runner mold

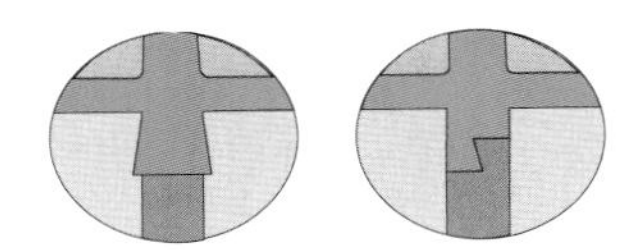

Figure 8.4 Common sprue puller designs which also act as a cold slug well

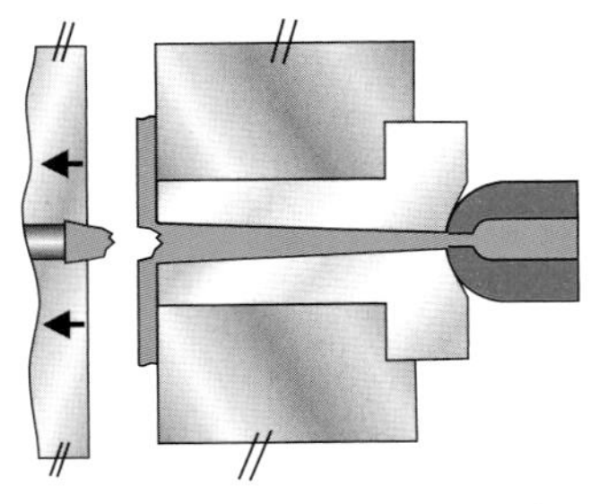

Figure 8.5 The large mass at the intersection of the sprue and the runner can cause the sprue to break on mold opening if it is not sufficiently solidified. This will cause the molder to extend the cycle time in order to allow the sprue to solidify enough to provide for it to be pulled from the bushing

Figure 8.6 Common case of careless sprue rotation, restricting flow and potentially affecting mold filling balance

intersection with the cold runner. This is also the location of the sprue puller, which is required to pull the cold sprue from the sprue bushing during mold opening. If this thick region is not frozen sufficiently at the time of mold opening, it will break and the sprue will remain lodged in the sprue bushing (see Fig. 8.5). When this happens, the cooling cycle must be extended so that the sprue is sufficiently solidified to allow for it to be extracted from the sprue bushing without breaking. This additional time may far exceed the time required to cool the part.

The sprue bushing is the source of a number of problems that are commonly ignored. Two problems are the issue of sprue cooling (discussed earlier) and the lack of positive positioning of the sprue. As the sprue bushing is generally a slip-fit through the top clamp and the cavity plates, it receives very little cooling. In addition, a portion of the primary runner is generally machined in the top of the sprue bushing, linking the sprue to the main runner. Rarely is any positive anti-rotation provided on the bushing to assure that the runners align (see Fig. 8.6). The result is the common case in which these two sections of the runner become misaligned during application, creating changes in product during the production run. Figure 8.7 shows a modified sprue design, in which cooling is provided. Gaskets are required and a retaining screw in the base of the sprue is required so that the sprue does not back out and allow water leakage. In addition, a locating pin is located in the shoulder of the sprue bushing, ensuring runner alignment. The pin can be welded to the mold to ensure positioning and use of the proper sprue during mold maintenance.

Following are some general guidelines to consider when designing or specifying a sprue:

- The sprue must not freeze before any other section of the melt delivery system. This is necessary to permit sufficient transmission of holding pressure to the part-forming cavity.
- The sprue must de-mold easily and reliably.
- The base of the sprue should be recessed so as to act as a cold slug well.

The dimensions of the sprue depend primarily on the dimensions of the molded part and especially on the part's wall thickness. Some general dimensions are given below:

- The sprue "O" dimension should be at least 0.794 mm (1/32 in.) larger than the opening in the nozzle.
- The sprue length must be tapered (between 1° and 4°) and be smooth around its circumference in order for the sprue to be pulled out of the orifice when the mold is opened (in rare occasions a zero draft can be used with very high shrink materials). Ejection of the sprue also requires that there must not be any flash where the sprue interfaces with the machine's nozzle.

- The standard sprue taper is approximately 2.4°, which is equivalent to ½ in. per foot.
- Sprue bushing radii should be slightly larger than the radius of the nozzle face.
- The outlet sprue diameter should be at least 1.5 mm larger than the wall thickness of the part being molded. This helps to ensure that the sprue freezes after the part and thereby remains open for the holding pressure.
- The sprue bushing should have positioning pin to prevent rotation and misalignment of runners

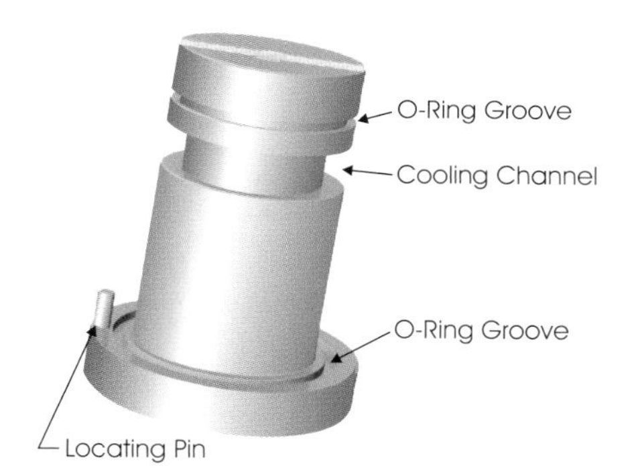

Figure 8.7 Sprue bushing design providing for cooling and a locating pin to ensure runner alignment

With high shrink semi-crystalline materials, the taper in the sprue can be reduced. This is desirable when the sprue is controlling the molding cycle. In some cases a sprue with zero draft may be used with high shrink materials such as POM.

The sprue shown in Fig. 8.8 is from a single-cavity prototype. The angled sprue allowed the part to be centered in the mold and run on a smaller machine. This type of sprue requires that the part and sprue be ejected before they are fully frozen. At a high temperature, the sprue will deflect and allow for ejection.

Figure 8.8 In this application, the sprue was angled to allow the prototype mold to be run in a smaller machine. Retraction of the sprue was enabled by the flexibility of the warm sprue during ejection

8.1.2 Hot Sprue

A hot sprue can often be used as an alternative to a cold sprue. The hot sprue directs the melt from the machine nozzle directly into a single cavity, or into a cold runner that will either feed multiple cavities or multiple gates feeding a single cavity. In a single-cavity mold, the hot sprue eliminates the need to cut the more common cold sprue from the molded part manually as well as the need to deal with the left over sprue.

In a case where the large diameter of a cold sprue would control cycle time, the hot sprue provides an excellent alternative. The hot sprue has few of the drawbacks of a hot runner system. They are relatively inexpensive, easy to install, operate, and maintain. They generally require only one additional temperature controller and act much like an extended injection nozzle.

8.2 The Cold Runner

The cold runner is one of the most influential components of a successful molding, but is often grossly misunderstood and its impact underestimated.

As discussed in Chapter 5, the ideal runner has a full round cross sectional shape which provides the optimum ratio of the perimeter of the runner geometry to cross-sectional area of the runner. However, the full round

$$\text{Shape Factor} = \frac{\text{Perimeter}}{\text{Cross-sectional Area}}$$

Figure 8.9 Shape factor used in a study of the efficiency of various runner cross sections

Figure 8.10 Results of a study on efficiency of various runner cross section designs. Those with the lowest shape factor for a runner with a given cross-sectional area were the most efficient

runner has the disadvantage of requiring the two halves of the runner to be machined in each of the two halves of the mold. These two half runners must then closely match up to form the full round runner when the mold is closed. Due to the added cost of the full round runner and the potential for misalignment, alternatives to full round runners are often used.

Alternative runner shapes include: trapezoidal, parabolic, and half round. These alternatives are often much easier to machine because there is no concern of matching up runner halves as with the full round runner. However, these alternative runners are less efficient as they will result in a higher pressure drop per runner volume than the full round. The higher pressure drop per runner volume results from the larger ratio of the perimeter of the runner cross-section to its cross-sectional area (i.e., has a larger shape factor as described in Fig. 8.9).

There is a predictable relationship between the pressure drop through a runner and its shape factor. In a study by Neely and Hennebicque [3], runners with the same cross-sectional areas but increasing perimeters were found to increase pressure during injection by over 33%. Figure 8.10 compares pressures developed in a test mold, which evaluated the effect of changing runner cross-sectional shapes. The flow through five different

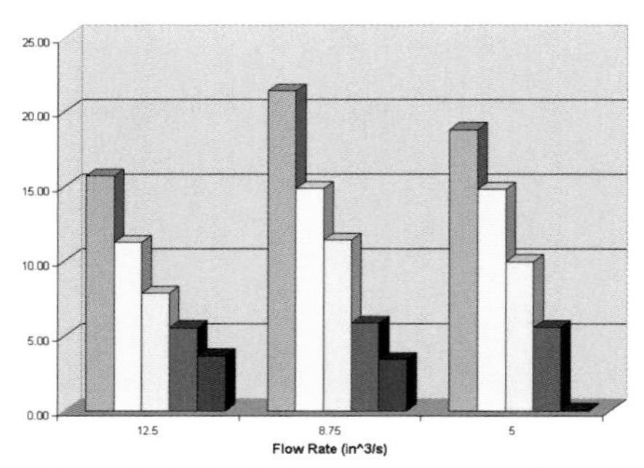

PMMA-Variation in Pressure vs. a Full Round At Each Flow Rate

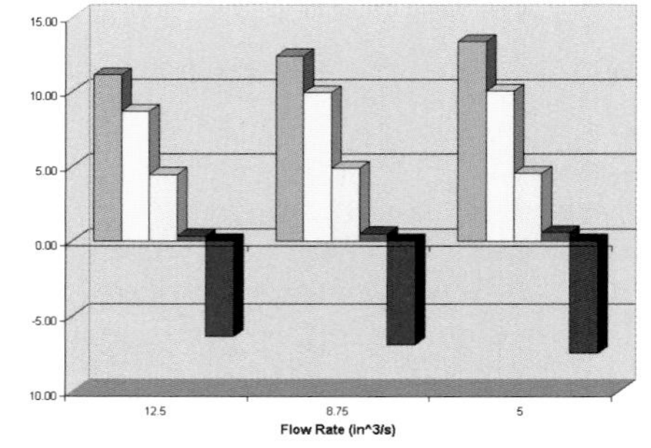

PP-Variation in Pressure vs. a Full Round At Each Flow Rate

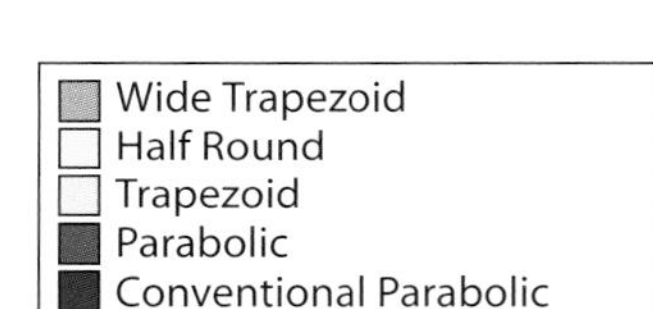

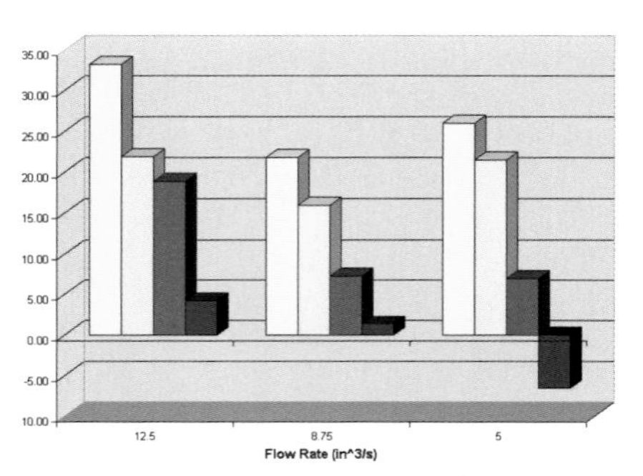

ABS-Variation in Pressure vs. a Full Round At Each Flow Rate

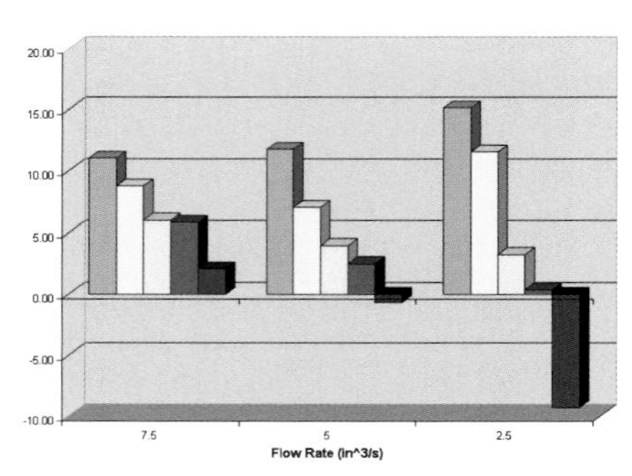

PBT-Variation in Pressure vs. a Full Round At Each Flow Rate

cross sectional shapes were contrasted to the ideal full round runner. Flow pressures through the runners were evaluated at three flow rates with multiple materials. The top left graph in Figure 8.10 shows that at an injection rate of 12.5 in.^{3}/s, the PMMA had a 15% higher pressure flowing through a wide trapezoid shape than when flowing through the full round. It can be seen that for each injection rate and each of the materials tested, the degree of variation between geometries differs; however, a general trend is clear. The least efficient geometry, in terms of pressure drop, is the wide trapezoid followed by the half round, trapezoid, and, finally, the parabolic. The conventional parabolic geometry appears to result in the lowest pressure. However, the conventional parabolic has a larger cross-sectional area than all the other geometries, because it was based on a full round in which the sides were given a 10° taper to form the parabolic shape (see Fig. 8.11).

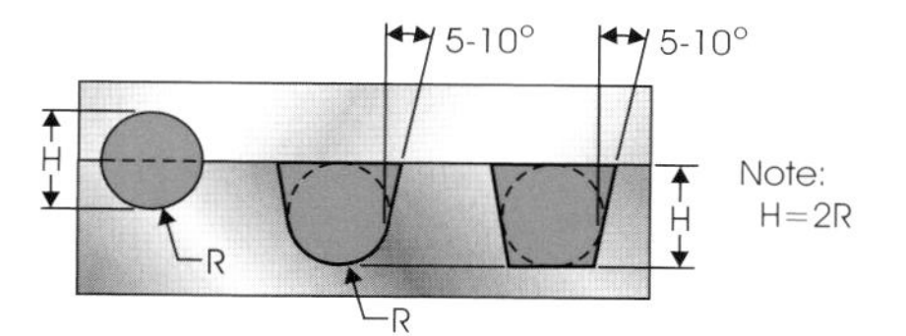

Figure 8.11 Recommended runner designs; left to right – full round; parabolic; trapezoidal

The only non-full round shapes that should be considered for injection molding are correctly designed parabolic and trapezoidal (see Fig. 8.11 for correct designs). The designs should be based on the diameter of the desired full round runner. The radius used on the parabolic runner should be identical to the radius of the desired full round runner. The draft on the two sides of the two designs is based on ejection requirements. The 10° draft is easier to eject and is recommended for most low shrink materials. The 5° draft may be used with good runner ejection and high shrink materials.

8.2.1 Important Machining Considerations

The importance of proper machining of a cold runner is commonly underestimated. Sloppy machining and lack of tolerances on runners and gates are a major cause of mold reworks and process problems. Runner cross-sections should either be full round or parabolic. The two halves of a full round runner should be toleranced to match up to within 0.05 mm (0.002 in.). This tolerance should also apply to the diameter. Sprue bushings should be keyed to prevent rotation and to assure alignment with the runner to within the same tolerance.

Neely and Hennebicque also evaluated the effect of offset full round runners, which would result from poorly matched assembled mold halves. The study contrasted the flow through two halves of a branching runner fed by a common sprue. They found that an offset of only 0.254 mm (0.010 in.) in one branch of a 3.125 mm (0.125 in.) full round runner created flow variations of 59.14% compared to the other runner branch that was fully aligned [3]. Increasing the offset to 1.016 mm (0.040 in.) created flow variations of 83.42%. Further studies measured pressure drop through steadily increasing offsets of the two halves of a full round runner. It was found that the pressure drop through the runner increased by approximately 3–7% when offset by 0.254 mm (0.010 in.) and 10–23% when offset by 0.762 mm (0.030 in.), depending on material and process.

This study dramatically illustrates the significant impact of only minor imperfections commonly found in runners.

8.2.2 Sizing of Runners

Unfortunately, cold runners are generally sized based simply on the availability of a standard cutter size or by tradition.

When sizing a runner, the following three factors should be considered:

- The runner must be large enough to allow the part to fill. This can be determined using injection molding simulation software
- The runner must be large enough to provide for compensating flow throughout the required packing stage. This is generally determined using rules of thumb.
- The runner diameter should be minimized in order that it does not control the cooling time nor add excessive regrind.

To help assure compensating flow, the runner diameter should be approximately 1.5 times the thickness of the part wall to where the gate is attached. Conservative recommendations suggest that the runner should be 1.5 times the thickness of the thickest region of the part. This assumes that the gate is placed in the thickest location of the part. If there are thicker regions away from the gate region, such as flanges or bosses, the runner need not be any thicker than the wall into which it is gated. Under these conditions, packing will be limited by the wall thickness between the gate and the thick regions – not by the runner diameter.

Many designers vary the runner diameter along its length (this is discussed in detail in Chapter 5.5.4. If this is the case, the runner should be sized from the part to the sprue. The smallest diameter runner section will be attached to the gate and should be sized based on both filling *and packing requirements.*

In a study by Knepper [1], it was found that it was possible to fill and maintain good pack control of a molded part with runners having diameters no larger than the parts wall thickness. Though the study did not evaluate runners smaller than this, results seemed to indicate that this would be possible. Of the materials tested (PP, ABS, Acrylic, Acetal, PC and Nylon) only the PC clearly had a problem packing out the test mold with the smaller runner diameters. It is theorized that the small diameter runners are able to sustain packing due to the relatively high flow rate, and shear rate, that results during compensation/packing phase. This is the primary reason that pack control can often be maintained even with small restrictive gates. Though the study was somewhat limited in scope, it did provide some interesting insight and demonstrated a need for further investigation. The potential benefit resulting from this study is that in some cases a molder could use a smaller than expected runner diameter with a thick-walled part. This would reduce excess runner scrap and cycle time.

8.2.3 Venting

In most molds it is wise to begin venting at the sprue and in the runner region. This eliminates much of the air trapped in the runner path from being pushed into the cavity. This early venting may also allow the melt to be injected more quickly and with less injection pressure.

Vents can be placed at the intersection of the primary and secondary runners. Additional venting will be beneficial, but is often omitted owing to the added expense. An additional strategic position for vents would be just prior to entering the gates of the part.

Vents can be at least 4 mm wide by 1.5 mm long. The depth of the vent depends on the material but is generally less than 0.025 mm. This vent should be relieved into a 1 mm deep vent relief that purges to atmosphere. It is critical that if vents are to be used in the runner, that they be placed symmetrically along the runner. If a branch or gate is to be vented, then all like branches and gates should be vented in the same fashion. Unbalanced venting can contribute to unbalanced mold filling.

8.2.4 Runner Ejection

8.2.4.1 Sprue Puller

When designing cold runner molds, the ejection of the runner and sprue should be taken into consideration. To ensure that the runner system does not stick to the cavity half of the mold a sprue puller is used. There are many designs of sprue pullers, and some commonly used ones are shown in Fig. 8.12. These designs work fairly well, keeping the sprue from sticking to the cavity half of the mold; however, the design in Fig. 8.12E can result in obstructions to melt flow.

In order to maximize the functionality of the sprue puller without creating a restricted flow path it is recommended that a recessed sprue puller design be used. Designs A, B, and C are recommended. Design D is recommended for applications in which it is desirable for a robot to pick the runner from the mold. Design E results in a flow restriction and therefore should be avoided. An additional advantage of using the recessed designs is that the recess also acts as a built-in cold slug well. Studies show that the vast majority of cold slugs are trapped in the recessed sprue puller/cold slug well [2].

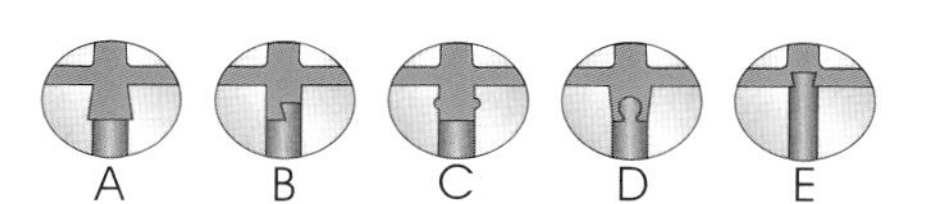

Figure 8.12 Variations in sprue puller designs. Design E should be avoided as it restricts the melt flow and requires a striper or an automated sprue picker

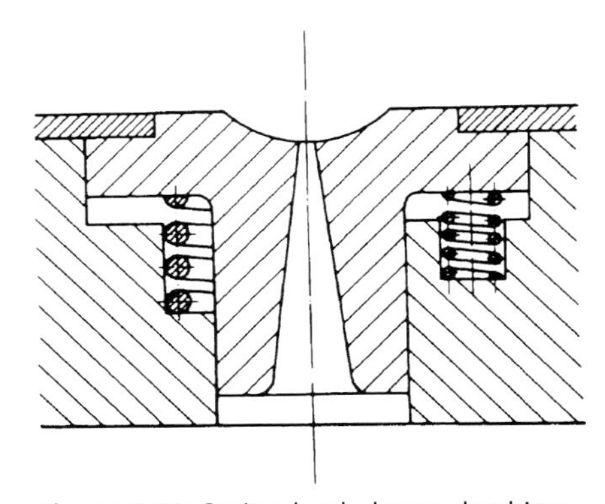

Figure 8.13 Spring-loaded sprue bushing

Another, less common option for removing the sprue from the bushing is shown in Fig. 8.13. The sprue bushing is spring-loaded. After the mold has been filled and the nozzle is retracted from the bushing, springs push back the bushing and loosen the frozen sprue. This reduces the problem of a sprue breaking at the sprue puller.

8.2.4.2 Secondary Sprue/Cold Drop

Flow restriction problems may be even more apparent in the secondary sprue, or cold drop, of three-plate molds. Here, the sprue puller must be capable of holding the cold drop section in place while the gate is torn from the part. To assure the cold drop and gate are held securely, a protruding sprue puller is often used (see Fig. 8.14A). As the second parting line opens, the gate is torn from the part generally in the first millimeter of movement. During this one millimeter movement, the base of the sprue is encapsulated by the surrounding mold sprue bushing, thus preventing it from expanding and releasing from the protruding sprue puller. Once the sprue is fully retracted, it is easily stripped from the protruding sprue puller as it is now free to deform as it is stripped off.

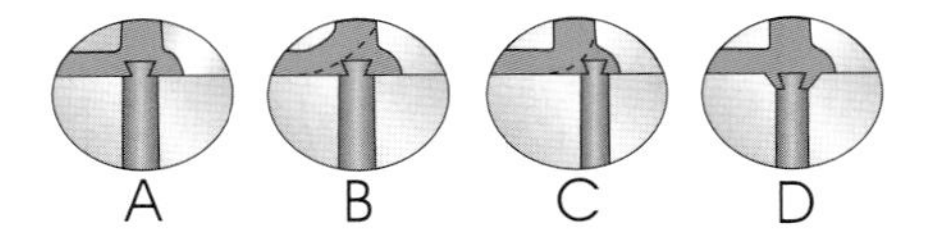

Figure 8.14 Sprue puller designs for secondary sprues in 3-plate cold runner molds. It is critical that the sprue position be consistent and that it does not restrict flow to the cavity

A major flaw of this approach is its negative effect on mold filling and packing. To assure its mechanical operation, the protruding sprue puller must fill much of the flow channel at the base of the sprue. This can result in a significant pressure drop and any inconsistencies will affect the melt filling balance. Figure 8.14B, C, and D provide alternative designs to minimize, or eliminate, the restriction of the sprue puller. Design B is the same as design A, but modifies the base of the secondary sprue to open up the melt flow path. This path should be limited so as not to increase the mass of plastic, which will take longer to solidify. However, it should provide a flow path that is no smaller than the runner cross section. With design C, the sprue puller is offset such that it does not interfere with filling or packing of the part. The shortcoming of this design is that the sprue will be pulled at an angle rather than straight back. For this, the base of the cold drop should be radiused to reduce the chance of the sprue breaking away. In addition, the sprue should not be offset any more than required to allow a free flow path for the material. This method also has the added advantage that the protruding sprue puller will stay cooler, allowing the plastic to freeze onto it faster, as it is not in the main flow stream. Design D is another successfully used design that does not restrict flow in the runner. Whatever sprue puller design is used, it should not restrict flow and should provide for a runner path that is no smaller than the runner. It is critical in multi-cavity molds that the sprue puller design and positioning be consistent so that they do not negatively effect the filling balance to each cavity.

8.2.4.3 Runner

The cold runner should be reviewed for potential ejection problems. If ejector pins are to be used, they must not extend into the flow channel. Pins should either be flush with the bottom edge of the runner or recessed. A slightly recessed pin will not create any mold filling problems.

A full round runner generally requires little assistance for ejection. The sprue puller can often be enough (assuming all gates and the part cavities have adequate ejection). If a trapezoid or parabolic runner is to be used, ejector pins should be used along the runner. Low shrink materials, such as polycarbonate, can be a particular problem. Here, the draft on the sides of these parabolic runners should be 10 degrees and ejection should be placed at the intersections of the primary and secondary runners. Ejectors may be placed at every intersection depending on the potential for hang-up. Recessing the ejector pin assures extended contact with the ejector pin during ejection. A recess of 0.5 to 3 mm (0.020 to 0.120 in.) is generally adequate (similar as used with tunnel gates, see Fig. 8.32). Again, be sure that the ejector pin does not protrude into the runner in any way, which would restrict flow.

8.2.5 Cold Slug Wells

Cold slug wells are used extensively on cold runner molds to prevent problems associated with cold slugs. Cold slugs are solid, or partially solid, pieces of plastic material that form in the nozzle tip as a result of its contact with the cold mold. During injection, the cold slug is injected into the sprue and travels with the melt stream. It may become lodged in the runner causing a blockage or disturbance in the melt flow.

Cavity balance could potentially be jeopardized by a cold slug trapped in the runner system. In the case of multi-cavity molds, one cavity may become partially or completely blocked, which will affect the filling of the other cavities as well. To take it one step further, even if all cavities fill, the flow variations could affect the amount of shrinkage and warpage of each of the parts that are formed.

In order to address these issues, cold slug wells are often employed. A cold slug well is simply an extension of the feed runner beyond its intersection with branching runners. This extension provides a place for a cold slug to become lodged as it travels at the tip of the flow front, before it reaches the part cavity (see Fig. 8.15). Standard practice is to place cold slug wells at each intersection in a runner system. However, studies have found that nearly all cold slugs are trapped in the recessed base of a cold sprue, which acts as the sprue puller. If the recessed sprue is used, it should capture most cold slugs [2].

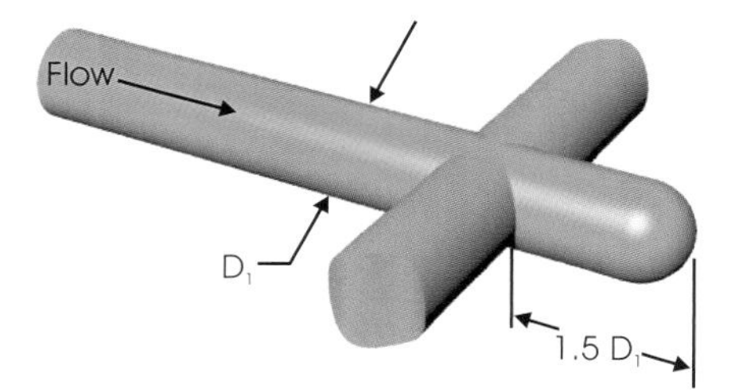

Figure 8.15 Recommended design of a cold slug well. Excessive length to the cold slug only wastes material

An alternate strategy is *not* to attempt to capture the cold slug beyond the sprue. Once a cold slug enters a primary runner it can be expected to affect the filling balance between cavities and thereby create a variation in the formation of the parts in the different cavities. If the cold sprue is not caught at a downstream branch, it is likely to sufficiently restrict, or block off, flow to a cavity (or group of cavities) such that it is obvious that defective parts were molded. If the cold slug well catches the slug, the mold will most likely fill and the resultant product defects will be masked. It may be best to know that cold slugs are occurring and address the problem at its source.

8.3 Runners for Three-Plate Cold Runner Molds

The primary advantage of the three-plate cold runner mold (Fig. 8.16) over the two-plate cold runner mold is that gating is no longer limited to the perimeter of the part cavity. Compared to hot runner systems, three-plate molds are low cost, relatively easy to operate, have fast startups, require less skill to operate, and provide for easy color changes.

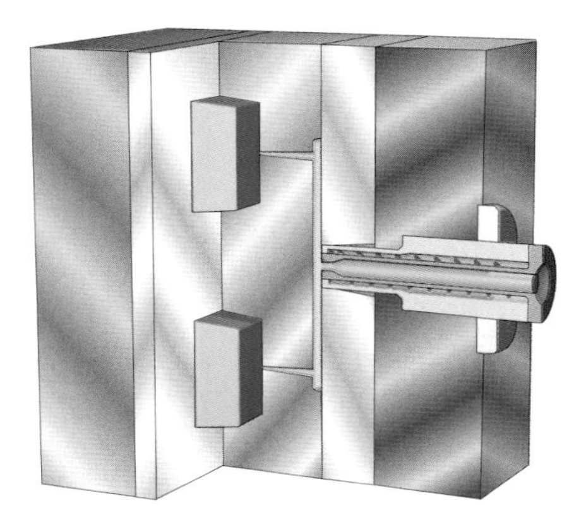

Figure 8.16 Typical 3-plate cold runner mold illustrating the delivery of the melt to the cavity. Use of a hot sprue reduces the required opening stroke and runner ejection problems

The three-plate cold runner mold has a second parting plane located behind the cavity plate. The second parting plane, between the cavity plate and top clamp plate, provides for a runner to travel under the mold cavity to any position relative to the part cavity. A secondary sprue then transfers the melt from the runner, through the mold cavity insert, to a desired location on the part cavity. The secondary sprue is attached to the part by a small-diameter pin gate. Owing to the increased flexibility in gating locations, the three-plate cold runner mold might be used in multi-cavity molds producing parts such as a cup, where gating in the center of the cavity is desirable.

Figure 8.17 illustrates one variation of a three-plate cold runner mold. Here, the ejection of the part and runner begins with the mold first opening at a primary parting line, defined between the core and cavity plates (Step 2). At a position where the part has been fully retracted from the cavity, a pull rod (A) will begin to pull a floating cavity plate to open the mold at a second parting line (Step 3). As the secondary parting line opens, a stationary sprue puller with an undercut holds the base of the secondary sprue, or cold drop, such that the mold opens sufficiently for the secondary sprue to be fully relieved. The runner stripper plate moves forward to strip the secondary sprues from the stationary sprue puller, allowing the runner to be ejected (Step 4). In the design shown in Fig. 8.17, both the molded part and the runner are ejected using a stripper plate. The stripper plate, which ejects the molded part is triggered by the action of the ejector plate (not shown) acting on a push rod (C). The stripper plate ejecting the runner is activated by a second puller pin (D), which pulls the plate forward as the mold is opening.

The disadvantages of the three-plate mold compared to the two-plate cold runner mold include the added complexity of the mold, which potentially

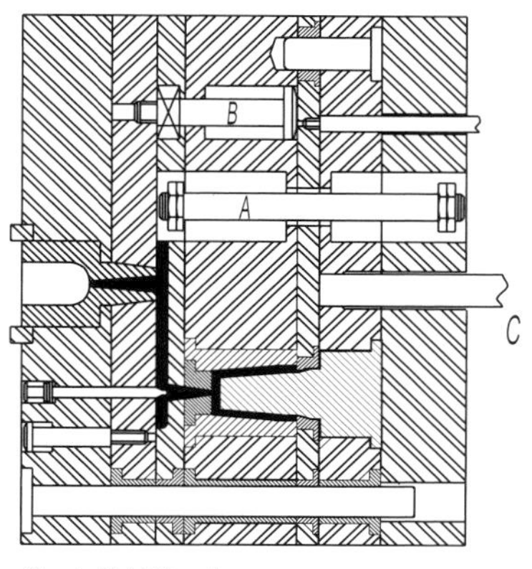

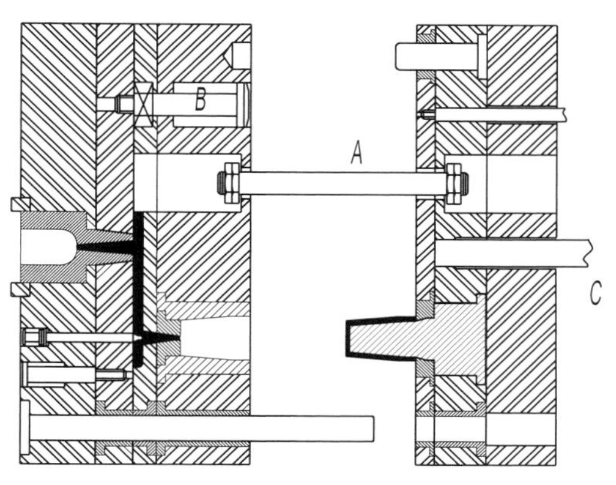

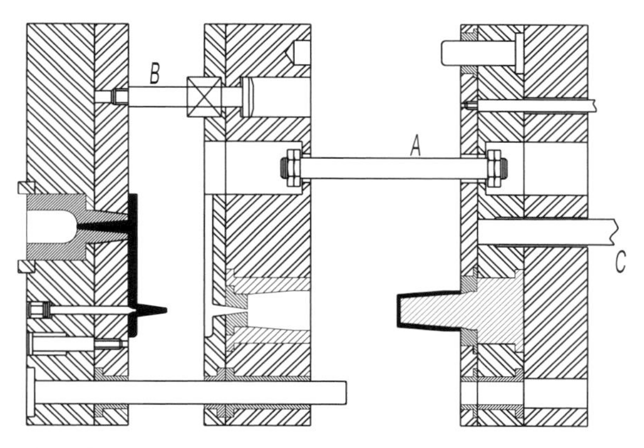

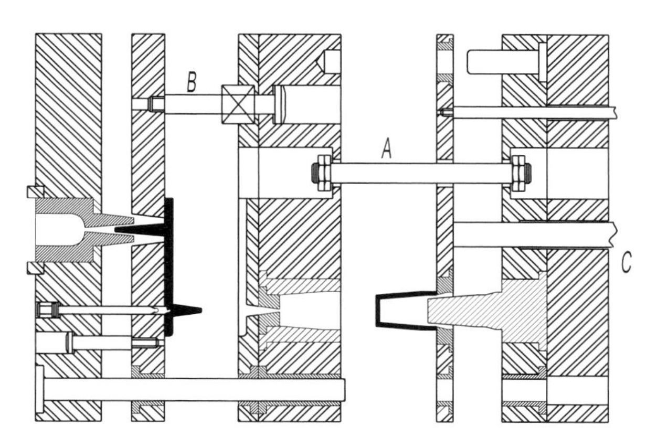

Figure 8.17 Opening and ejection action of a three plate mold with a hot sprue

creates increased maintenance and operations problems. Molds must provide the proper opening, and the sequencing of the opening of the primary and secondary parting lines, as well as the activation of ejection of both the runner and the molded parts. The cavities and cavity retainer plates must now float on guide pins as the mold opens and closes. This increased number of moving parts augments wear-related problems. Opening and closing is more difficult to set up and control. The addition of the second parting line requires that the mold opens further, which will increase cycle time and mold opening requirements. There is also increased potential for runners hanging up during ejection. As the opening distance of the secondary parting line is limited, the runner may bridge across the mold halves as it attempts to fall free during ejection. Hot sprues are generally used with three-plate cold runner molds to reduce the opening stroke required to eject the runner (see Fig. 8.18).

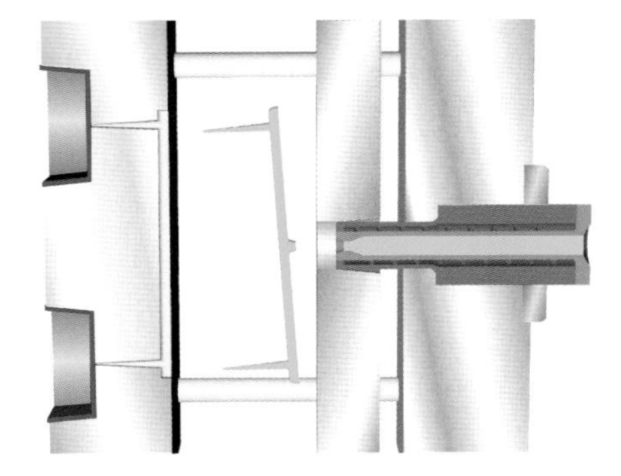

Figure 8.18 Illustration of hot sprue utilized in 3-plate mold

The three-plate cold runner offers the same advantage of providing flexibility in gating locations as a hot runner. However, the three-plate cold runner has the same disadvantage as the two-plate cold runner in dealing with the frozen runner. In addition, the three-plate cold runner will normally have a higher pressure drop during mold filling than any of the other mold types, particularly hot runner molds. The fact that the runner must pass under the cavity plate and then travel up through the plate usually results in a longer runner than in the two-plate cold runner mold. However, this additional length of runner will not usually increase the required clamp force because much of the runner generally travels under the cavities, along a separate plain, and thereby does not significantly increase the projected area of the molding. In contrast to a hot runner mold, this runner must have a relatively small cross-section to conserve on regrind and to ensure that the runner does not extend the cooling cycle of the process. This is not a concern with a hot runner mold; therefore, the runner can be a larger, more freely flowing channel. Finally, gate designs in a three-plate cold runner mold are limited to small, restricted gates, which must be designed to allow them to be torn away from the part during mold opening.

8.4 Gate Designs

The gate is the link between the part and the runner system. It is normally a restricted area that facilitates separation of the runner from the part. The size, shape, and placement of the gate can significantly affect the ability to successfully mold a product. The key feature of the gate is to allow for easy, potentially automatic, separation of the part from the runner system, while allowing for filling and packing of the part.

It is desirable that the gate be designed to allow for easy removal from the part. This implies that the cross-section of the gate needs to be relatively small. However, too small a gate can restrict the packing of the part, cause over-shearing of the material, jetting, and other gate-related defects. It is commonly recommended that the thickness of a gate, or the diameter of a gate, be 40% to 70% of the wall thickness of the part to which it is attached. The smaller cross-section allows for easier gate removal but increases the potential for the molding problems mentioned earlier. Gate lengths should be as short as possible to reduce pressure drop. A shorter gate will also tend to improve packing in cold runner molds. Gate lengths of 0.75 to 1 mm are desirable.

Evaluating gates for excessive shear:

Determine mold fill rate (Q), volume of all cavities (V_1), volume of the cold runner (V_2) and mold fill time (t). If the mold is not yet in production, estimate fill time or use mold filling analysis to determine optimum fill.

$$Q = \frac{V_1 + V_2}{t} \tag{8.1}$$

The shear rate through a gate with a round cross section can be found by:

$$\dot{\gamma} = \frac{32\left(Q/(number\ of\ gates)\right)}{\pi d^3} \tag{8.2}$$

The shear rate through a gate with a rectangular cross section can be found by:

$$\dot{\gamma} = \frac{6\left(Q/(number\ of\ gates)\right)}{wh^2} \tag{8.3}$$

where d = gate diameter, w = gate width, and h = gate height.

Read Section 4.2.13 and reference Table 4.2 for shear rate limits.

NOTE: The above assumes that fill rate to all gates is balanced. Unbalanced filling can cause shear rate spikes of well over 100% in the last filling cavities.

A part should be gated in a location that does not impair the part appearance. Gates also should be located at the thickest area of the part allowing material to flow from the thickest areas to the thinnest areas maintaining flow and packing paths. Gating into thin sections of a part will result in premature freeze off of the thin region, thereby preventing thicker sections from being packed out. This can result in warpage, sinks, and/or voids. To prevent gas traps, gates should be positioned to ensure that the last area to be filled can be vented. Gating into regions of sudden changes in thickness can also cause hesitation, sink marks, and voids.

There are a wide variety of gate designs that have been developed to address various needs.

8.4.1 Sprue Gate

Sprue gating refers to cases where there is no traditional runner system or conventional gate. It is different from any other type of gate because the part is gated directly from the sprue (see Fig. 8.19). Sprue gates are used in single-cavity molds and provide for the melt to be delivered to the center of the cavity. This is ideal for many cylindrical or symmetrically shaped parts, such as buckets, tubs, helmets, cups, disk-shaped parts, and so forth.

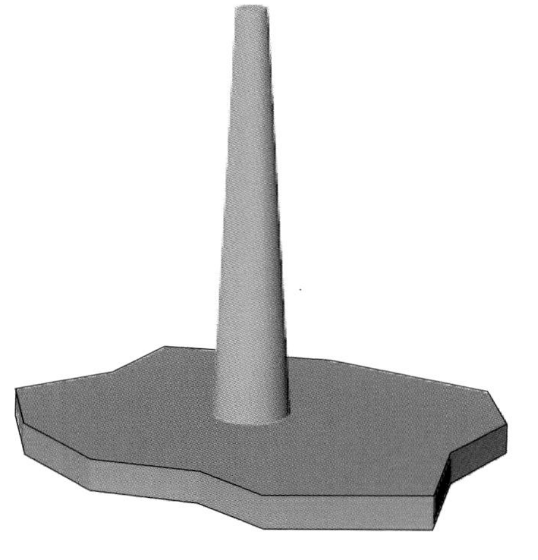

Figure 8.19 Simple sprue gates can be used in single cavity molds

Sprue gates should be located in the thickest section of the part. The sprue is tapered to facilitate ejection with the molded part. For cold sprues, the sprue must be manually removed from the molded part after molding. With a hot sprue, a small diameter, or valve gate, can be used and will result in automatic degating of the part from the sprue. The sprue gate has the tendency to increase the probability of gate blush, particularly when used with glass-filled materials. Direct sprue gating does not allow for cold slug wells. Sprue designs are presented in Section 8.1.

8.4.2 Common Edge Gate

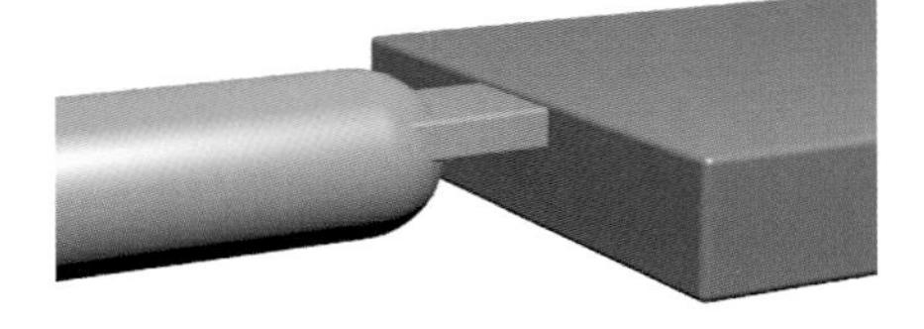

Figure 8.20 Common edge gate

Common edge gates are the most basic type of gate. They are normally rectangular in cross-section and attach to the part, along its perimeter, at the parting line of the mold (see Fig. 8.20). They are used when automatic degating is impractical or undesirable. An edge gate would be preferable in a multi-cavity mold where parts are to be positioned for automated post-molding assembly. The edge gate will remain with the part maintaining the molded part's position and orientation on the runner, which will provide for easy post-mold handling, such as assembly, decoration, or inspection. The primary disadvantage of the edge gate is the need for manual degating when control of post-molding positioning is not required.

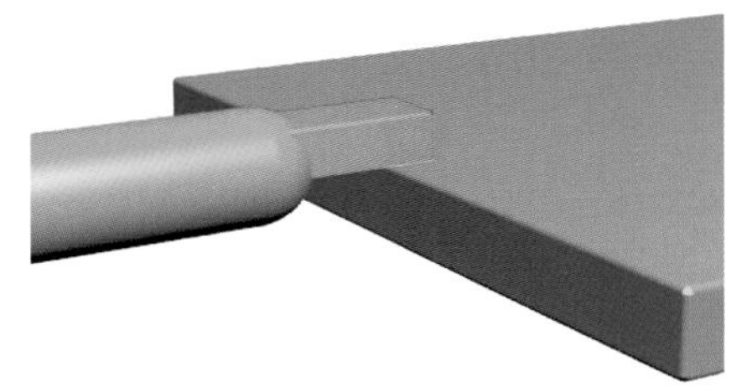

Figure 8.21 Lapped edge gate

Variations of the edge gate include the lapped edge gate (Fig. 8.21) and the notched edge gate (Fig. 8.22). The lapped edge gate reduces the possibility of jetting. The melt traveling through the gate is directed downward opposite the gate onto an opposing wall. The impingement on the wall creates a flow resistance, which inhibits jetting. The major drawback of this design is that the lapped gate cannot easily be cut off in a single operation without leaving a cosmetic blemish, and may require machining to be removed cleanly. To avoid degating problems, this type of gate should be placed in a location where the lapped region will not need to be removed; however, the resulting flow through the cavity from this location should be considered.

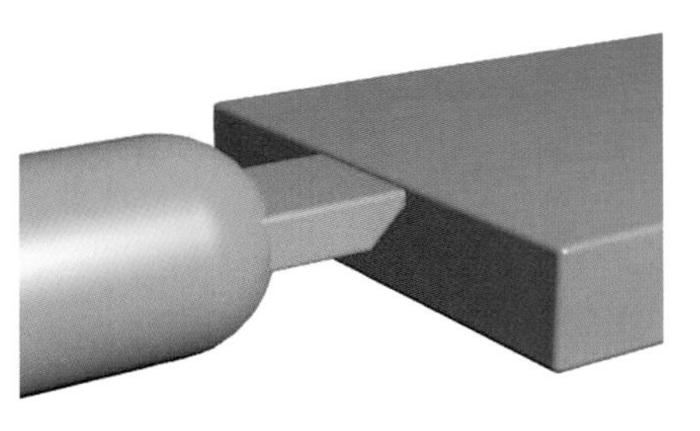

Figure 8.22 Notched edge gate

The notched edge gate is much like a standard edge gate but can simplify post-mold degating and possibly provide for automatic degating when brittle materials are being molded. The notch creates a focused fracture point, which will break in a brittle material when flexed. This flexing could be performed after demolding or while still in a mold that features two-stage ejection. With two-stage ejection, an ejector pin first acts on the runner and gate, raising it at an angle, relative to the stationary part, that is sufficient to break it at the notch. Following this, a second stage in the ejection pushes the part from the mold.

Dimensions:

Edge gates should have a thickness of 0.5 to 0.7 times the part wall thickness. Thicker gates are preferred because they improve packing. Land length should be kept as short as possible, 0.5 to 1 mm is recommended. The short land reduces fill pressure and improves packing. The width of the gate is dependent on ease of degating and melt flow conditions. A narrower gate allows for easier degating, but it also increases flow rate through the gate and increases the possibility of jetting and other gate flaws.

8.4.3 Fan Gate

Fan gates are similar to a basic edge gate in that they are attached to the part at the parting line and require manual degating. The difference is that the fan gate expands out from the runner in the shape of a fan with its widest end opening to the cavity (see Fig. 8.23). The fan region can be relatively thick and feeds a thin gate land, which is attached directly to the part. This design spreads and slows the melt as it enters the cavity. The benefits of the slower flow and the broad uniform melt flow front include improved melt orientation, reduced chance of jetting, reduced stresses in the gate region of the part, and reduced shear rates through the gate. Therefore the fan gate is used to create a uniform flow front into wide parts where warpage and dimensional stability are main concerns.

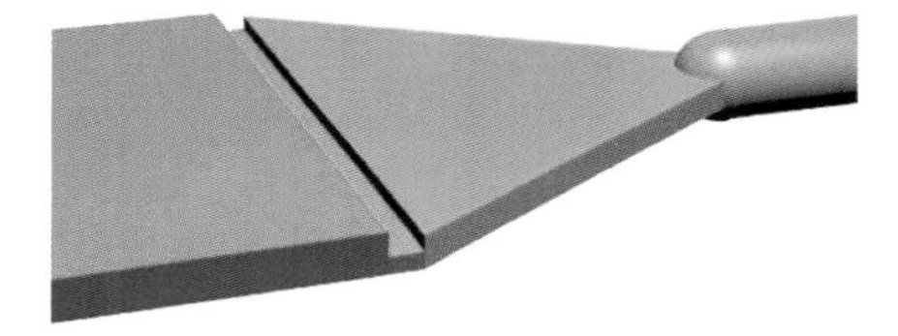

Figure 8.23 Fan gate

Another advantage of the fan gate is its ability to replace several more restrictive gates on a part, which eliminates the formation of weld lines between these gates. The major disadvantage is that its width causes a problem with degating. Manual degating may require fixtures, shearing devices, or machining.

Variations of fan gate designs are illustrated in Figs. 8.24. These designs are used to help improve the distribution of the melt across the thin gate land prior to entering the part forming cavity. The improved melt distribution will reduce hesitation that may otherwise cause the flow front to be less uniform.

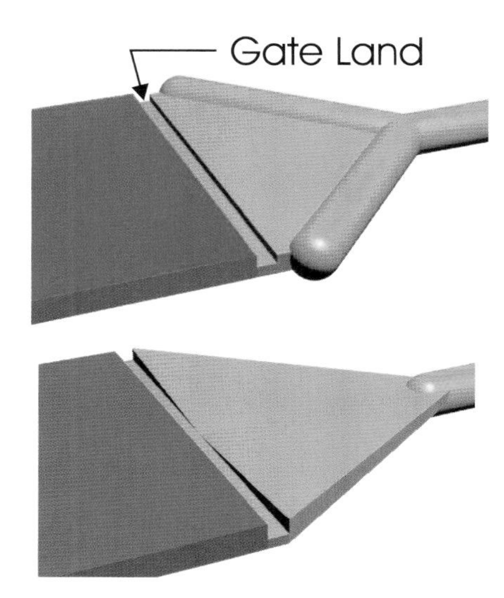

Figure 8.24 Variations of fan gates that helps improve melt distribution at the flow front

Dimensions:

As with other manually trimmed gates, the thickness of the fan gates land should be approximately 50 to 70% of the part thickness where it is attached. A thicker gate land is desirable as the relatively slow flow across the wide gate during compensation will freeze off sooner than a narrow gate. The surface feeding the gate land should be at least the thickness of the wall of the part where it is attached. The gate width is dependent on the application. Width can range from a few millimeters to the entire width of the part. Fan gates over 800 mm have been used in automotive door panels (see Fig. 4.14). Fan gates, particularly wider fan gates, are best designed with the use of injection molding simulation.

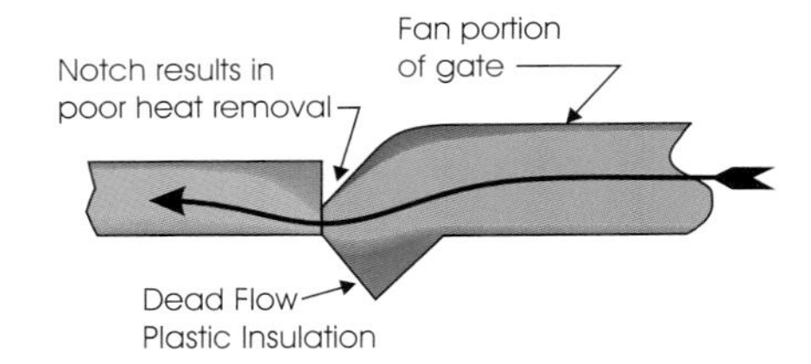

Figure 8.25 Gate designs to improve packing

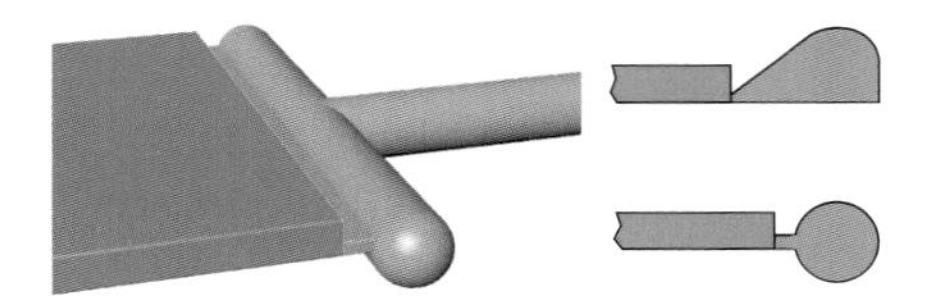

Figure 8.26 Typical film gate designs

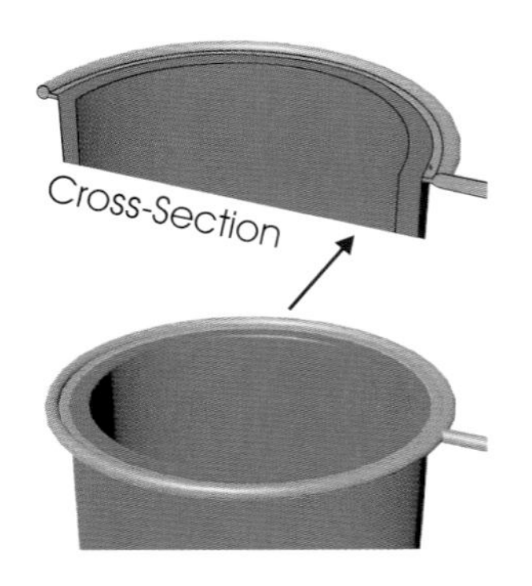

Figure 8.27 Typical ring gate

Wide Gates (fan, film and diaphragm): Wide gates reduce the melt velocity through the gate during both the filling and packing/compensation phase. This will have a positive effect during filling but a potentially negative effect during packing. The low melt flow rate during compensation will cause the gate to freeze more quickly than a narrow gate with the same thickness. As a result, the wider the gate, the thicker it needs to be. If a thinner gate is required, a wedge design with an insulator will improve packing (see Fig. 8.25).

8.4.4 Film Gate or Flash Gate

The film, or flash gate, attempts to capture the advantages of the fan gate, while it uses less space and material. In this design, the runner attaches to a gate manifold that distributes the melt along a broad thin gate land attached directly to the part (see Fig. 8.26). The disadvantage of this type of gate is the fact that the flow distribution across the gate and the flow rate through the gate is less predictable than in the fan gate. The melt has the tendency to hesitate at the thin gate land position closest to where it is fed by the runner. The melt will then race down the gate manifold and enter the cavity at positions away from where the runner feeds it. Increasing injection rate can potentially reverse this scenario. The resulting filling patterns are sensitive to process variations. Film, or flash, gates work best at fast fill rates where hesitation is minimized.

Dimensions:

Entrance of the gate to the part should be approximately 50 to 70% of the wall thickness of the part wall. The gate land can be a wedge shape where the gate tapers down from the larger runner manifold to the part. This will improve packing and allow a thinner gate land at its intersection with the part. The manifold diameter of the film gate is dependent on the length of the gate, the part's thickness, the flow length across the part, and material flow characteristics. The size is best determined using simulation.

8.4.5 Ring Gate

Ring gates are essentially film gates that have been wrapped around a cavity (see Fig. 8.27). They are generally used for gating into cylindrical parts in two-plate cold runner molds. The objective of a ring gate is to eliminate weld lines, provide uniform flow, and resist core deflection. Among the drawbacks of this type of gate are difficulties in degating and highly unpredictable and unbalanced flow, similar to that found in film gates. Use of mold filling analysis, with carefully modeled gates, may help if this gate is required.

Variations of the ring gate include the use of edge or tunnel gates if automatic degating is required; however, this benefit does not come without a price. Flow balance is not always achieved and this could result in poor concentricity or core deflection. Figure 8.28 is a design commonly seen, but is not recommended as it will result in unbalanced filling. Figure 8.29 illustrates a preferred gating arrangement when attempting to achieve a balanced filling. Optimum balance will require the use of melt rotation technology. The obvious downside of using multiple gates is that they will result in multiple weld lines.

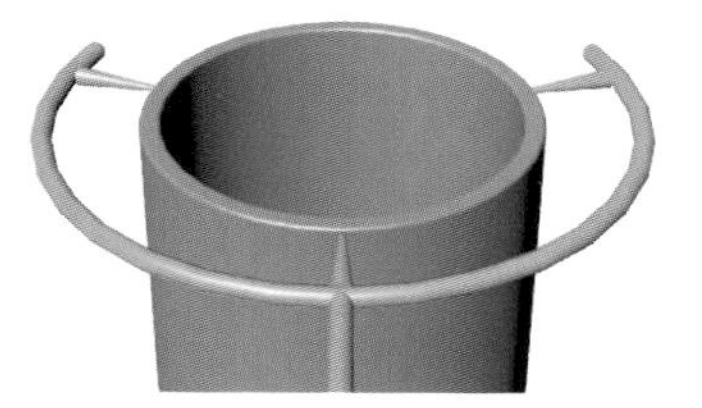

Figure 8.28 Poorly balanced variation of ring gate designed to aid in degating

Dimensions:

Ring gates follow the same general guidelines as film gates. The manifold diameter of a ring gate depends on the type of plastics to be molded, the weight and dimensions of the molded part, and the flow length.

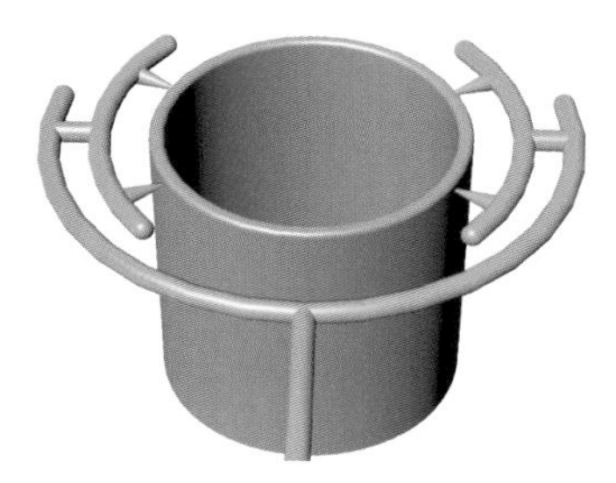

Figure 8.29 Improved balance in variation of ring gate designed to aid in degating

8.4.6 Diaphragm Gate

Diaphragm gates, or disk gates, are used for cylindrical shaped parts that are normally open on either end (like a tube). They are used in three-plate cold runner, hot runner, or single-cavity sprue-gated molds (see Fig. 8.30). Diaphragm gates are generally used when concentricity is an important dimensional requirement. The advantages of this type of gate include the elimination of weld lines, minimal potential for core deflection, and development of an ideal flow pattern for predictable shrinkage and minimal unwanted warpage. From the standpoint of uniform flow, this is the ideal gate for most cylindrical parts. The primary drawback is gate removal difficulties.

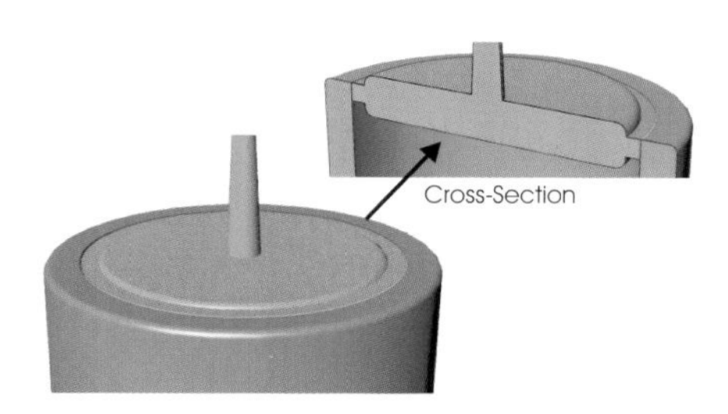

Figure 8.30 Typical diaphragm gate and cross-section

Modifications of the diaphragm, or disk, gate are the spoke, also known as the spider, multipoint, or cross gate (see Fig. 8.31). The advantage of this type of gate compared to the diaphragm/disk gates is its ease of degating. Disadvantages associated with implementing spoke gates are the formation of welds.

Dimensions:

The diaphragm gate should initially be at least 1.25 times the thickness of the part wall to which it is attached. This should feed into a gate land that can either have a straight land length of about 0.5 mm or be wedge shaped (see Figure 8.26). The gate opening to the part should be approximately 50 to 70% of the part's wall thickness. The runner section of a spoke gate should follow the design of a cold runner and the actual gate should follow the guidelines for the particular gate used.

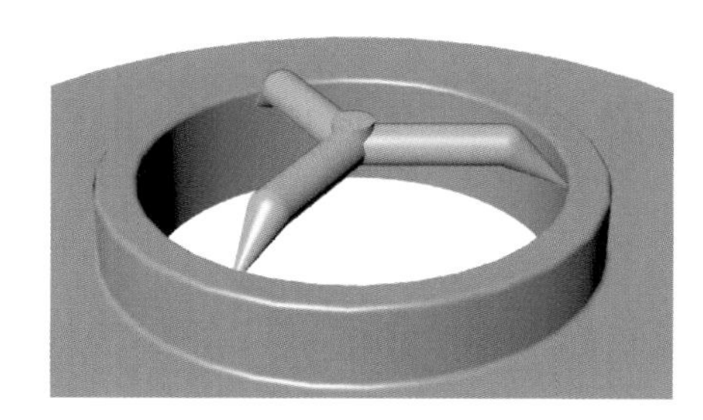

Figure 8.31 Spider type gate that replaces a rig gate. The spider allows for simpler, or automatic, degating

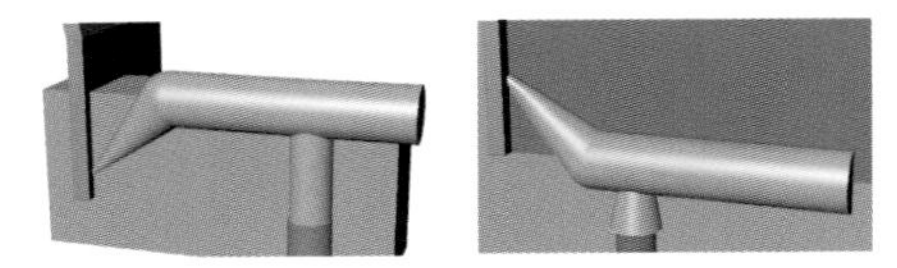

Figure 8.32 Typical tunnel gate designs

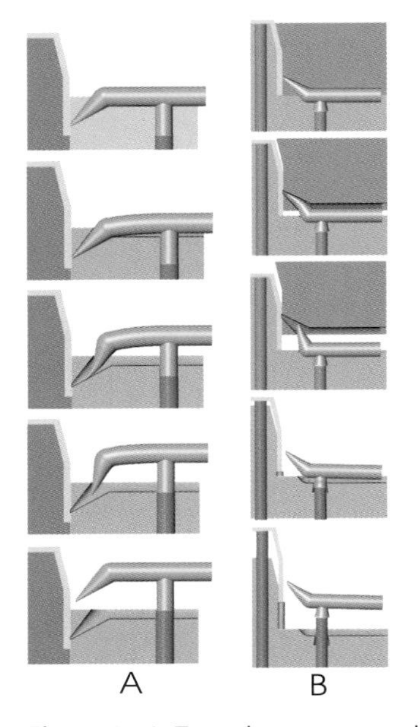

Figure 8.33 Tunnel gates are used when automatic degating is desired. Two methods of extracting the gate from the mold are shown here

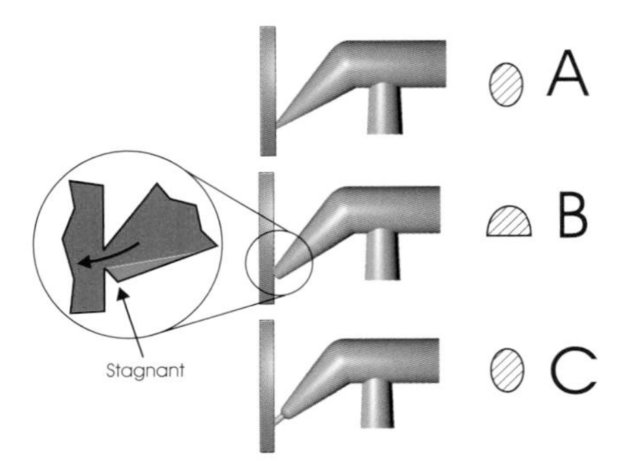

Figure 8.34 Variations of tunnel gate designs. Design "B" is preferred for improved packing and reduced gate vestige

8.4.7 Tunnel Gate

A tunnel gate, also known as a submarine gate, is typically conical in shape, with the smallest end of the cone attached to the part. The gate is cut in the mold such that it tunnels from the runner to the cavity below the parting line of the mold (Fig. 8.32). During ejection, the gate is sheared, or torn away, from the part as it is pulled through the tunnel. Figure 8.33 shows two common variations of tunnel gate ejection. These gates are commonly used to provide automatic degating in standard two-plate cold runner molds; however, tunnel gates can be used in combination with a variety of mold and runner types and configurations providing automatic degating. Its primary advantage is automatic degating in cold runner systems.

The objective of automatic degating reduces the flexibility in design of the tunnel gate. The gate opening must be kept to a minimum to assure that it breaks away from the part during ejection without damaging the part or leaving an excessive gate vestige. However, if the gate diameter is too small, it will result in over-shearing the material and premature freeze-off, preventing proper packing of the cavity. The gates are tapered so that they can be ejected from the mold. The gate-tip diameters typically vary between 40 and 70% of the wall thickness of the part, but they are material- and application-dependent. It is preferred that the plastic material being used is not too brittle because the gate and runner must distort enough to be retracted from the tunnel. The angle at which the gate tunnels toward the part and the taper on the gate will affect how much it is distorted during ejection. Increasing the thickness of the main body of the gate will cause the gate to be warmer and more flexible during ejection, allowing use of brittle materials.

Variations in gate tip design can affect the gate vestige, shearing, packing, and pressure drop experienced by the melt. Design A in Fig. 8.34 is the most common tip design. Design B shown in Fig. 8.34 is less common and has a D-shaped opening to the cavity; however, this design can improve packing, particularly when used with semi-crystalline materials that have sharp melt temperature transitions. As melt passes through the gate, material will be trapped at the base of the gate, insulating the flowing material from the cold steel at the bottom side of the gate tip. This gate can also reduce gate vestige. The drawback of this design is that it is more difficult to maintain consistent gate sizes between gates. In addition, with the D-shaped tip, small amounts of wear will more quickly increase the opening size compared with the more traditional, full conical tunnel gate. This will cause gate-size variations from one cavity to another in multi-cavity molds.

Dimensions:

Gate openings should be 30 to 70% of the wall thickness. However, it is highly likely that parts formed with the smaller gate opening will have poor control of packing and part shrinkage. The length of the gate should be kept as short as possible. The shorter gate will reduce pressure loss, improve packing and decrease ejection problems. Figure 8.35 shows some typical dimensions of a tunnel gate. Figure 8.35B shows a short gate fed by a parabolic runner. The distance between the gate opening and the KO pin provides room for the gate to flex during ejection. The extension of the runner down into the KO pin hole provides positive control of the runner movement during ejection. The design shown in Fig. 8.35B is highly recommended for semi-crystalline materials.

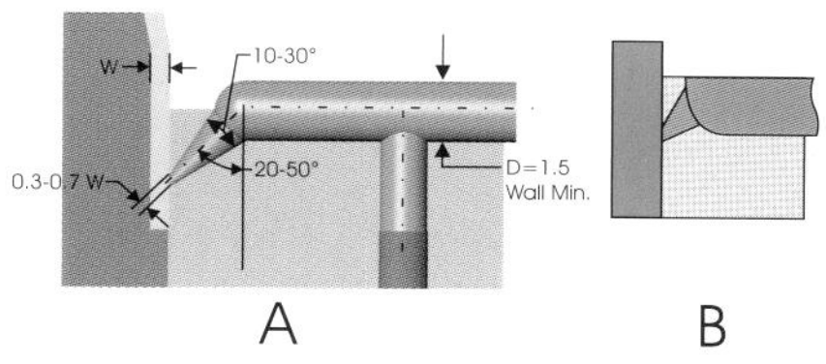

Figure 8.35 Tunnel gates design details

8.4.8 Cashew or Banana Gate

Cashew, or banana, gates are a variation of tunnel gates, except that they can provide gating in regions that cannot be reached by the standard tunnel gate (see Fig. 8.36). The primary limitation is that the curved shape requires that the material in the gate goes through considerable distortion during ejection. This requires that the material have good ductility at time of ejection. If a non-ductile material is used, its ductility can be increased by increasing the cross-section of the body of the gate so that it is hotter and more flexible at the time of ejection. This method of increasing the cross sectional area of the gate body works best with amorphous materials, which have a broad solidification temperature.

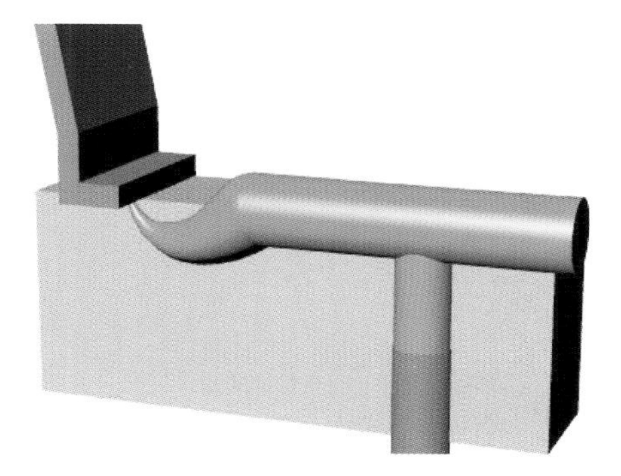

Figure 8.36 Cashew gate

Due to the complex geometry of the cashew gate some companies prefer to purchase specially made gate inserts. Figure 8.37 illustrates a standardized cashew gate insert offered by EXAflow [5]. These powder injection molded gates include a relief along their side walls, which will help thermally insulate the gate from the cooled mold. The warmer gate will remain more flexible and thereby help with ejection. Table 8.1 provides additional data for Fig. 8.37.

Table 8.1 Different ExaFlow Insert Designs for Various Common Plastic Materials

Plastic material	MINIFLOW	GTR/GTE 10 mm	GTR/GTE 12 mm	GTR/GTE 14 mm
Polyolefin Polyamide (PE, PP, PA, etc.)	L = 17–20	L = 20–25	L = 22–27	L = 24–30
Styrenic based materials (ABS, ASA, etc.)	L = 22–27	L = 25–27	L = 27–32	L = 30–35
Thermoplastic Elastomer (TPE) Polyurethane (TPU)	L = 15–20	L = 15–25	L = 17–27	L = 20–30
PA + PF POM	L = 25–30	L = 30–35	L = 32–37	L = 35–40

(Courtesy: Exaflow)

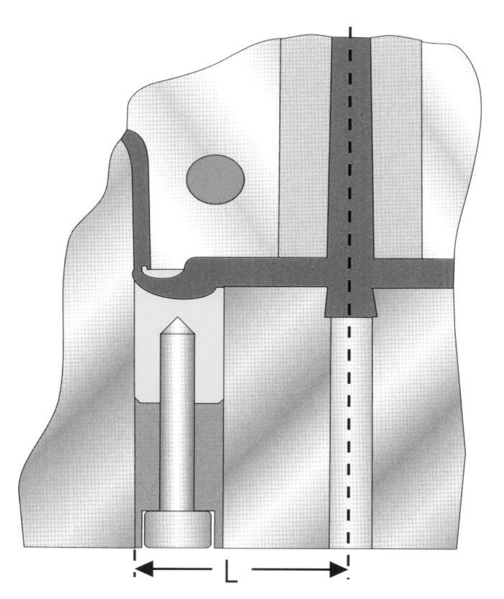

Figure 8.37 Cashew gate mold insert marketed by EXAflow

Dimensions:

Gate openings should be 30 to 70% of the wall thickness. It is highly likely that parts formed with the smaller gate opening will have poor control of packing and part shrinkage.

8.4.9 Jump Gate

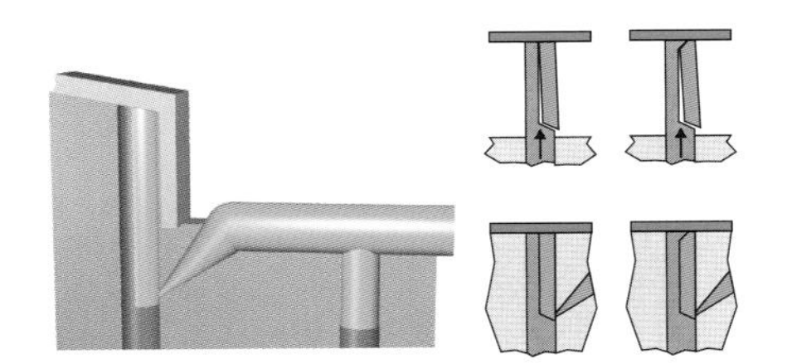

Figure 8.38 Jump gate for gating into interior of a part

Jump gates can be used to eliminate gate vestige from the outside of a part. The jump gate is another variation of a tunnel gate, but this type of gate jumps past the outer wall into some internal feature of the part (see Fig. 8.38). It is used when automatic degating is required in a two-plate cold runner mold, but there is a cosmetic or physical concern with the gate being attached to the exterior of the part. The feature to which the tunnel gate jumps is often added to the part solely for the purpose of providing a connection to the gate. This feature can either be manually removed after ejection, or designed such that it will not interfere with the normal operation or the assembly of the molded part. The design shown in Fig. 8.38 shows the internal part feature being formed by a modified ejector pin. The notch created at the top of the pin allows the feature to be easily broken from the part when a notch-sensitive plastic material is being used.

8.4.10 Pin Point Gate

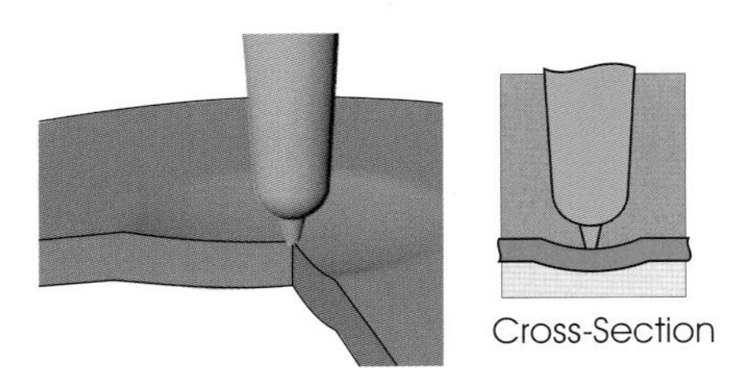

Figure 8.39 Pin point gate as used in 3-plate cold runner molds

A pin point gate is a restricted gate used in three-plate cold runner molds (Fig. 8.39), where the runner system is located on the secondary mold parting line and the part cavity is on the primary parting line. This gate must be small enough to be torn away from the part wall without damaging the wall. It should also be tapered so that it breaks away at its smallest cross-section at the part in order to minimize gate vestage. The wall is often dimpled as shown to recess any gate vestage.

Dimensions:

Like all restrictive tear gates, the diameter of the pin point gate must be kept relatively small. Diameters should be approx. 40 to 50% of the wall thickness and the length should range from 0.5 to 1 mm.

8.4.11 Chisel Gate

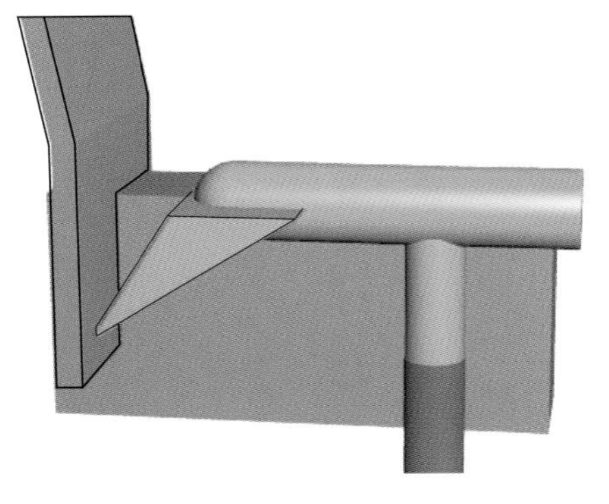

Figure 8.40 Chisel gates are a cross between a tunnel gate and a fan gate

A chisel gate is a cross between a tunnel gate and a fan gate (see Fig. 8.40). Like a tunnel gate, the chisel gate tunnels into the part and is torn, or sheared, off during ejection. Instead of having a circular cross-section, however, the chisel gate has a flat profile. To eliminate an undercut, the chisel gate is widest where it attaches to the runner. It then tapers with a decreasing width and thickness as it tunnels toward the cavity wall.

8.4.12 Tab Gate

A tab, or runner overflow, gate can sometimes be used to reduce jetting or other gate-related problems by slowing down the velocity of the melt as it exits the gate. The melt approaching the gate is split between the gate and the overflow (see Fig. 8.41). This reduces the flow front velocity as the melt initially emerges from the gate. Through careful sizing of the overflow, it is possible to control the melt flow front velocity and minimize problems such as jetting. This method is not fool proof, and in some cases even increases the gate problem by creating excessive hesitation at the gate. Unfortunately, determining whether a tab gate will work is often left to trial and error [6].

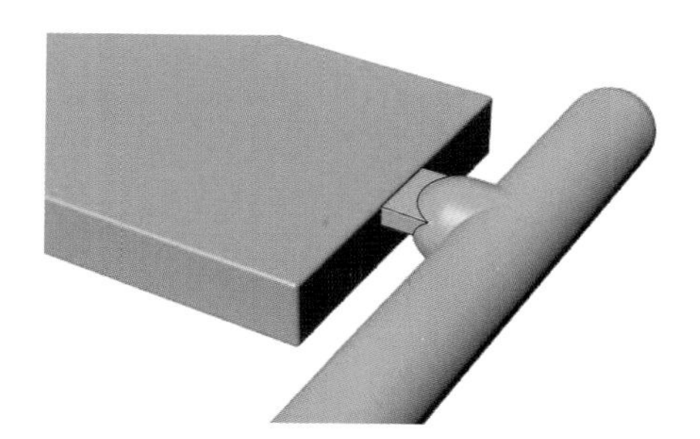

Figure 8.41 Tab gates are used to reduce risk of jetting

References

[1] P. C. Knepper, "The Effect of Runner Diameter on Packing of a Plastic Part with Injection Molding," SPE Annual Technical Conference, Chicago (2004)

[2] P. Auell, B. Martonik, "Cold Slug Wells in Injection Molding," SPE Annual Technical Conference, San Francisco (2002), p. 3561

[3] A. M. Neely, M. E. Hennebicque, "The Effect of Runner Shape on Mold Filling and Product Variation," SPE Annual Technical Conference, Nashville (2003), p. 3368

[4] G. Miller, T. Lacey, "Design and Material Issues Effecting Jetting During Injection Molding," SPE Annual Technical Conference, Nashville (2003), p. 3383

[5] Exaflow GmbH & Co. KG, Steinschönauer Str. 4c, 64823 Groß Umstadt, Germay, http://www.exaflow.de

[6] J. A. Haury, "Predicting and Eliminating Jetting when Using Injection Molding Analysis Software," SPE Annual Technical Conference, Montreal (1991), p. 2659

9 Hot Runner Molds

9.1 Overview

During the past several decades, the use of hot runners has increased to approximately 30% of all new molds. Though there are many advantages, this chapter will point out a number of problems that the molder must be aware of. Most hot runners are purchased from a company that specializes in their design and manufacture. The hot runner is then assembled into a mold.

Some of the challenges the hot runner must overcome include the following:

- The runner must withstand potential internal melt pressures of over 30,000 psi. This is a particular challenge at assembly points within the system and between the system and the mold. The high pressure can create leakage through the smallest of openings, or it can create openings where none previously existed. Additionally, design and construction must be able to resist distortion or bursting at these high pressures. Hot runners today are normally fully enclosed in a housing, which both isolates the system thermally and protects personnel from any risk of a component bursting under the high pressure.
- The runners are required to control deflection and to coexist within a mold, which must also resist deflection under the extreme pressures and forces generated by the melt and the molding machine's clamp. Structural rigidity of the assembled hot runner and mold must be sufficient to allow it to maintain tolerances of the assembled components. These tolerances can be as small as 0.005 mm (0.0002 in.).
- The hot runner must coexist with a cold mold. This requires that:
  - the hot runner and cooled mold cavities be well isolated from each other such that their opposing thermal requirements do not have a negative impact on each other.
  - the heat from the hot runner system be isolated from the molding machine platens.
- The hot nozzles must provide a gate, which will allow plastic to be molten on one side and frozen on the other without clogging, stringing, or drooling. The hot gate must minimize the negative effect of localized heating from the gate region, because this could create localized stress in the molded part.
- The system must provide for thermal expansion of the hot runner relative to the cold mold with which it interfaces. Temperature difference, which can be well over 200 °C, will cause significant relative movement between the hot and cold components during heating and cooling.

- The hot runner must have a flow channel that provides controlled and balanced flow to each of the gates in a multi-drop system.
- The flow channel must be smooth, free flowing, and free of dead flow areas. Potential for flow stagnation exists in corners, junction areas of assembled components, and around invasive flow channel components such as heaters, valve pins, gate inserts, and static mixers.

9.1.1 Advantages and Disadvantage of Hot Runner Systems

The advantages and disadvantages of hot runners compared to cold runners are normally provided by companies that produce hot runners. But because nobody specifically sells cold runners, there is generally little marketing in their favor.

9.1.1.1 Advantages of Hot Runners

One distinct advantage of a hot runner system is that it eliminates the need to deal with the left over cold runner. Other advantages commonly claimed need to be questioned, based on the particular circumstance of the mold. Below is a summary list these claims, followed by a discussion of their validity:

- Faster cycle time (e.g., no sprue and runner cooling time, no injection time for filling runner, no plastification time required for runner material, reduced clamp stroke)
  The claim of faster cycle time is generally based on the expectations that the cycle of a cold runner system is determined by the cooling of the sprue and runner, the additional time needed to fill the runner during mold filling, and the additional time needed for plasticating the melt required to fill the cold runner. The fill time advantage of the hot runner is generally only a fraction of a second. However, in high speed applications, this may be significant. Eliminating the additional cooling time required for the cold sprue is potentially the biggest advantage. This can generally be addressed with a hot sprue, costing about $ 800 and a temperature controller. The cold runner is less likely to effect cooling time if it was properly sized. Even though a cold runner will nearly always take longer to cool than the part, it should be realized that the runner does not have to be solidified to the same extent as the part and can be ejected at a higher temperature. This is particularly the case with semi-crystalline materials. However, there are cases where high cavity filling pressure may require a large runner cross-section to assure filling. The reduced plastification time will not affect most products with the exception of thin-walled high-speed molding.
  The hot runner may be of advantage in clamp movement and time required for ejection of the runner. This is particularly true when

compared to a three-plate cold runner, which must open along two parting lines to the distance required to eject both the part and the cold runner. In addition, the speed of opening and closing the three-plate mold generally must be slower than a hot runner because movement of the additional plates and ejection of the runner requires more control.

- Reduced clamp stroke vs. three-plate
  The hot runner should definitely reduce the required clamp stroke because the mold only needs to open enough for the part to be ejected. Ejection of a runner in a three-plate cold runner mold can easily double the distance required for part/runner ejection
- Reduced energy (plastification, filling pressure, granulators)
  Even when considering the added energy of heating the hot runner, it is normally expected that the hot runner will cause a reduction in total energy used. However, the difference may be minimal. The hot runner will require less material to be plasticated (no runner to fill) and eliminates the use of granulators. However, the hot runner must be heated and additional cooling must be provided to the mold to counter the hot runner's effect on mold cooling and to protect the molding machines platens from heating.
- Consistent melt temperature directly to the cavity
  In most cases, this is not true. The only thermal variation created in a cold runner would be shear-induced. Surrounding mold temperature and its effect on the melt is negligible if the mold has reasonable cooling. Therefore, other than shear-induced melt variations, the thermal history to any given gate will be consistent throughout filling and packing. The hot runner adds potential thermal variations as material is traveling through and residing in regions of the runner that have their own individual temperature control. When material is not flowing (injection and compensation phases), it is sitting in contact with the heated runner. Variations along, and between, heated regions will create variations in the melt temperature. Some molders even purposely vary temperatures across a manifold and between drops to try to compensate for other problems. This can include observed imbalances, which are most likely resulting from shear-induced melt variations.
  In addition, if the melt temperature from the injection unit is not controlled to be the same as in the manifold, a temperature variation can be created within a given shot (beginning to end of shot). This can be further influenced by the volume of material in the shot relative to the volume of material residing in the hot runner. The larger the volume of the runner, the longer the thermal history of the material.
- Increased shot size capability
  It is unusual for the volume of the runner to cause the mold to exceed the shot capacity of the molding machine. However, it is worth reviewing.

- Reduced clamp tonnage
  The hot runner will almost always reduce the required clamp tonnage as there is no reactive force on the clamp provided by a hot runner.
- Improved automation
  A hot runner eliminates the need to design, build, operate, and maintain additional automation equipment that would be required to handle a cold runner.
- Cleaner work environment
  Eliminating the need to handle the runners and the resulting regrind will have this effect.
- Reduced injection pressure
  This is often the case with externally heated manifolds and drops whose flow channels can often be sized larger than cold runners. This advantage becomes more questionable when annular flow channels are used such as in internally heated components and valve gates.
- Elimination of welds with the help of valve gates
  Valve gates have the distinct advantage of sequencing their injection and potentially eliminating the flaws created by a weld line.
- Use of stack molds
  Though stack molds can be produced with cold runners, this is rarely done and cold runner designs would be very limiting.
- Ability to address mold filling imbalances by individually controlling drop temperatures.
  As discussed in Chapter 7, this is a false and fickle solution to imbalance. Balance of filling, packing, and melt conditions are the objective. Creating a filling balance by varying drop temperatures will increase variations in melt conditions and packing, which will create product variations. In addition, it corrupts the critical thermal design of the hot runner system, resulting in a narrow and sensitive process window and can extend cycle time because the drops – requiring extra heat for attempting a fill balance – will dictate the cooling time (see Section 7.3.1).

9.1.1.2 Disadvantages of Hot Runners

In addition to the many successes, companies have found in using hot runners, there are also a number of horror stories. Most systems are built by one of the dozens of companies located worldwide that specialize exclusively in hot runner design and production. Despite the specialization of these companies, troubles with startup and operation of these systems are commonplace. Many of the problems are caused by the lack of operator skill when implementing a hot runner or by the unique features of the system they are utilizing which they have not been trained to use. In addition, the market for hot runner systems is very competitive, which sometimes causes a company to over-sell their capabilities, and under-advise their customers of potential problems. Many of these companies are relatively small, with only a minimum research budget. At least some parts of all hot runner systems are custom made; however, competitive pricing limits the engineering that can go into a system. Where injection molding simulation should be used with most applications, rules of thumb and generalized guidelines are used to economize time and money. This generally is not the fault of the hot runner supplier, but rather caused by the customer. Management paradigms dictate that we drive the purchase price of molds down and demand a fast delivery. This results in initial hot runner costs being lowered, while the cost for start-up and operation is increased. This also results in longer startup times and lower productivity. Low cost and fast delivery drives down quality in design and manufacture, leading to compromised capabilities.

The most significant fault of a hot runner is often the purchaser of the system. Short-sighted buyers drive costs down by reducing engineering and ignoring need for training. The result is long-term problems and costs.

Below is a summary of a number of negative features of using hot runner molds. Many of these can be minimized by providing for increased system engineering and increased operator training.

- Requirement for operators, process technicians, process engineers and maintenance with a higher skill level
- More process sensitive regarding issues such as gate freeze, gate drool, balancing of flow and melt conditions to each gate, over-packing, thermal control, and foreign impurities in the system resulting from stagnant or slow flow areas
- More difficult to achieve a consistent gate seal than can be achieved with cold runners
- High cost of hot runner system including electrical controllers and hydraulic or pneumatic controls required for valve gating
- Cost of maintenance for hot runner and control systems
- Unscheduled cycle interruptions caused by clogged gates, damaged electrical components, and leakage
- Does not provide for moving of gate locations without designing and building a new hot runner as well as significant rework to mold
- Potential for extended down time caused by system leakage. Leakage can cause severe damage to the electrical system and valve gate components
- Potential for severe damage from heater voltage error resulting from mismatched heater and controller
- Little flexibility in changing plastic materials as assembly seal-off of most runner components is temperature specific. Materials with significantly different operating temperatures would be most limiting. In addition, flow channel geometry is not tolerant of material changes and gate tips are commonly material specific
- Extended time required for color change. A correctly designed system and proper process procedure can minimize this problem

- Some temperature-sensitive materials are difficult or, depending on the system chosen, impossible to run in hot runner systems
- Cooling of cavities in the stationary, hot half, is more complex. The molding machine and cavities must be insulated from the heat of the hot runner. In addition, gate cooling is sensitive and specific to the given material run. Further, the presence of hot drops reduces available cooling locations and creates differential cooling across the wall of the part where the gate tip reaches the part
- Shot control can be more difficult owing to melt compressibility in the large molten hot runner channel
- If filling is pressure limited, increasing runner diameter in the manifold, if possible, would be very expensive. Hot nozzles would generally require replacement.

9.1.1.3 Summary of Attributes of Different Runner Systems

Table 9.1 provides a limited contrast of some of the more distinct advantages and features of the various types of hot runner systems and cold runner systems.

Table 9.1 Summary of Attributes of Different Runner Systems

Mold Design Systems	Distinct Advantages/Features
Two-plate cold runner	Easy color changes Easy to operate and maintain Low cost Tolerant of all plastic materials Consistent operation
Three-plate cold runner	Gate location flexibility Lower cost compared to hot runner Fast start-ups Easy color changes Tolerant of all plastic materials Consistent operation
Hot runners in general	Gate location flexibility Less runner scrap Decreased cycle time with thin-walled parts Reduced clamp tonnage
Externally heated manifold and drops	Low pressure drop
Externally heated manifold with internally heated drops/nozzles	Reduces potential for leakage

Mold Design Systems	Distinct Advantages/Features
Internally heated manifold with internally heated drops/nozzles	Eliminates most leakage problems Potential for good gate tip control Improves on isolating heat from the cavity
Insulated hot runner	Low cost compared to other hot runner systems Low pressure loss Less complicated design

9.2 Overview of Multi-Cavity Hot Runner Systems (Contrasting Systems)

A multi-cavity hot runner system is similar to that of the cold runner in a three-plate mold in so far as the runner travels behind the cavity plate, providing flexibility in gating locations. Flexible gating is achieved without the downsides of cold runners. The designs of hot runner molds are quite complex relative to the simplicity of cold runner molds. The hot runner system is comprised of the sprue, manifold, and the drop. The sprue introduces the melt to the manifold from the machine nozzle. The manifold delivers the melt to selected positions behind the cavity plate. The manifold, contained within the mold body, normally travels parallel to the molding machine's platens. Hot drops then provide passage for the melt from the manifold either directly to the part-forming cavity or indirectly by connecting to a cold runner, which might feed multiple part-forming cavities (indirect feed). The drop is generally, but not always, positioned at 90° to the manifold and travels through the cavity plate.

The manifold in a hot runner system is internally heated, externally heated, or insulated. The drops are internally heated, externally heated, heat conducting, or insulated. Each heated manifold and drop has its own individual heater and temperature controller. A heat-conducting drop is made of a highly thermally conductive material, such as beryllium copper, and conducts its heat from the manifold. An insulated hot runner system has no direct heat source other than the molten plastic flowing through it. Each of these systems is described in more detail in the following sections.

There are several combinations of manifolds and drops used to create the various hot runner systems. Each combination, rather than each separate type of manifold and drop, will be discussed in the following sections. Figure 9.1 illustrates the more common combinations:

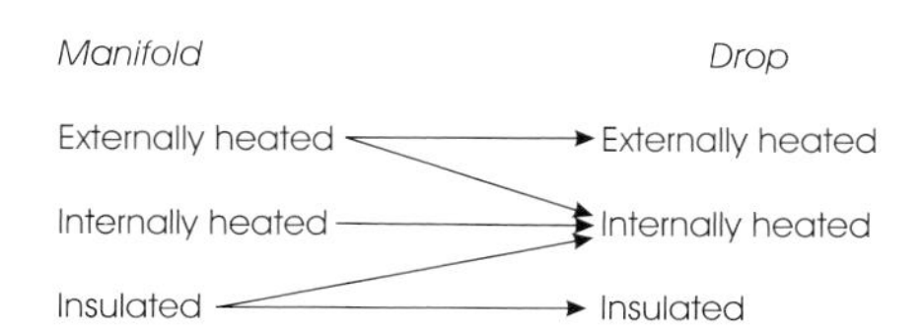

Figure 9.1 Externally heated hot manifold and nozzle combination

Owing to the complexity of the hot runner systems, most are produced by companies specializing in their design and manufacture. A company building a hot runner mold normally purchases standard components, or an entire hot half runner system, from one of these specialized companies.

Because there are numerous competing companies, there are numerous variations between the systems offered. The following examples present some of these variations.

9.2.1 Externally Heated Manifold and Drops/Nozzles

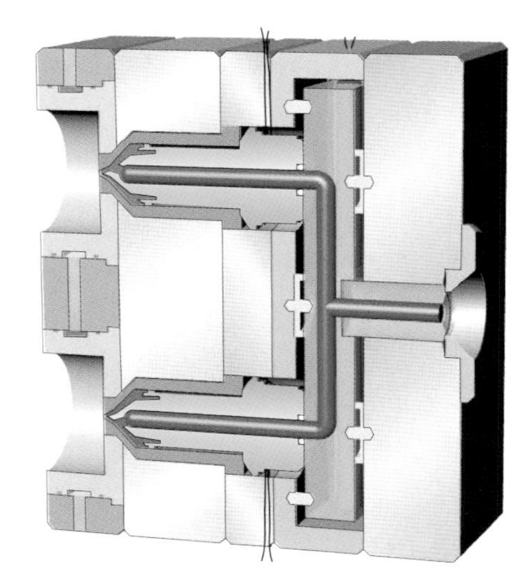

Figure 9.2 Externally heated manifold and internally heated drops

With the possible exception of insulated hot runners, externally heated systems (see Fig. 9.2) have the ability to provide the lowest pressure drop of all mold types. The flow channels are cylindrical in cross-section and generally have a larger diameter than a cold runner mold. The cylindrical flow channel is the most efficient shape for flow. Because cooling of the runner and regrind are not issues in the case of hot runner, a larger diameter is permissible. Both the larger diameter and the fact that there is no growing frozen layer in the runner system contribute to the relatively low pressure drop in these types of molds.

Of the various hot runner molds, externally heated manifolds and drops are recommended for thermally-sensitive and high-viscosity materials owing to their free-flowing open channel. This type of system may also provide the most homogeneous melt temperature of all hot runner systems. Melt injected at a temperature of 260 °C (500 °F) can be surrounded by a flow channel that is heated to 260 °C (500 °F). Therefore the temperature gradient across the flow channel can be reduced, in contrast to an internally heated system, which is surrounded by a cold wall. In addition, because the heating of the manifold and drop is external, a frozen layer of plastic along the outer channel does not develop as with internally heated or insulated systems. Without this frozen layer, a larger flow channel will be available and color changes can be achieved more easily than with an internally heated hot runner system.

Disadvantages of an externally heated runner system include the potential for leaking of molten plastic and the difficulty of isolating its heat from the cold mold cavity. Improper design, assembly, or operation can result in plastic leaking between the drop and the manifold. This leaking plastic can engulf the manifold, entombing it thus destroying heaters, wires, and thermocouples. In addition, the external heat source is in direct conflict with the cooling of the mold. The drops are comprised of a heated nozzle, which must be surrounded by an air space to help insulate it from the cooled cavity. This air space will require additional space in the mold. The proximity of nozzles approaching a cavity not only introduces localized hot spots but also limits positioning of cooling lines. Externally heated drops also have more limited control of the critical gate tip. Heat conducting gate tips are often used and, when properly designed, can minimize these problems. The tip conducts heat from the heated outer nozzle body. However, it should be realized that the same heaters controlling the nozzle body temperature are controlling the gate tip temperature. This restricts the control and may result in a compromise of temperatures for each of these two regions of the drop.

9.2.2 Externally Heated Manifold with Internally Heated Drops

The internally heated drop (see Fig. 9.3) can potentially provide better gate tip control (less freezing, drooling, stringing, etc.) because heat can be applied more directly to the tip. This potential advantage of gate tip control is highly dependent on the design. In addition, as the heater is surrounded by plastic, the heat from the probe is more naturally isolated from the mold cavity, where the part is trying to freeze. Furthermore, as the heater is located in the midstream of the plastic, a frozen layer will develop around the perimeter of the cooled flow channel. This frozen layer helps act as a seal around the flow channel and reduces the potential of leaking at the juncture of the manifold and the drop. In addition, as the internally heated drop does not include a nozzle requiring an air gap, it eliminates the leakage potential at its gate end.

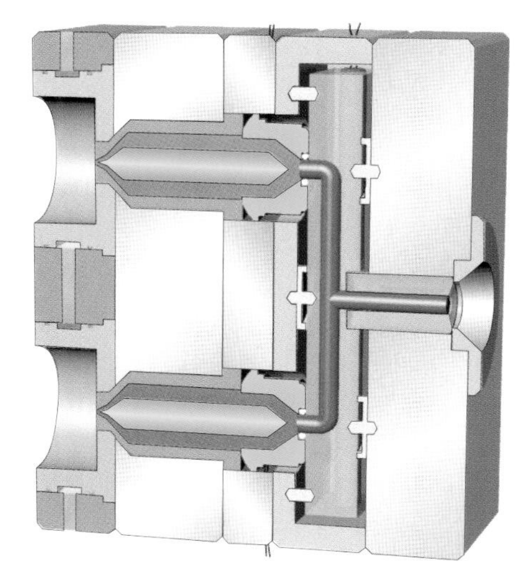

Figure 9.3 Example of an externally heated manifold with internally heated drops

Negative aspects of the internally heated drop include the fact that, because the heater is located in the middle of the flow channel; it causes a restriction to flow. This location can also result in stagnant flow regions depending on how the heater is located in the melt stream. These factors combine to increase pressure loss during mold filling, the degradation of material, problems with color changes, and undesirable temperature variations in the melt (created between the heated probe and the cold mold wall).

The internal heater creates an inefficient annular-shaped flow channel. The result is that the pressure drop through an internally heated drop can be significantly higher than through a full round drop with an equivalent amount of plastic material. This may not have much significance if there is sufficient pressure to fill the mold cavities. Although the diameters of the flow channels can be increased to decrease filling pressure, one must be cautious of the resulting increase in material residence time and poor flush rate.

During normal color changes, the material in the frozen layer will never be flushed and can continue to bleed into the cavity. Acceptable color changes can often be achieved by first running hot, fast cycles to melt and reduce the thickness of the frozen layer of the colored material. The second colored material is then added as the process is returned to normal. The result is that the frozen original color is covered by a frozen layer of the new color. If no color changes are required, this is not an issue.

Due to the non-ideal flow channel of an internally heated drop, it generally cannot be used with thermally-sensitive materials. It should also be used with caution when molding clear and light-colored materials. If stagnation points exist, with time many materials will degrade in these locations, break away, and undesirable streaking or black specks will develop in lighter colored parts.

9.2.3 Internally Heated Manifold and Internally Heated Drops

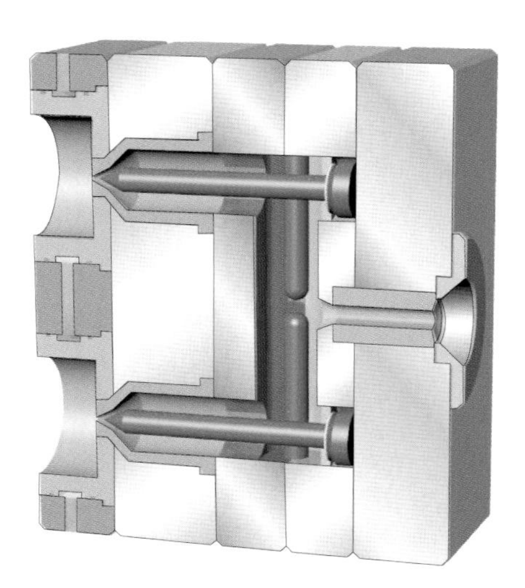

Figure 9.4 Internally heated manifold with internally heated drops

The combination of internally heated manifold and drop eliminates most leaking problems, naturally provides good isolation of the heater from the surrounding mold, and potentially provides for improved gate tip control (see Fig. 9.4). As heaters are internal, there is no need for a separate manifold block or nozzle, which would have to be heated and insulated from the surrounding mold with an air space. Plastic around the perimeter of the flow channel freezes against the colder mold, solidifies, and provides a thermally insulating boundary of plastic. This reduces the challenge that exists with an externally heated hot runner system of insulating the heat from the part-forming cavity where the plastic is trying to freeze. In addition, without the air gap surrounding the manifold in an internally heated system, leakage concerns are virtually eliminated.

Many of the concerns with the internally heated manifold and drop are similar to those discussed previously regarding the use of the internally heated hot drops. Some of the problems are further complicated by the increased use of internal heaters and the required crossing arrangement of these heaters (see Fig. 9.4). This crossing arrangement of internal heaters significantly increases the opportunity for stagnant flow regions to develop in which material can hang up and degrade. Therefore, these systems are not recommended for use with thermally-sensitive materials. In addition, the pressure drop during mold filling will be the highest of any of the hot runner molds per amount of material in the runner. The flow cross-section can be increased to reduce the pressure but will result in increased residence time and reduced flush rate for the molten material. The development of the frozen layer around the perimeter of the flow channel continues to create a challenge with regard to color changes.

9.2.4 Insulated Manifold and Drops

In this type of runner (Fig. 9.5), no heat is introduced other than the heat from the molten plastic flowing through it. Therefore, the insulated system must be run with fast and regular cycles to ensure that the melt in the manifold and drops does not solidify. A very large diameter flow channel is generally used [often 20 to over 30 mm (0.79 to 1.2 in.) diameter]. The first plastic injected into the mold fills the runner system. The plastic material along the perimeter of the cooler flow channel freezes, acting as a thermal insulator. As the mold cycle continues, new material passes between these frozen outer insulating layers. Keeping this flow channel open depends on material moving through the runner fast enough so that the molten material in the center is continually replaced and does not have a chance to freeze completely.

The main advantages of the insulated systems over the heated systems is low cost, the ability to have thorough color changes, low pressure loss due to the large diameter runners, and less introduction of heat to the part-forming cavities.

The molds are designed such that a parting line is located along the melt's flow channel. This parting line can easily be opened while the mold is still in the molding machine. Once the melt has frozen, this parting plane can be opened and the runner completely removed. This technique provides for very thorough color changes as well as for easy servicing the mold if the runner inadvertently freezes off.

Despite its advantages, this type of hot runner is rarely used owing to the many drawbacks related to its lack of temperature control. Any interruption or variation in the cycle can cause significant variations in flow and melt conditions as well as the potential for the gate or runner to freeze off completely. In addition, there is an inherent variation in melt temperature across the flow channel, which can cause unacceptable results with most materials. This system is normally limited to low-tolerance parts and commodity plastics such as polyethylenes, polypropylenes, and polystyrenes. Often, internally heated drops are used in conjunction with an insulated manifold to increase temperature control of the gate tips (see Fig. 9.5). The top half of Fig. 9.5 illustrates the use of a heated probe, while the lower half illustrates a fully insulated drop. Insulated runner systems are not sold by companies that specialize in designing and building hot runners. The designs, and use, of these systems are more of an art developed by a handful of companies. The analysis of insulated runners is not addressed very well by the commercial injection molding computer-aided engineering (CAE) programs.

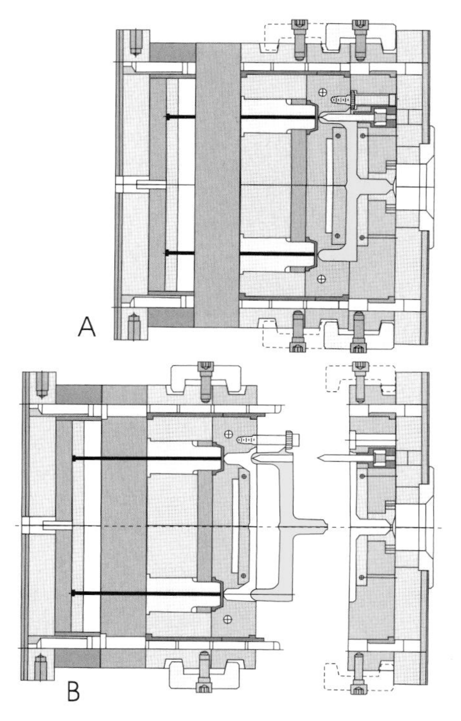

Figure 9.5 Insulated hot runner system during molding (Fig A) showing an example of an internally heated drop (top half) and a fully insulated drop (no heater – bottom half). Figure B shows the mold open for removal of the runner when frozen

9.3 Stack Molds

Stack molds are specially designed molds that allow doubling, tripling, or even quadrupling the number of cavities in a mold with very little increase in clamp tonnage or platen size. This is accomplished by adding multiple parting lines, where cavities along a given parting line are positioned back-to-back relative to cavities in the additional parting lines. This design ensures that any cavity positioned back-to-back does not increase the projected area and therefore does not increase the reactive force on the clamp. Though two-level stacks are most common (Fig. 9.6). Stack molds are also made with three and four levels (see Fig. 9.7) The additional total mold height of a stack mold generally requires a specially designed clamp with increased daylight and increased clamp movement.

Most stack molds are fed using a hot runner system (see Fig. 9.6). The melt is fed from the machine nozzle through a "sprue bar" to a hot manifold positioned between parting lines. In a two-level stack, the sprue bar passes across a first parting line to the center section of the mold where the hot manifold is located. The manifold then delivers the melt to drops feeding the cavities along both parting lines of the mold. During mold opening, the

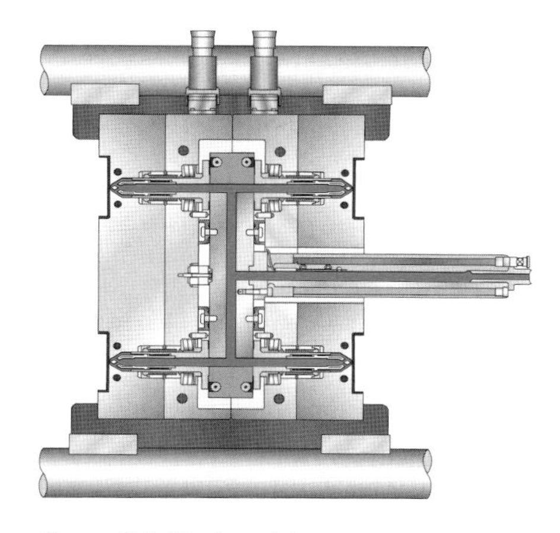

Figure 9.6 Stack mold

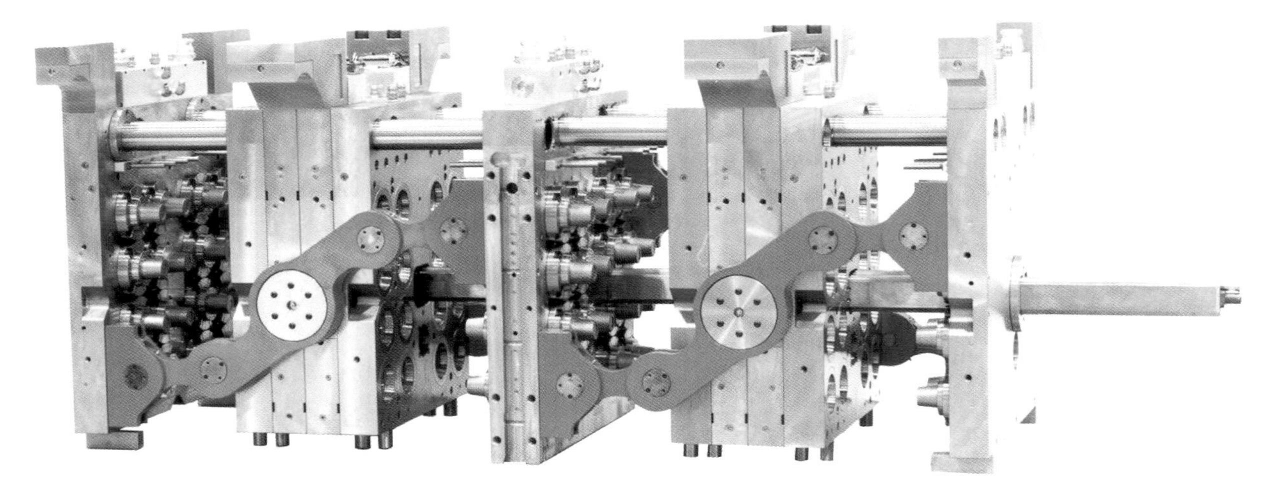

Figure 9.7 Multi-level 4 × 16 stack mold (Husky)

sprue bar is pulled away from the injection molding machine's nozzle as it remains attached to the central hot manifold.

Figure 9.8 shows a 128-cavity, two level stack mold hot runner system prior to assembly into the mold.

In some cases it is desirable to mold large parts in a stack mold. These may be positioned as 2 × 1 stacks, which locate two cavities back-to-back, central on the mold. Cavities located in the center of the mold do not allow a sprue bar to go through the center of the mold. In such a case, the central manifold can be fed by a flow channel located external to the mold (see Figs. 9.9).

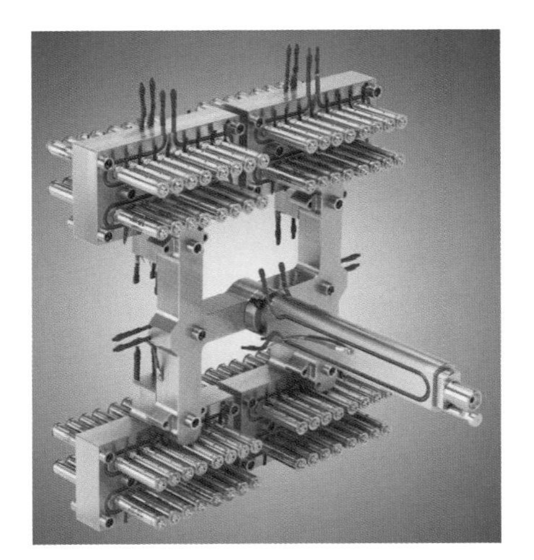

Figure 9.8 128 drop 2 × 64 hot mold manifold (INCOE)

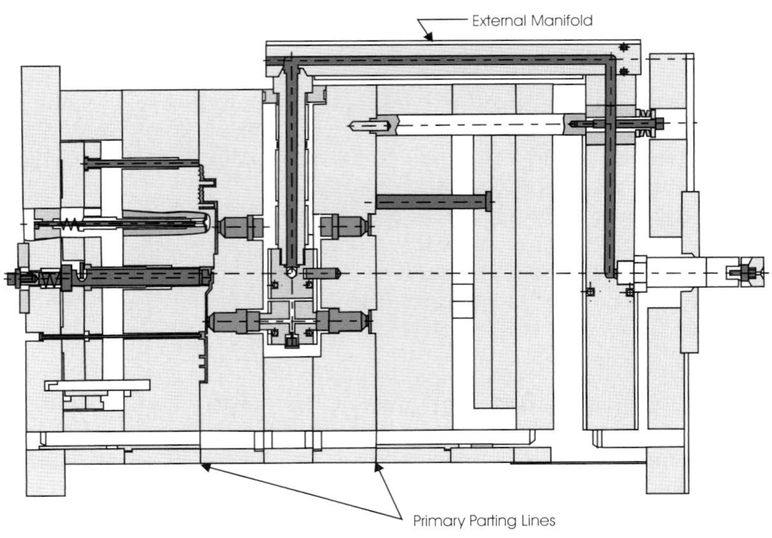

Figure 9.9 Hot runners allow the melt to be routed external to the mold. This method is sometimes used in stack molds molding a large part positioned in the center of each of the two parting planes

10 Hot Runner Flow Channel Design

The flow channel design of a hot runner is more critical than that of a cold runner. Problems with a cold runner are relatively easy to address. Relocating, resizing, or reshaping of a cold runner is relatively simple and inexpensive. Similar actions in a hot runner are much more complex and expensive. Modification in runner layout would require the replacement of the entire runner system. In addition, because the cold runner is ejected after every cycle, it eliminates any concerns of stagnant flow that must be addressed in hot runners.

The hot runner channel is normally gun-drilled as opposed to a cold runner, which is milled onto the surface of a plate. The drilling operation dictates much of the design of the hot runner. It limits the channels to straight sections. Changes in direction are accomplished through cross-drilling. There are a couple of companies such as P.E.T.S. that cast their hot runner, which provides them opportunities not available in drilled manifolds. Despite some of the limitations, the fact that the hot runner does not have to be ejected between cycles provides additional opportunities for runner layout and design that could not be achieved in cold runners. This includes improved geometrical layouts through the use of level changes and the use of internal features such as static mixers and valve gates.

It is required that the hot runner flow channel be smooth and free of any areas that will prevent material from being freely flushed. This creates a challenge in locations where channel components are assembled, with internal to flow channel features, or at corners and branches. Even small misalignments of a tenth of a millimeter between the manifold and the drop channels can create a problem with some materials and processes. To help with smoothing of channels, after machining, hand working of components, and assembly, the manifold is sometimes treated to an extended flush with an abrasive slurry. Extrude Hone's ECPolishing™ is one method that takes rough surfaces from drilling, milling, and turning and reduces the surface roughness. A fairly complex geometry can be improved from a 300 microinch R_a internal finish to a 10–20 microinch R_a finish in less than 10 minutes [1]. The ExtrudeHone processes use a polymer carrier mixed with abrasives. The viscosity of the carrier and the abrasive grain size and concentration can be altered to achieve various finishing results.

10.1 Layout for Balanced Molding

Though the layout of most hot runners are somewhat limited due to the need of straight drilled sections, they do allow for runner sections to be layered and manifolds to be stacked, which improve the opportunities for

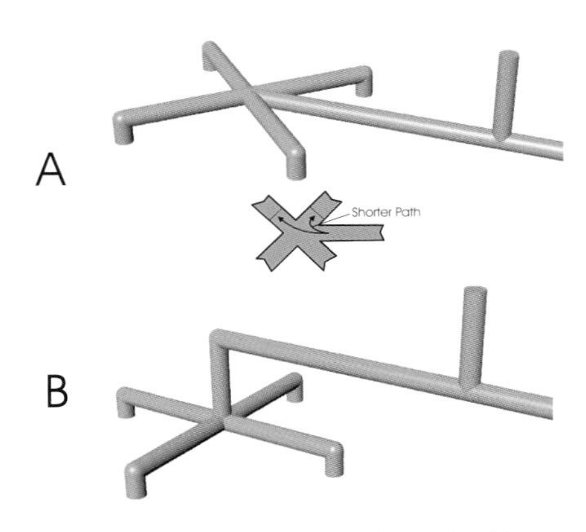

Figure 10.1 Single level "X" branched hot runners (A) result in variations in flow length between inner and outer branches. Use of a level change (B) improves the geometrical balance

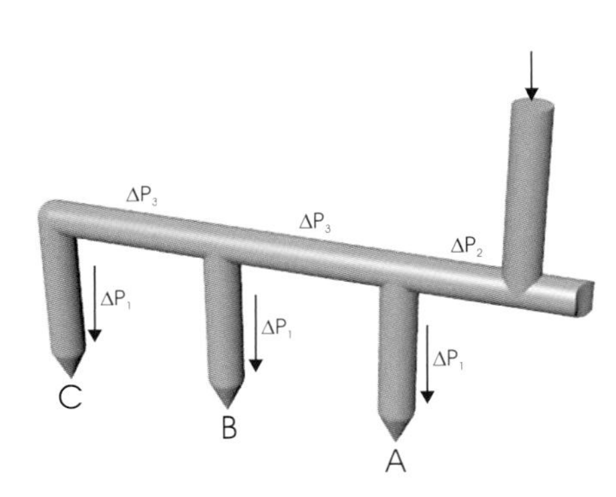

Figure 10.2 In this 6-cavity fishbone runner layout, pressure to fill cavity A is $P_1 + P_2$; for cavity B it is $P_1 + P_2 + P_3$; for cavity C it is $P_1 + P_2 + 2P_3$. Increasing the primary runner diameter will improve filling balance

achieving geometrically balanced runner layouts. This is particularly the case when "X" branched runners are used (see Fig. 10.1A–B) On close examination of the intersection of a feed runner to the branching "X," Fig. 10.1A shows that flow lengths to the inside vs. the outside branching runners are different. In addition, any additional flow effects, such as elastic flow conditions or inertia (normally considered negligible) may cause variations between flows. The alternative design shown in Fig. 10.1B places the feed runner at a different level and a bridging runner that approaches the "X" from above, rather than the side. This provides an ideal geometric balance and eliminates the variations in the two flow conditions mentioned earlier. In addition, channels can even be routed externally to the main body of the mold. This approach is sometimes used in 2 × 1 stack molds of large parts as shown in Fig. 9.9.

A geometrically balanced runner layout is suggested in order to improve potential for uniform filling to each gate in the hot runner system. The geometric balance should increase the process window and improve the melt consistency between parts. Geometric balances were previously discussed in Chapter 5.

As with cold runners, hot runners are also sometimes laid out in fishbone arrangements. These will not provide as good a flow or melt balance as a geometrically designed layout. A difference in the melt's temperature, shear, and pressure history to each cavity along the branch is unavoidable. Therefore, it can be expected that parts molded in each of the cavities will form under somewhat different conditions.

If the fishbone design is absolutely required, it is suggested that the manifold feeding the drops be kept relatively large. This reduces the pressure drop between the progressive cavities along the main channel. In Fig. 10.2, pressure to fill cavity A is $P_1 + P_2$; cavity B is $P_1 + P_2 + P_3$; cavity C is $P_1 + P_2 + 2P_3$.

Elevation changes in hot runners have been used for a number of reasons. They not only increase the chances to achieve a geometric balance of the channel but they can help address alignment problems created from thermal expansion of the hot runner relative to the cold mold. In externally heated molds with high cavitation, there can be a concern with the alignment between the numerous hot nozzles and the manifold flow channels.

10.2 Cross-Sectional Shape

Flow channels in hot runners are either full round or annular. The only exception occurs in some insulated molds that have flow channels cut as trapezoids; however, as material around the perimeter freezes, the actual flow becomes closer to full round.

As most hot runner channels are drilled, their perimeters are nearly always circular. Use of internal heaters or valve gates will create an annular flow channel. The only exception to the full round, or annular shape, is in locations of the gate or when a static mixer might be used.

The full round channel is preferred as it has the minimum pressure drop per cross-sectional area of material. This allows smaller channels, which reduces the mass of material in the manifold and the relative residence time. The lower pressure drop allows for a smaller channel, which improves the flush of material through the channel and particularly along the flow channel walls. In addition, this channel shape has the least potential for slow or stagnant flow areas, which can lead to material degradation and color change problems.

Annular channels result from use of internally heated hot runners and drops as well as with valve-gated drops. Compromising by using a non-full round runner is typically done to achieve other desired effects. The most common application of annular flow channels is with the use of valve gates and internally heated drops. There are also some cases where internally heated manifolds are used along with the internally heated drops. Section 5.5.2 compared the relative pressure drop of a full round runner to that of an annular shaped runner. In the example provided, pressure drop was 1200% higher for the annular flow channel with an equivalent cross-sectional area to the full round runner. This can be significantly influenced by actual flow channel size, flow rate and material viscosity.

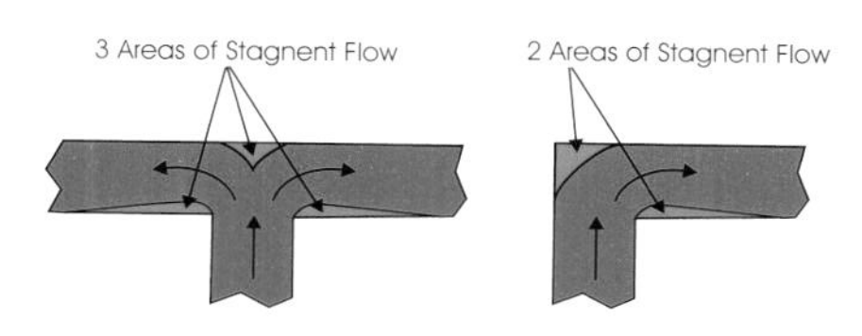

Figure 10.3 Low flow areas developed in corners and branches of hot runners

10.3 Corners

Corners are always a concern in a hot runner mold as they can result in slow flow or stagnant flow areas. Figure 10.3 illustrates expected slow flow and stagnant flow areas in both a "T" branch and a 90 degree corner. The corners in a hot manifold are initially formed by cross-drilling. To create a manifold for the four drop runner layout shown in Fig. 10.4A, the primary runner is first gun-drilled as a through hole (Fig. 10.4B). Next, the two secondary runners are cross-drilled, intersecting the center of the primary runner (Fig. 10.4C). The alignment of these crossing holes is typically within 0.1 mm (0.004 in.). The vertical entrance (sprue) and exit channels feeding the hot drops are then drilled at 90 degrees to the previous channels (Fig. 10.4D). Each of the through holes is then plugged at the six locations shown in Fig. 10.4E. The plugs are designed to minimize dead flow. The plugs must be carefully machined and assembled to assure alignment of any channel walls in the insert and the manifold. In addition, the inserts must prevent leaking and potential blow out at pressures that can exceed 30,000 psi.

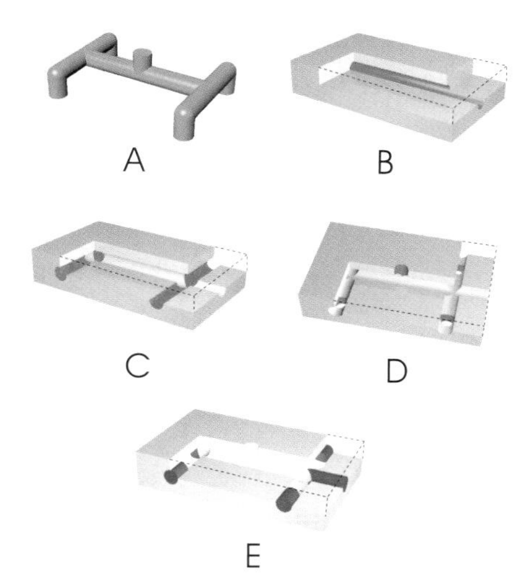

Figure 10.4 The 4-cavity H-pattern hot runner (A) manifold is created using the progression of drilling and insertion of plugs shown. The primary runner is drilled (B); secondary runners are then cross-drilled intersecting the primary runner (C); the 90 degree change in direction feeding the nozzle is then drilled, intersecting the main runners (D); all drilled holes are then plugged (E)

Common plug designs fit within the original gun-drilled channels and a counter bore of the same channel. The face of the plug, creating the corner of the flow channel, can be shaped with a ball mill, creating a fully radiused corner (Fig. 10.5). Prior to plugging the holes, the inside of the corners of the intersecting drilled flow channels are normally filed by hand to remove burs and to reduce the sharpness of the corner.

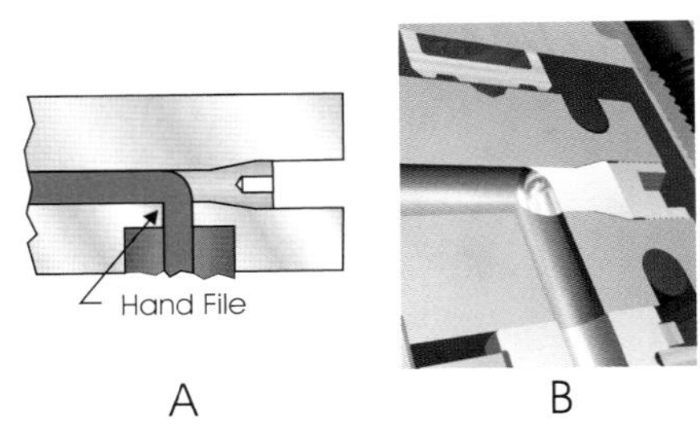

Figure 10.5 Various plug designs are used to cap the drilled holes. The outside corner is generally formed by the plug, which can be shaped to minimize stagnant or low flow regions. The inside corner is generally hand-worked to break the corner and remove burs left from machining (Fig. 10.5B – compliments of Husky)

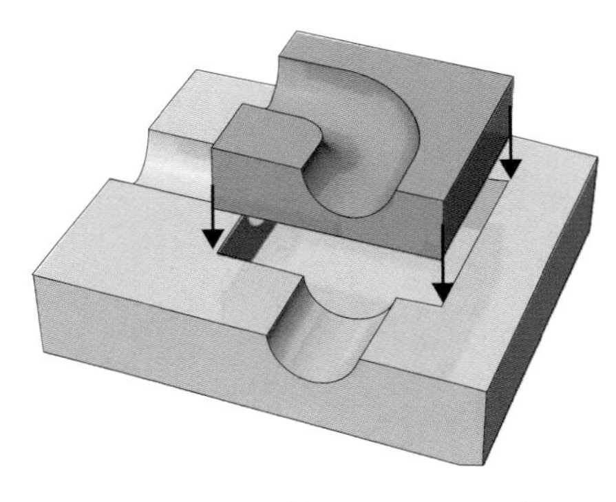

Figure 10.6 Special inserts are sometimes used when a liberal inside and outside radius is required at the corner. This is generally used for highly shear and thermally sensitive materials like PVC

In particularly demanding flow conditions, where stream-lining of the flow channel is a must, specialized inserts can be provided that allow for fully radiused flow channels. These are either machined in two halves or cast (see Fig. 10.6).

10.4 Effect of Diameter

10.4.1 Pressure

The most obvious consequence of changing the runner's cross-sectional size (diameter) is the effect on pressure drop through the channel. Pressure developed by a Newtonian fluid flowing through a round channel is inversely proportional to the radius to its fourth power. Therefore the impact of changing the diameter of a flow channel with a Newtonian fluid is quite significant. The effect on a non-Newtonian fluid is not nearly as straight forward as the viscosity is significantly affected by the change in shear rate that occurs with a change in runner diameter.

Contrasting the pressure drop through a round channel of a Newtonian fluid can be reduced to a direct ratio of the channel radius.

$$\Delta P = \frac{8Q\eta L}{\pi r^4} \tag{10.1}$$

If we assume injection flow rate (Q), viscosity (η), and flow length (L) are the same when comparing these runner diameters, then the following can be assumed.

$$\frac{8Q\eta L}{\pi} = K \tag{10.2}$$

Therefore, pressure is found to be a function of the fourth power of the runner radius and the constant K.

$$\Delta P = \frac{K}{r^4} \tag{10.3}$$

In reality, at the same flow rate, the shear rate increases with decreasing runner diameter, which will reduce the viscosity of the non-Newtonian material. Table 10.1 compares the effect of changing diameter and increasing cross-sectional area on the pressure through a 200 mm long round runner channel. MFG Pressure (column 3) are pressure predictions from Moldflow's 2D mold filling analysis program MFG, which accounts for viscosity changes from non-Newtonian shear thinning and melt temperature changes (this data is also plotted in Fig. 10.7, showing the non-linear relationship of pressure increase with decreasing channel diameter). Theoretical pressure (column 4) is based on the above formula. Here, K is determined by the value at which its calculated pressure will match that of MFG when a 25 mm flow channel is analyzed. Note the increasing difference in calculated pressures as the flow channel diameter is progressively decreased. As the theoretical pressure calculations assume viscosity to be constant, differences in pressure between the two methods build to over 3800% in the 4 mm diameter runner in which shear rates are highest. This contrast accentuates the need to consider the non-Newtonian and temperature effect on a plastic's viscosity.

Table 10.1 Effect of Changing Diameter and Increasing Cross-sectional Area on the Pressure through a 200 mm Long Round Runner Channel

Diameter (mm)	X-sectional area (mm²)	MFG pressure (MPa)	Theoretical pressure (MPa)	Temperature rise (°C)
25.00	19.63	1.66	1.66	– 0.29
24.00	18.85	1.81	1.95	– 0.22
23.00	18.06	1.99	2.32	– 0.12
22.00	17.28	2.18	2.77	– 0.01
21.00	16.49	2.41	3.33	0.11
20.00	15.71	2.68	4.05	0.26
19.00	14.92	2.99	4.97	0.43
18.00	14.14	3.36	6.17	0.63
17.00	13.35	3.80	7.76	0.86
16.00	12.57	4.32	9.89	1.14
15.00	11.78	4.96	12.80	1.48
14.00	11.00	5.74	16.87	1.91
13.00	10.21	6.71	22.69	2.43
12.00	9.42	7.93	31.25	3.08
11.00	8.64	9.51	44.26	3.92
10.00	7.85	11.58	64.80	5.03
9.00	7.07	14.37	98.77	6.51
8.00	6.28	18.21	158.20	8.55
7.00	5.50	23.69	269.89	11.44
6.00	4.71	31.84	500.00	15.72
5.00	3.93	44.56	1036.80	22.35
4.00	3.14	65.88	2531.25	33.31

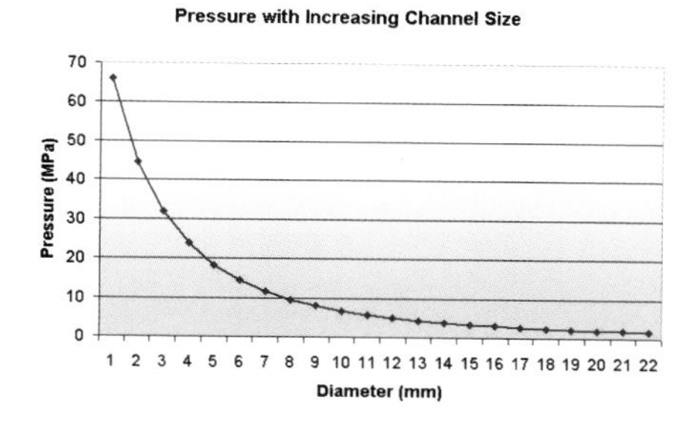

Figure 10.7 Non-linear effect of runner channel diameter on pressure drop (pressure is material-dependent)

Because there is no concern with regrind or lengthy cooling time, as with a cold runner, a hot runner can have a larger diameter flow channel. The larger diameter will reduce the pressure required to fill the mold. This can be important in applications with part-forming cavities that will require a high pressure to fill. However, increasing the diameter has a number of negative effects that must be considered. Increasing the runner diameter will increase residence time of the melt in the heated manifold and reduce flow rate through it. The increased thermal exposure in the manifold may be detrimental to some materials. The resulting slower melt velocity will reduce the ability to flush materials along the channel walls and any other low-flow regions. This can increase the potential for degradation of materials and complicate color change.

It is critical that the channel diameters be properly sized. With most materials it is desirable to minimize the diameter, while assuring that the cavity can be filled at the desired melt temperature and injection rate. A small diameter improves channel flush, unless for some shear- and thermally-sensitive materials for which excessive shear may be detrimental. However, be aware that increasing diameter may decrease shear, but will increase material residence time and reduce flush at slow flow areas.

If there is any concern with mold filling, a computer mold filing simulation and analysis should be performed on a model, which includes details of the runner system and a representation of the cavity. Analyses run without representation of the cavity geometry will not provide the required pressure information.

10.4.2 Shot Control

Because molten plastic is compressible at the high pressures used during injection molding, the increased volume of a large-diameter runner will act like a spring. As the melt fills a mold under high pressure, it becomes compressed, with the highest compression occurring in the injection unit and the hot runner, where pressures are the highest. As the injection screw moves forward at a constant rate, the melt flow rate in the cavity will accelerate as the melt is compressed in the screw barrel and manifold. While the pressure rapidly builds, the molten material in the manifold will act more rigidly as it continues to drive the material forward. If the screw stops moving prior to the cavity being filled, the melt front will continue to move forward for some distance because the melt expands as it decompresses. The larger the volume of material in the barrel and the hot runner, the further the melt may advance on its own during this decompression. This is one of the reasons why it is commonly recommended that the screw and barrel of the molding machine be sized such that about 70 % of the shot is used for high precision molding. Similarly, a smaller diameter runner will reduce runner volume and potentially improve shot control.

As a result of this behavior of plastic materials, excessive material in the hot runner can reduce shot control. A 40 % increase in the channel diameter, from 0.625 in. to 0.875 in., nearly doubles the volume of material in the manifold (that is a 100% increase). A given single hot runner manifold section, 254 mm (10.0 in.) long by 22.225 mm (0.875 in.) in diameter, contains a volume of 98.5 cm^3 (6.00 $in.^3$) of molten plastic material. If the material were a Montell, PRO-FAX 6231 polypropylene, according to the PVT data from Fig. 3.7, the molten material being processed will compress by more than 10% at a pressure of 120 MPa (17,404 psi). Therefore, with an average manifold pressure of 120 MPa (17404 psi), as pressure is relieved, a volume of 9.85 cm^3 (0.60 $in.^3$) of material could expand forward into the cavity (pressure loss from the expanding melt and freezing of the advancing melt would reduce this somewhat). If the melt from the larger diameter runner were to decompress the full 9.85 cm^3 into a part cavity with a wall thickness of 1.5 mm (0.060 in.), the melt would advance and occupy an additional 65.7 cm^2 (10.2 $in.^2$) of the cavity. In a screw with a 3.175 cm (1.25 in.) diameter, this volume is equivalent to 1.245 cm (0.49 in.) of screw travel. Given a second section of the same hot runner manifold feeding a second cavity, the additional volume of material would result in screw travel of 2.49 cm (0.98 in.). The flow behavior of the melt is like a spring, where the first end is attached to a driving force and the second end to a movable object. When the driving force is quickly applied to the first end, the spring will move the object (flow front) forward. When the driving force is removed, the first end of the spring will stop while the second end will continue to move the object forward as it decompresses. Note, the second end may not fully decompress as the object, the melt, is still experiencing a resistance.

10.4.3 Color Change

Color changes are most difficult in systems that experience annular flow. A full round channel is generally preferred. However, color changes can often be achieved as long as they are considered during the design stage and proper process procedure is used.

Successful color change with all hot runner systems requires colors to be purged not only from low flow regions, such as corners, branches, and mismatched assemblies, but also from the primary channel walls. In fact, removal of material from the channel wall may be the controlling factor in many hot runner systems.

It is generally accepted that a plastic's flow velocity is maximum near the middle of the flow channel and zero at a channel wall. This implies that the material at the channel walls will remain there forever. At slower flow rates, the velocity gradient across the channel is parabolic (see Fig. 10.8), which indicates that the flow velocity gradually increases from zero at the wall to maximum at the center. Material near the wall in the low velocity region is the last to be replaced. As flow rate is increased, the velocity gradient becomes more plug-like. The plug-like velocity distribution is not only a result of high flow rates, but causes higher flow velocity nearer the

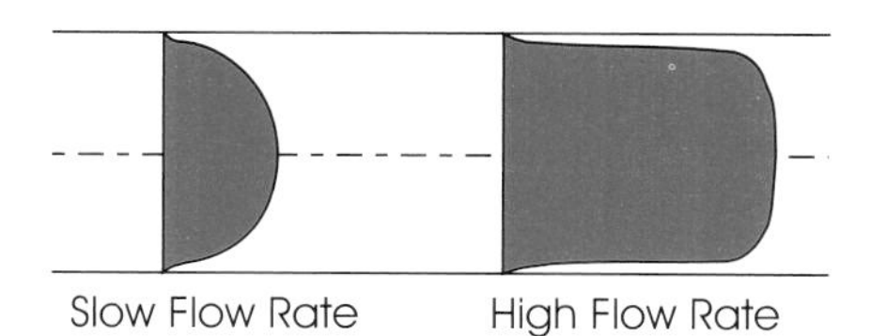

Figure 10.8 Increasing flow rate increases flow velocity near the channel wall, improving material flush rate and color change

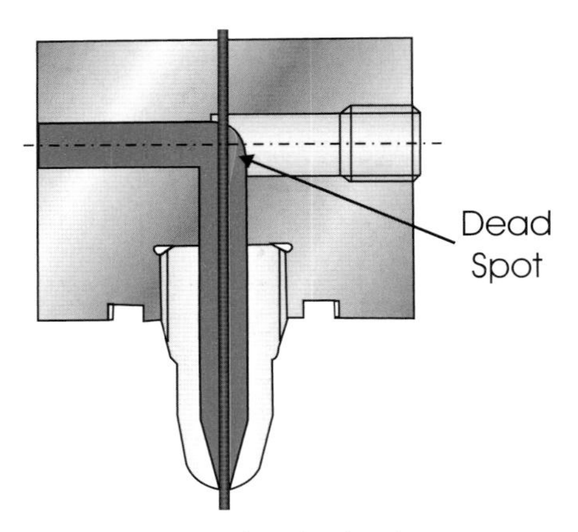

Figure 10.9 A valve pin often has an inherent dead spot on the far side of the pin where material can stagnate

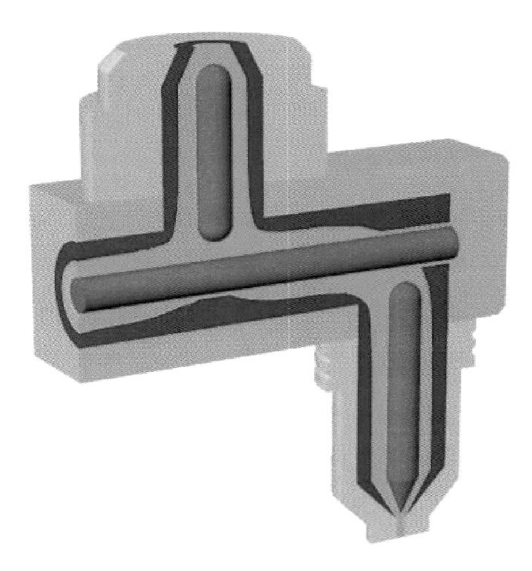

Figure 10.10 Internally heated hot runner systems result in a frozen layer along the channel perimeter (blue) and more dead/stagnant flow areas on the downstream side of the internal heaters

channel wall. This and the increased shear stress on the channel wall improve a material's flush and the rate at which a color change can take place.

With annular flow systems, such as valve-gated or internally heated systems, there are almost always low or stagnant flow areas from which it is difficult to flush a first material or material color. Figure 10.9 shows the slow flow region on the downstream side of a valve pin that can result in valve-gated nozzles. Figure 10.10 illustrates the melt conditions that commonly exist in an internally heated hot runner system. Internally heated systems include both slow and dead flow areas on the downstream side of the internal heaters and frozen material around the flow channel's perimeter. The first material injected into the runner will never be removed from the channel wall without disassembling the entire system and physically machining it out, or flushing it in a fluidized bed cleaning system. However, often a color change can take place if the temperature of the system is increased and the runner purged at high flow rates with the new material. The raised temperature and fast flow rates will allow the new material to eat away at the frozen first color. The temperature is then lowered and a normal process is resumed. This allows the new colored material to cover the first color.

Externally heated systems also create issues with color change. Any place of slow or stagnant flow will be slow to purge. This includes corners and misaligned components of the runner. Of course, these issues can be minimized with careful design and processing to designed conditions. This includes processing at temperatures that aid in proper alignment of assembled components.

In a study performed by Bouti, an "older 2 × 16 hot runner stack mold" not optimized for color changes was compared to a runner in which the flow channel diameters had been decreased to improve flush rate. The results of the study can be seen in Fig. 10.11. The smaller channel size reduced residence time to almost 1/3 of that of the original channel and increased the shear rate through the channel by nearly 400%. This improved design also reduced time for color change to nearly one third of the original from 6.75 to 2.42 hours [2]. This study was in good agreement with a previous study by Hume, which found that the time for a color change was approximately proportional to the material's residence time in the manifold, that is, reducing the channel volume by half will reduce color change time by approximately 50% [3]. The study also found that the channel diameter was a more significant contributor to reducing time for color change than corner radiusing or even a 0.40 mm mismatch between the nozzle and manifold flow channel. Though a majority of the part had changed color in only 20 cycles, elements of the original colored material remained in the immediate gate region of the part after as many as 400 cycles. This material was concluded to be coming from material remaining at the wall of the channel because of its zero flow during injection.

Key factors increasing time for color change in a hot runner:

- Insufficient flow rate (flush rate of the flow channel):
  - The higher the flow rate, the higher the velocity of the melt near the flow channel. At high flow rates, the velocity distribution across a flow channel begins to become less parabolic and more plug-like. Therefore, not only does the overall velocity of the material increase with a smaller channel, but the highest shear rates move closer to the channel wall, thereby more quickly removing a previous colored material.

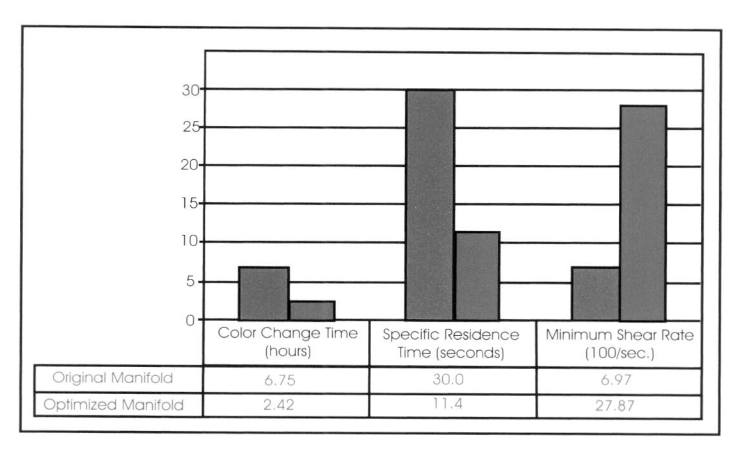

Figure 10.11 Results of study showing the positive effects of decreasing the runner diameter (Optimized Manifold) in a hot runner. Color change time and residence time are significantly reduced. Shear rate increases indicating higher flush rate [2]

- Long residence time:
  - The fewer the shots in the manifold, the more frequently the material is replaced during a color change.
- Cold spots along the flow channel:
  - The plastic material will solidify at cold spots in the flow channel. These can result from poorly isolating the system from the surrounding cold mold. Cold spots are expected at the gate region, where the nozzle must interface with the solidifying molded part. These regions often require that the melt be heated higher than its normal process temperature and quickly cycled in order for this material to be removed.
- Mismatched assembled components (particularly in the region of manifold-to-nozzle and nozzle-to-mold):
  - Mismatched components create dead or slow flow locations in the flow channel. Incorrectly calculating the thermal expansion of the manifold, relative to the hot nozzle, or not operating at the specified temperatures can result in mismatched components.
- Melt channel turns and branches:
  - Inside and outside radii of a bend in the runner will create low flow regions, which are more difficult to flush during a color change.
- Flow obstructions (internal heaters, valve pins, nozzle tips, etc.):
  - The downstream side of an obstruction in the flow channel will create a low flow area, which can create difficulties during a color change. This is typical of a valve gate or and internally heated runner system

NOTE: See Section 13.4 for color change procedure.

10.4.4 Material Change

Material changes are not normally recommended. The hot runner is designed for a specific material. Sizing of the channel is dependent on the material's flow characteristics and flush requirements. Thermal expansion fittings and gate tip position of most hot runners are designed to the specific temperature at which a material is to be run. This is particularly important with externally heated systems, where seal and alignment between the drops and manifold are designed to a specific temperature. Increasing the temperature will create undue stress in the compression assembly of the drop and manifold, because they go through excessive expansion. Decreasing the temperature can result in too little stress at these junctions, resulting in leakage. In both cases, the alignment between flow channels will be affected.

If a material change is required, attempt to process the new material at the same, or close to the same, temperature as the material for which the system was designed.

References

[1] Extrude Hone, 1 Industry Blvd., Irwin, PA 15642. www.extrudehone.com

[2] A. Bouti, "A New Hot Runner Nozzle Speeds Color Change and Eliminates Flowlines," SPE Annual Technical Conference, San Francisco (2002)

[3] W. Hume, "Advanced Hot Runner System Design for Enhanced Color Change Capabilities," SPE Annual Technical Conference, San Francisco (1994), p. 1160

[4] N.N.: Modern Hot Runner Technology (G), Company Brochure, Hutter Plastic, Altach.

[5] Menges, Mohren, "How to Make Injection Molds" (1986) Hanser Publishers, Munich

11 Hot Runner Drops, Nozzles and Gates

The hot drop delivers the melt to a gate tip, which is the hot runner's interface to the part-forming cavity. The critical challenge of the gate tip is to allow for the plastic part to freeze on its outlet side and provide for the plastic material to remain molten on its inlet side. The distance between inlet and outlet side is often a fraction of a millimeter.

Hot drops and gate tips are arguably the most critical elements of the hot runner system. Some of the fundamental requirements of the drops are to:

- Conduct heat to the gate (prevent gate from freezing)
- Provide thermal separation between drop and cavity
- Minimize flow restrictions or areas of material hang-up
- Provide good temperature control to the melt
- Provide for a continuous flow channel between the manifold and the gate tip
- Minimize pressure loss
- Prevent leaking

Some of the fundamental requirements of the hot gate include:

- Resisting freezing and blocking further melt from entering the cavity
- Resisting stringing or drooling
- Minimizing localized heat being transferred to the part
- Minimizing pressure loss
- Minimizing dead flow areas
- Minimizing gate vestige (provide clean separation of melt and solid part)
- Resisting wear – particularly from abrasive materials

In an attempt to achieve these objectives, most suppliers of hot runner systems offer dozens of gate designs. Each design is matched to a particular material and the requirements of the particular mold and of the part being molded. Most gates are designed to be removable, which allows for easy replacement if worn or if a particular gate does not provide the desired results. Figure 11.1 shows a variety of standard nozzle tip designs offered by Husky, which are designed for various materials and applications.

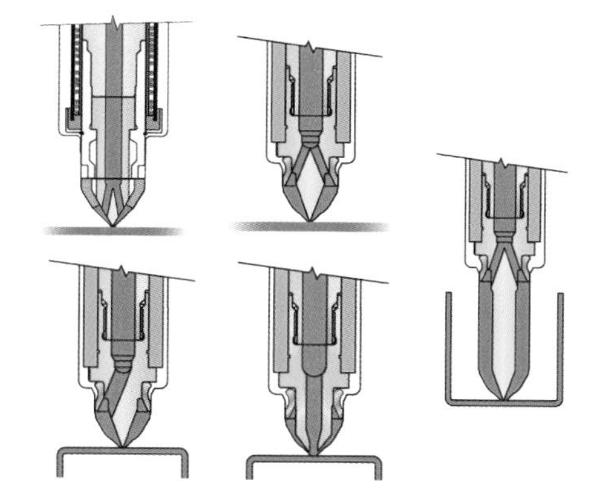

Figure 11.1 Various removable hot gate designs (Courtesy: Husky)

Hot gates are either open or shut-off. Open gates are generally restrictive, while mechanical shut-off gates allow for a larger gate orifice.

11.1 Hot Drops

The hot drop is integral to the gate and can significantly influence its design and performance. Hot drops are either internally or externally heated. They are either directly heated via their own heaters and temperature controllers or they are indirectly heated. Indirectly heated drops are made of high thermal conductivity materials, such as beryllium copper, and conduct their heat from the heated manifolds. These types of drops are relatively simple and of low cost, but do not provide for individual temperature control. With externally heated systems the body of the hot drop is generally referred to as a nozzle. The heater in an internally heated drop is generally referred to as a probe.

11.1.1 Externally Heated Hot Drops (Nozzles)

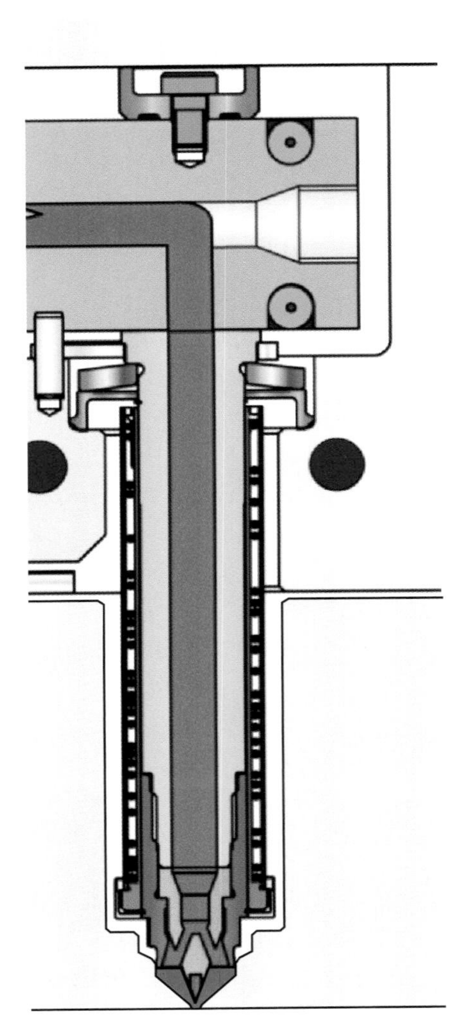

Figure 11.2 Externally heated hot nozzle with conductive gate tip heater (Courtesy: Husky)

Externally heated drops provide for the ideal full round flow channel. Radially from the center, they are normally comprised of the full round flow channel, surrounded by a bushing (see Fig. 11.2). The bushing is generally surrounded by a band-type heater, or in some cases the heater may be cast into the bushing. This heated bushing, or nozzle, is then surrounded by an air gap, which helps insulate it from the cold mold. A thermocouple is placed into the nozzle to provide feedback to the nozzle temperature controller.

The nozzle is attached to the manifold in a variety of different methods. The nozzle is generally screwed into, bolted onto, or pressed against the manifold. The assembly of the nozzle and the manifold provides for a flow channel to deliver the melt to the gate tip. One major challenge for the externally heated nozzle is that it must prevent leaking of the high-pressure melt at its base, where it attaches to the manifold, and at its tip, where it delivers melt to the cavity.

The tip of the hot nozzle is often integral to the design and operation of the gate. In gates that feed directly to a part it is generally required that the gate vestige be kept as small as possible. This demands that the gate opening be kept to a minimum. In order to keep the small gate open, a heated *gate tip* is often used. This gate tip is generally an insert, mid stream of the gate orifice, and gets its heat indirectly (via conduction) from the heated nozzle. This indirect heating limits temperature control of the gate tip. Any adjustment to the nozzle temperature affects the gate tip, and visa versa. With some designs, the heated nozzle may have two distinctly different heaters and temperature controls. One heater is for the mid and lower body of the nozzle and a second is for the top of the nozzle and the gate tip. This two zone heater design provides improved control of these different regions.

11.1.2 Internally Heated Hot Drops

Internally heated hot drops are generally comprised of a flow channel, which may flow through a dedicated bushing (Fig. 11.3A) or may be machined directly into the "A" half mold plates (see Fig. 11.3B). A heated *probe* positioned in the center of the channel forms an annular flow space. A thermocouple is placed within the heated probe to provide feedback to a temperature controller. As molten plastic is injected into the mold, it freezes along the outer walls of the channel while remaining molten near the heated probe. The frozen layer helps seal the channel from leaking and insulate the heat from the hot drop from the cooled mold. If a bushing is used, an additional insulating air gap can be included. This design results in a high pressure drop per volume of material.

The center heater commonly is a cartridge heater, though other methods are also used. Heat is normally conducted to the gate tip, which is integral to the center heating probe. This continuity between heater probe and gate tip appears to have more potential for improved gate tip control, relative to the more isolated gate tips in externally heated nozzles. However, this is very much dependent on the designs. Seiki Corporation was the first company to develop a separate heater in the tip of their probe, which could be controlled independently of the main drop, or nozzle. This provided unique capabilities for gate tip control (see Section 11.3.5).

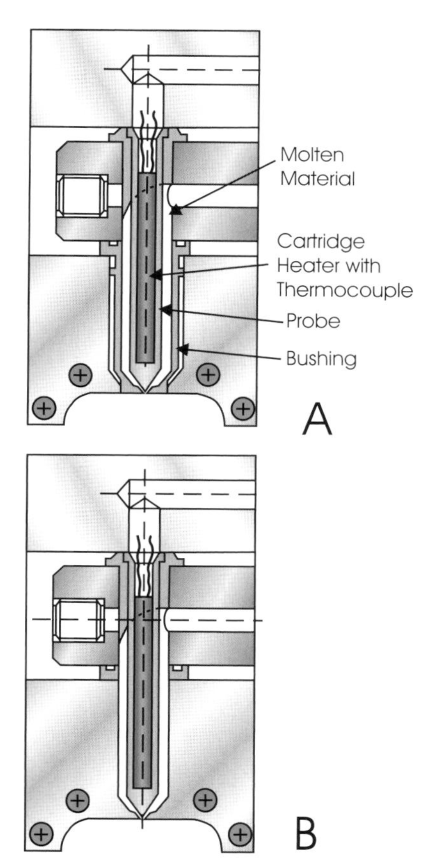

Figure 11.3 Internally heated drop designs. Fig. A includes an insulating bushing; Fig. B excludes a bushing

11.1.3 Heat Conducting Nozzles

Other variations of hot runner systems include externally heated manifolds with heat conducting hot nozzles or drops, which are more common with homemade designs. The heat conducting nozzles are normally made of a high thermal conductivity material such as a beryllium copper. All heat to the drop is conducted from the heated manifold; therefore, these drops do not have individual heat control. In Fig. 11.4, the nozzle, made of a high thermal conductivity material, is screwed into the manifold to maximize heat transfer. The nozzle is surrounded by an external bushing. During start up, plastic will spill out into the air gap between the conductive nozzle and the surrounding bushing. This material will remain there and act as an insulator. Further insulation from the mold is provided by an air gap created between the bushing and the cold steel wall. Figure 11.5 is a more compact variation as it does not have an external bushing.

Advantage of this system:

- Low cost
- No temperature controls or heaters to deals with

Disadvantage of this system:

- The drop lengths are limited because they must conduct their heat from the manifold.

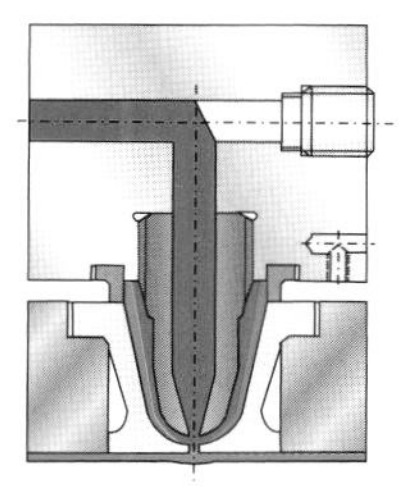

Figure 11.4 Heat conducting nozzle/drop with external insulating bushing

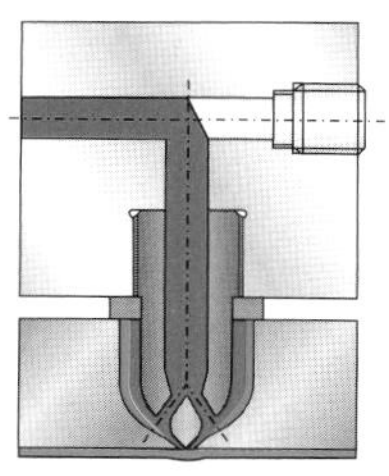

Figure 11.5 Head conducting nozzle/drop with conductive gate heater and no insulating bushing

- No individual temperature control of drops.
- Drop-to-drop thermal control is dependent on the uniform temperature distribution in the manifold. Some molders will argue that if the manifold is melt-balanced; there is no need for individual temperature control of the drops.

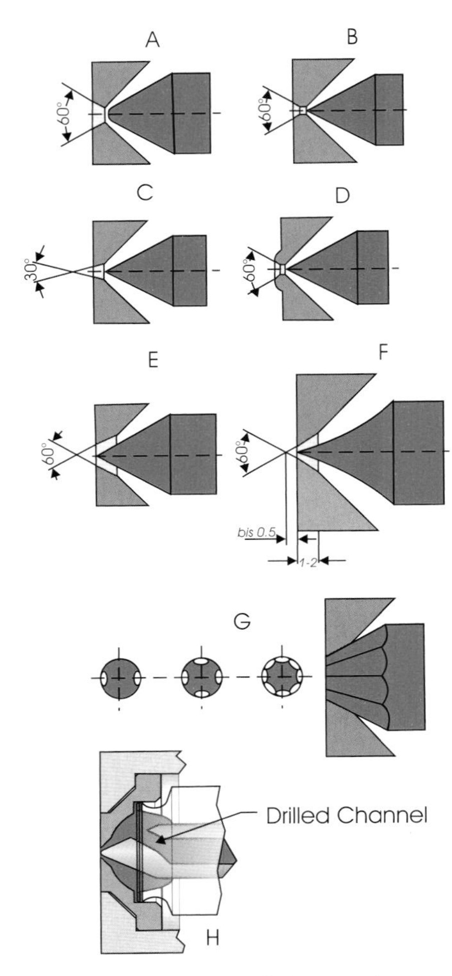

Figure 11.6 Tip design for hot nozzles. A and B are standard pinpoint designs, resulting in small, partly melted remains on molding; C is a flat pinpoint gate, resulting in little protruding tip; D is a recessed pinpoint gate, resulting in non-protruding tip; E is a small annular gate, resulting in small protrusion; F is a small pinpoint gate, resulting in protruding tip; G is a profile gate for discs, gears, rings, etc. [4]; H illustrates a common gate tip assembly

11.2 Restrictive/Pin Point Gates

Open pin point gates are the most commonly used when gating directly into a part. As illustrated in Fig. 11.6, there are various types of gates that, depending on their design, leave different gate vestiges on a molded part. A gate design should be chosen satisfying the specified gate vestige requirements while at the same time resulting in the lowest pressure loss and potential for freeze-off. A minimum gate vestige is generally achieved by extending a heated nozzle or probe tip into the orifice of a small restrictive gate opening. The heater increases the control of gate freeze but results in a small annular flow channel, which increases pressure drop during injection as well as the shear on the material being molded. This small opening also increases the potential for the gates to clog, particularly when running regrind.

The open pin point gate uses a small orifice to minimize gate vestige. The gate orifice can range from 3 mm in diameter to less than 1 mm, depending on the material and gate vestige requirements. A 0.75 mm diameter is common. The gate orifice is either machined directly into the cavity or is included in a *gate insert*. The gate insert can provide for easy replacement of worn or damaged gates. They can also be used to help insulate cooling of the cavity steel from the gate orifice. This helps prevent premature gate freeze with semi-crystalline materials. Gate inserts are not as commonly used with most amorphous materials for which increased cooling in the gate is desirable to prevent stringing.

A heated gate tip is used to help keep the gate region from freezing off. Heat is normally conducted to the gate tip from the nozzle body or, with internally heated systems, the heated probe. The gate tip assists with preventing gate freeze; however, the resulting annular flow channel can be highly restrictive to flow. The gate tip of an externally heated system is normally attached to the nozzle body by either being directly threaded into or onto the body, or by being trapped by a threaded shoulder nut against the nozzle body. The threaded bushing is said to help prevent unthreading of the tip, which can result from the high melt pressures developed during injection and packing. Melt passage through the gate tip insert is provided by drilling or forming holes (see Fig. 11.6H).

Consistent temperature of drops is critical to gate tip positioning and the resultant flow restriction. Varying drop temperature to address filling imbalances can create conflicting results. A molder may increase nozzle/

drop temperature to try to reduce melt viscosity and improve flow. However, the resulting thermal expansion can force the tip deeper into the gate orifice. This will corrupt the design, affecting the flow restriction, pressure drop, shear rate, and gate vestige. Figure 11.7 illustrates how increasing nozzle temperature will cause the nozzle to grow in length and cause a gate tip heater to restrict the gate opening.

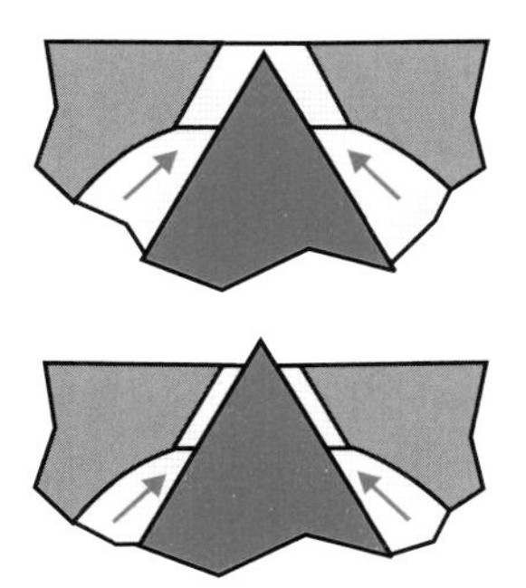

Figure 11.7 Effect of thermal expansion on gate tip heater positioning

At least some gate vestige is expected with pin point gates. Therefore, the gate is normally recessed in the part so that any gate vestige will not protrude beyond the part's primary surface (see Fig. 11.8). This requires a small protrusion of the cavity steel and matching recess in the core in the vicinity of the gate. Increasing the depth and decreasing the diameter of this gate dimple reduces the amount of steel immediately surrounding the gate. This reduced steel restricts the heat transfer required for good cooling of the gate orifice. The reduced diameter is commonly required on small parts. Figure 11.8 shows two gate designs that can be used to create the recessed gate dimple. The design shown in Fig. 11.8A results in poor cooling of the gate, which can cause stringing. The design in Fig. 11.8B increases the amount of steel immediately surrounding the gate, thereby improving cooling [2].

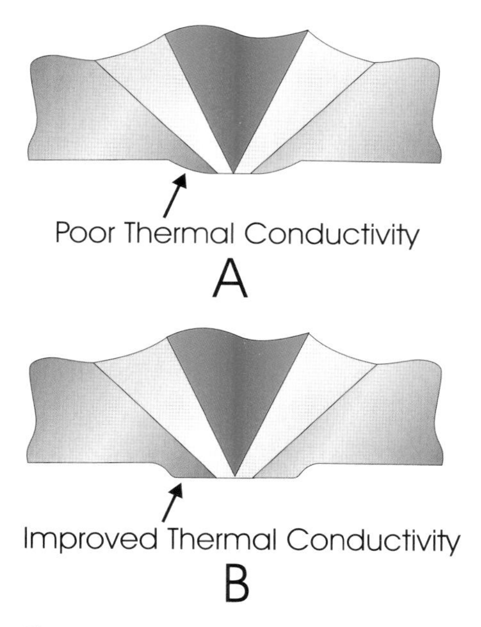

Figure 11.8 Dimples often used in direct gating from hot drops to position gate vestige below the part surface. Thickening the wall at the gate dimple reduces effects of high shear stress

11.3 Gate Design Considerations

11.3.1 Gate Freeze-Off

Premature gate freeze-off is much more likely to occur with semi-crystalline materials, such as nylon or POM, than with amorphous materials owing to their more distinct solidification temperatures. This can be a particular problem with many of the small restrictive gates designed to minimize gate vestige. In the worst case, if frozen, the reverse inlet taper of a gate orifice can trap the frozen material and prevent a further inlet shot. This problem can commonly be resolved by carefully applying a flame to the clogged gate tip (though this is not a desirable technique it is often required). Most often the frozen, or partially frozen, material in the gate can be dislodged with the high melt pressures developed during injection. However, this can result in unbalanced part filling because dislodging the material will often be inconsistent between the multiple gates in a multi-drop hot runner system. In addition, inconsistencies in packing resulting from variations in gate freeze will result in inconsistencies in the molded product.

To help reduce gate freeze, a heat conducting gate tip is required. In addition, the gate orifice is commonly located into a gate insert. This insert can include a small air gap around its perimeter to help thermally insulate it from the surrounding cold mold. The resulting higher temperature gate orifice will help prevent gate freeze (see Section 12.5 for more detail).

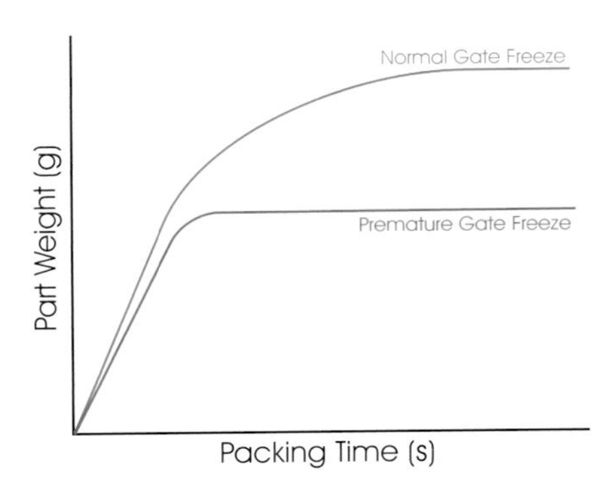

Figure 11.9 Graph of normal and premature gate freeze times relating part weight to packing time

A gate freeze study can be used to help to determine if gate freeze is preventing proper packing of a part. With a reasonable pack pressure set, begin molding parts using a very short pack time. Gradually increase pack time while decreasing cooling time proportionally, that is, increase pack time by 1 second and decrease cooling time by 1 second. All other process conditions should be kept the same. Save and weigh a few shots from each of the progressive pack times. Part weight vs. pack time should be plotted. If the gate is open during the entire pack process, the curve is more likely to show a gradual change in weight through its transition to a frozen part (see Fig. 11.9). If the gate freezes prematurely, the curve will more abruptly change as shown in Fig. 11.9. This is particularly true with semi-crystalline materials. It is best to know the approximate freeze time of the material in the part, which may be estimated using mold filling and/or mold cooling analysis. Some of the plastic material suppliers, such as Dupont, can provide excellent information as to the estimated freeze time of their materials. Dupont's technical representatives use a program they developed called CAMDO® to provide data to customers upon request.

11.3.2 Stringing/Drooling

Stringing occurs during the ejection of a part and is most likely to occur with amorphous materials rather than with semi-crystalline materials. During ejection, cooled high-viscosity material in the gate is pulled from the gate orifice, stretching into a thin string. This typically happens when the gate is not cooled sufficiently to prevent the material from stringing or when there is insufficient screw suck-back following packing. Stringing can be reduced by using more restrictive gates, increasing cooling in the region of the gate orifice or increasing screw suck-back. Cooled gate inserts can be used to minimize this effect, while the use of mechanical shut-off nozzles will virtually eliminate stringing.

Drooling can occur during mold opening/ejection and generally results when there is insufficient suck back from the screw. If a mechanical or thermal shut-off gate is not used, the hot runner must provide for the material in the hot runner to be decompressed.

11.3.3 Packing

With amorphous materials, as the runner, and often the gate, do not normally freeze during the molding cycle, pack pressure can be higher and will be more consistent through the pack stage of the molding cycle than with cold runner systems. In a cold runner, the freezing runner and gate will increase the pressure loss with time as material is fed into the cavity during the pack, or compensation, stage. This will create a degenerating pressure profile until the gate freezes. The degenerating pressure can sometimes be beneficial, because it acts to naturally profile the pressure and reduce over-pack of regions near the gate relative to regions away from the gate. To understand the benefits of profiled packing one must first understand what is occurring during the packing phase.

During packing, the regions near the gate will normally freeze last. This results from the fact that, during packing, material is flowing from the gate region to replenish shrinking material throughout the part. This is called "compensating flow," and this phase is referred to as the "compensation phase" of the injection molding cycle. Therefore, the highest volume of flow is nearest the gate, with nearly zero flow occurring in the extremities. The high flow in the gate region results in a greater amount of fresh material being fed though it, and a higher amount of frictional heating occurring here. This combination will cause the gate region of a part to remain at an elevated temperature longer than the extremities of the part.

During packing of a mold cavity, the flow of material will result in a pressure drop between the gate and regions further away. For example, if the pressure in the immediate vicinity of the gate is 60 MPa (8700 psi), pressure in regions away from the gate may not exceed 40MPa (5800 psi). If pack pressure is held at a constant 60 MPa until the regions near the gate freeze, the regions away from the gate would have been packed at 40 MPa, while those near the gate would have been packed at 60 MPa. This will result in variations in shrinkage and residual stresses in the molded parts, which are the fundamental causes of warpage. Therefore, it may be desirable to profile the packing pressure so that all regions of the part freeze at a pressure closer to 40 MPa. If the pressure near the gate is initially 60 MPa, and the pressure drop across the part is 20 MPa, it is desirable to gradually reduce the pressure as the part progressively freezes. If this is timed correctly, when the regions near the gate are freezing, the pressure near the gate should be only 40 MPa.

With a cold runner, a reduction of pack pressure will occur as the runner and gate progressively freeze. Therefore, it will naturally provide a degrading pressure profile. Figure 11.10A shows the degrading cavity pressure near the gate occurring with a cold runner mold, despite a constant hold/pack pressure. Figure 11.10B shows that the cavity pressure near the gate in a hot runner mold is constant with the constant hold/pack pressure [1]. Although the hot runner does not naturally accomplish a pack profile, it does provide a better potential capability to control the pack profile if desired. With a cold runner, if one wanted to control the pressure profile they must deal with the complication of the naturally degrading pressure profile present in the runner system. However, in a hot runner system, setting up a pressure profile in the part cavity would be easier because there is no freezing runner to complicate this process. This will allow a more direct and known control of the pressure profile.

Because the gate freeze in a hot runner system is less predictable than a cold runner, there is also the need to consider back flow in the gate region of the part. When pack pressure is terminated, any remaining unfrozen material

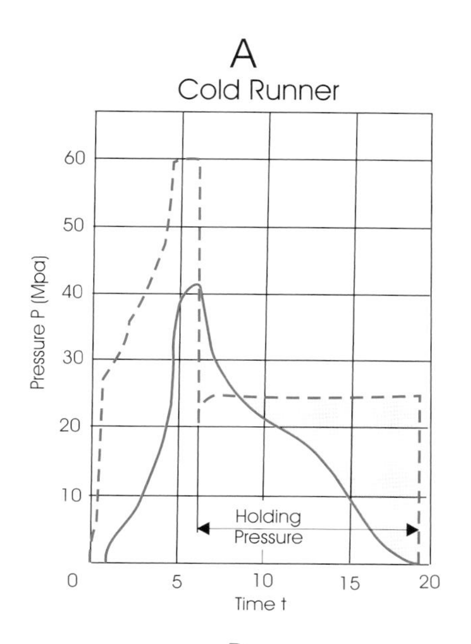

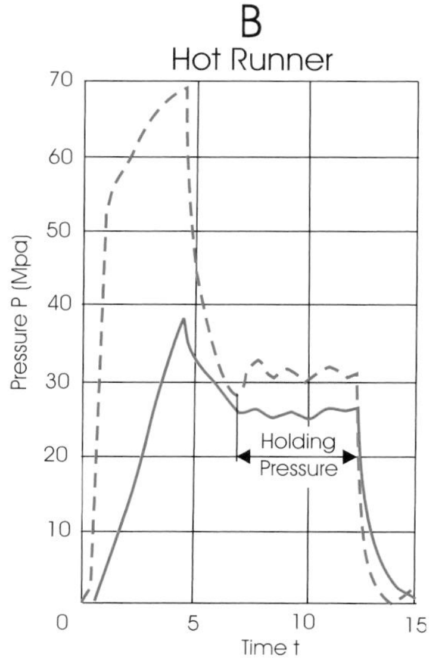

Figure 11.10 Comparison of pressure propagation during holding pressure in a conventional, solidifying cold runner system with an edge gate (top) to that in a hot manifold with open nozzle and direct gating (bottom): ——— cavity pressure close to gate, – – – – pressure in front of screw tip (calculated) [5]

mid-wall in the gate region of the part will be at a high pressure and will depressurize by driving material back through an unfrozen gate. This can increase sinks in the gate region, which will be relatively hot because of its close proximity to the heated drop.

11.3.4 Mechanical Valve Gates

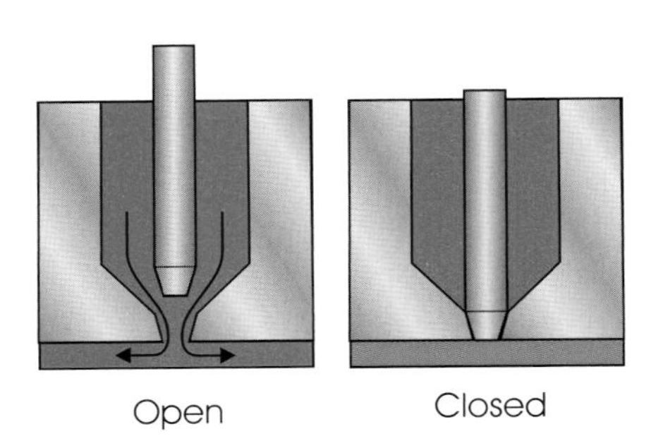

Figure 11.11 Opened and closed valve pins

Figure 11.11 illustrates a valve-gated system (left) and a more conventional open gate nozzle and tip (right). In this design the valve gate nozzle pin is hydraulically actuated. The hydraulics, drive piston, and cylinder are built into the top plate. Mechanical shut-off gates have spring, torsion bar, hydraulically, or pneumatically actuated valve pins. Before injection, the valve pin is mechanically retracted. This opens a large gate orifice to allow for the passage of the melt into the part-forming cavity. Before the melt at the gate has solidified completely, the pin moves forward into the closed position. Spring-actuated shut-off valve pins do not allow for controlled timing or positive movement of the valve. Therefore, most valve-gated systems are either pneumatically or hydraulically actuated. Pneumatically controlled systems utilize the plant's compressor generated air supply system, which is normally limited to less than 120 psi. The result is that the pneumatically controlled valve-gated systems require relatively large diameter air cylinders to operate and still do not provide the force, speed, and positive movement that can be had with a more compact hydraulically actuated valve gate. The hydraulically actuated systems normally use the injection molding machine's oil supply, which can provide pressures in excess of 2,000 psi. The negative aspect of these is the increased care and maintenance in setting up and operating a 2,000 psi hydraulic system.

The large gate orifice that results when the pin is retracted can lower injection pressure compared to the more restrictive open gate design. In high-speed molding applications the valve gate can also potentially reduce cycle time, as the valve can be closed allowing plastification during mold open stages.

One of the primary purposes of using a mechanical shut-off nozzle is to create a flawless gate with no drooling and stringing. These types of gates typically leave only a small witness ring rather than the standard gate vestige. Elimination of the standard gate vestige is often desirable for molded medical parts. A flawless surface will prevent the potential danger of surgical gloves and other protective clothing being torn by raged gate vestiges. Another advantage to the mechanical shut-off drop is that large gate openings are possible. This makes processing at high-flow rates much easier. Some disadvantages of mechanical shut-off drops are that their capital cost are high, that they are expensive to maintain, and that they increase the sophistication of the mold setup and operation (more functions to control). The addition of moving parts creates wear issues and the opportunity of leakage where the pin passes through the manifold into the melt stream.

The gate shut-offs in valve gates are either tapered or cylindrical (see Figure 11.12). With tapered shut-off pins, the tip of the pin is ground at an angle so that it seals to a matched angled surface in the gate orifice. Gate seal is a result of the tapered pin being driven against the matching taper in the orifice. The taper also can act as a guide when the gate is closing. Hydraulically actuated systems normally require that the gate's stroke be controlled at the base of the pin, in the drive piston (see Figure 11.13). If the forward stroke does not have this control, the pin could be bent at the high closing pressures, it could wear the gate orifice, and could progressively protrude into the cavity as the pin tip and gate orifice wear. Gate wear can result from both the pin tip – gate orifice contact and/or from the flowing plastic material. Filled materials can be particularly abrasive. The tapered pin will also develop a layer of frozen material around the inside tapered diameter of the gate orifice. This will limit the forward movement of the pin (see Fig. 11.14). An excessive buildup will result in a raised pin mark on the parts surface, which can also complicate ejection, and potentially stress the pin, which is now prematurely seating against the gate's tapered orifice.

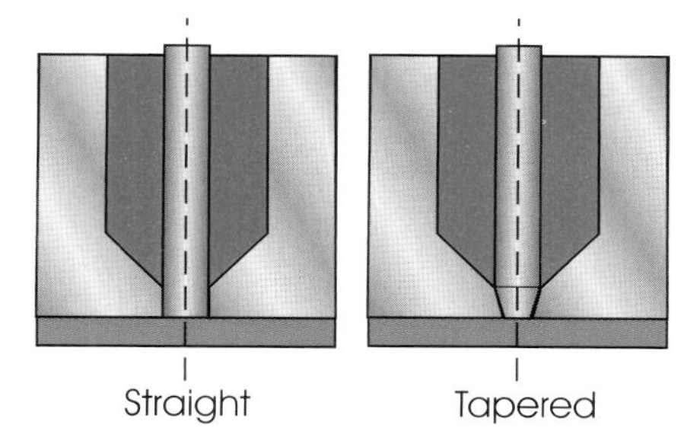

Figure 11.12 Straight and tapered designed valve gates used with hot runner molds

The straight pin tip eliminates some of the potential problems with the tapered pin as its closing movement does not result in it seating against any other surface. However, it introduces the increased demand for the pin being centered. An off-centered pin will cause wear on the pin tip and gate orifice. When the part ejects it could pull material from the resultant gap thereby creating a raised gate vestige. In addition, the lack of a tapered seat will result in an increasing gap between pin and orifice with continued wear. A tapered pin will be more forgiving with wear as its stroke can often be adjusted to maintain a seal. In addition, the frozen plastic material that formed around the straight gate's inside orifice will be driven forward into the part when the pin closes. This may create some problems with the part.

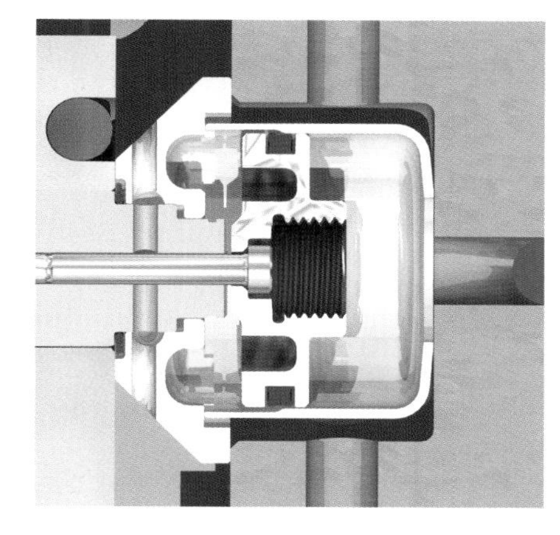

Figure 11.13 Close-up of valve pin stop utilized to control travel of the valve pin (Courtesy: Husky)

As the valve pin moves forward it compresses and pushes the material in the orifice into the cavity. To obtain a good seal, the tapered pin must be forced against the matching angled surface of the gate tip region.

The forward movement of the valve pin can be halted by a physical stop at the base of the valve pin (see Fig. 11.13). This, or some other means of controlling the stroke, can help prevent damage to the gate orifice and tip of the pin. When designing the gate insert, wear caused by cyclical loading by the valve pin must be a consideration. Problems with this design include freezing of material to the inside walls of the tapered gate orifice. This frozen material can prevent the pin from fully advancing. The result would be a slightly raised cylindrical surface on the part (see Fig. 11.14).

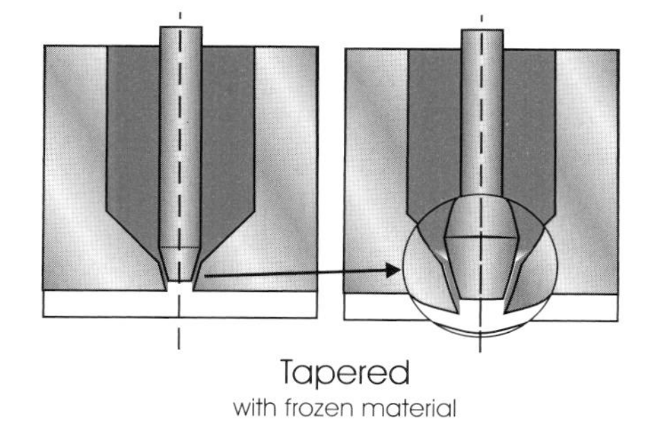

Figure 11.14 Closure of valve gate will be affected by the development of frozen plastic in the gate orifice

The cylindrical, non-tapered, designed shut-off valve pins eliminate this problem. The non-tapered pin seals around its perimeter using a tight tolerance slip fit in the cylindrical gate tip region. When the valve pin is actuated forward, it pushes the melt from the gate orifice into the cavity like a plunger. This design relies heavily on a tight tolerance between the outside diameter of the valve pin and the inside diameter of the gate orifice.

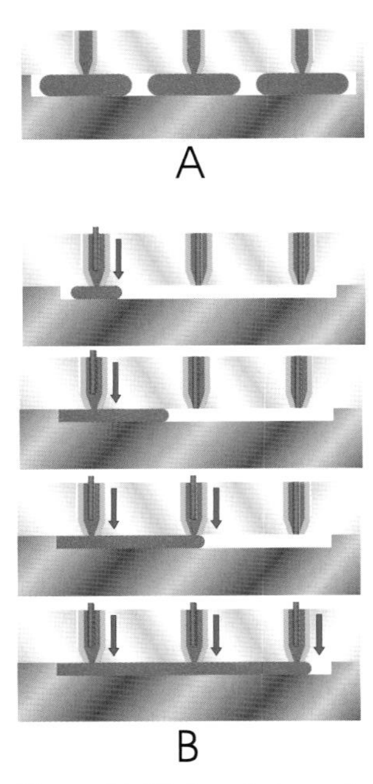

Figure 11.15 Part A shows the development of weld lines normally resulting from multiple gates; part B shows the use of sequential valve gating to eliminate weld lines in molds with multiple gates

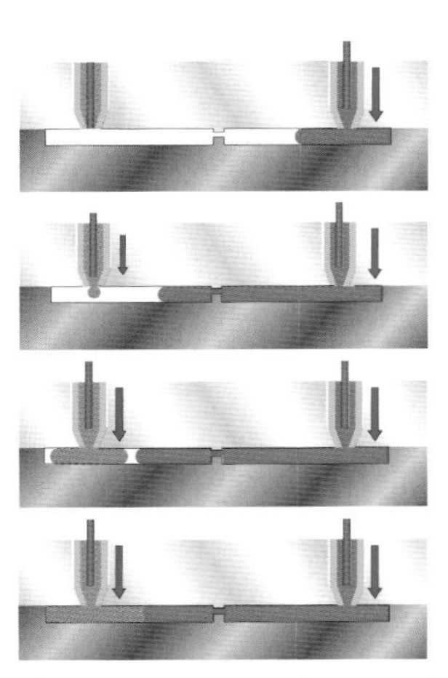

Figure 11.16 Use of sequential valve gating to assure good strength across the hinge and good packing on either side of the hinge

This type of shut-off is sensitive to wear, which would result in an annular gap forming between the valve pin and the gate orifice. An annular flash ring may form on the part as the wear between the valve pin and gate orifice increases.

Both types of valve pins work well with most resins, amorphous and semi-crystalline. The only limitation may be seen with resins that include abrasive fillers. A tapered design often has better success with abrasive resins because it is more tolerant of wear of the gate orifice.

Both valve gate designs require means to assure that the pin is centered on the gate orifice so that the pin and the orifice do not wear on each other during opening and closing. In addition, control of the pin stroke affects gate seal and whether the pin protrudes into the part or is recessed. Control of thermal expansion is also critical to the design and operation, which again accentuates the need to operate hot nozzles at designed temperatures.

11.3.4.1 Sequential Valve Gates

A unique capability of controlled mechanical valve gates is the ability to individually control the opening and closing of multiple gates within a single mold or cavity. Control of the timing of each valve provides a number of unique opportunities. Cascading injection has the potential of eliminating weld lines within a multi-gated part by controlling gate opening such that a single continuous flow front is developed. Figure 11.15A shows the cross section of a part being filled by three gates. This typical arrangement would result in two welds. Figure 11.15B shows the same part being molded with three valve gates that are sequentially opened as the melt front advances across the part. One of the three gates is initially opened and melt is injected. The melt is allowed to advance and pass a second nearby gate. As the melt passes over the second gate, this gate is opened and continues the filling to the third gate. The third gate then opens and fills the remainder of the part. This cascading effect eliminates the formation of a weld.

Another variation of sequentially opening the valve gates will result in a weld, but can be used to position the weld. Figure 11.16 illustrates the use on a part with an integral hinge. A gate on one side of the hinge may not provide the required packing control on the opposite side of the hinge. If two gates are used, an unacceptable weld may occur along the hinge. By sequencing the valve gates, one gate can feed the melt across the hinge before the second gate is opened. A variation of this would have the entire part filled by one of the valve gates and the second valve gate opened after the part is filled simply to provide pack control on both sides of the hinge.

11.3.5 Thermal Shut-Off Gates

A second type of shut-off gate is thermally actuated. Seiki Corporation first developed a thermal shut-off gate more than 20 years ago. Though originally patented and offered exclusively by Seiki, this gate tip heater concept is now available from other suppliers of hot runners. Seiki's first development was a separate gate tip heater positioned at the tip of its internally heated hot probe (see Fig. 11.17). The separate gate tip heater is individually controlled and can be cycled on during the injection and packing stages, and cycled off during the cooling phase to allow gate freeze to occur. The very small mass of the separate tip heater allows it to heat and cool very quickly. This gate tip design works best when excellent cooling is provided external to the gate orifice. In recent years, Seiki has also developed a similar approach using a separate gate tip heater in an externally heated nozzle design. Though this design is most commonly marketed to reduce the development of frozen material around the gate orifice in tapered mechanical valve pin systems, they also have a version used with open sprue gates (see Fig. 11.18).

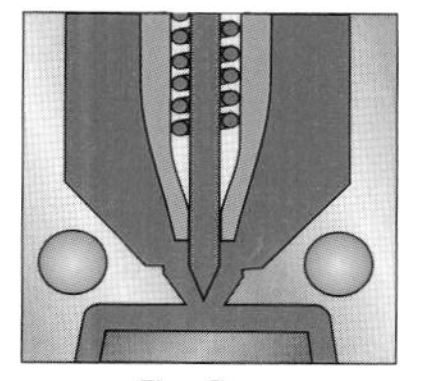

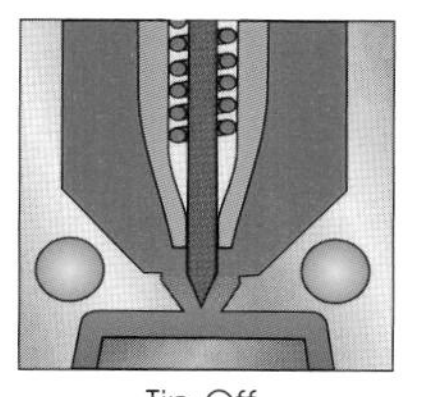

Figure 11.17 Thermal cycling of the spear gate tip heater allows increased control of the gate. The cycling of the separate tip heater can be controlled to close off and open the gate region during each injection cycle

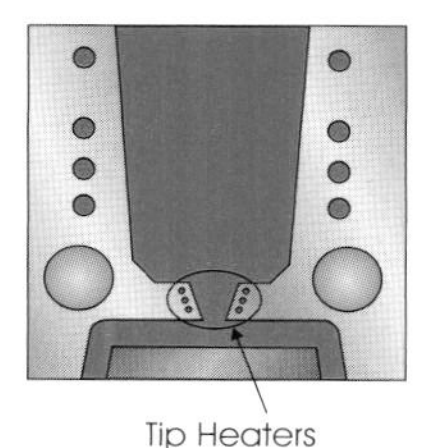

Figure 11.18 Seiki's separate gate tip control, used with valve gates, prevents buildup of frozen material in the gate orifice

11.3.6 Hot Edge Gates

Hot edge gates are another variety of hot gating. The only major difference is that the drop protrudes further into the mold and allows access to the part from the side. This also will allow for multi-cavity gating from the same hot drop or center-gating of a part with a cylindrical open center such as a fan. One example of a hot edge gate can be seen in Fig. 11.19.

Figure 11.20 is a multi-tip hot drop which includes conductive gate tip heaters for the multiple gates.

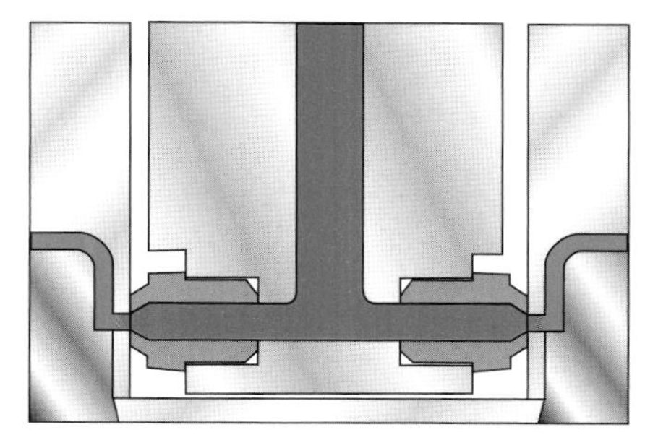

Figure 11.19 Edge gating to multiple parts with the same nozzle in a hot runner application

Figure 11.20 Multi-tip hot drop with heat conducting tip heaters utilized to create edge gating similar to the one in Fig. 11.19

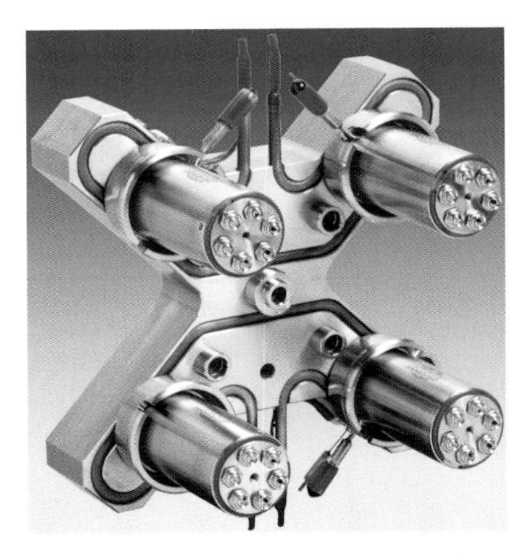

Figure 11.21 24 gate tips provided by four nozzles (Courtesy: INCOE)

11.3.7 Multi Tip Nozzles

Numerous variations of multi-tip nozzles are available from some of the hot runner suppliers. These are commonly used in multi-cavity molds molding small parts. The multi-tip eliminates the need to have individual nozzles feeding each cavity. A single zone heater is generally used to control the multi-tip nozzle. This can be considered both an advantage and a disadvantage. The argument against is that there is no longer individual temperature control to each cavity. However, this can also be an advantage if the melt being delivered to the multi-tip nozzle has symmetrical or homogeneous, conditions as discussed in Chapter 7. Figure 11.21 shows an INCOE 24-cavity hot runner system utilizing four multi-tip nozzles.

References

[1] Menges, Mohren, "How to Make Injection Molds", 1986, Hanser Publishers, Munich

[2] Klees, J., "Hot Runner Molds" seminar manual. John Klees Enterprise, Inc., www.johnklees.com

12 Thermal Issues of Hot Runner Systems

Other than insulated hot runner systems, hot runners require their own heating systems. This includes heaters, means to distribute the heat, heater controllers, and thermocouples to feed temperature information from the heated system to the temperature controllers.

In the design and operation of hot runner several issues have to be considered:

- The placement of heaters to provide uniform heating
- The amount of heat required for both startup and operation
- Maintenance and replacement of heaters
- Placement and replacement of thermocouples
- The heater controller
- Process for startup and operation

12.1 Heating

Proper selection of heaters, their design, placement, installation, and control are critical to the use of hot runner systems in order to avoid excessive downtime from failures and to achieve the desired control of the process. Heaters in hot runner applications are commonly custom-made to meet the requirements of the application. Customization can include wattage, watt density, heater coil distribution, and voltage that can range from 5 to 480 V.

There are four basic types of heaters used with most hot runner systems:

- Coil
- Band
- Tubular
- Cartridge

It is important to note that most heaters absorb moisture when they are not heated. This moisture can be picked up from the air and can easily ruin a heater. To avoid this, careful startup procedures must be followed (see also Section 12.2). Because heaters burn out and require replacement, their accessability and ease of replacement are important.

12.1.1 Coil (Cable) Heaters

Coil heaters are commonly used for externally heated sprues and nozzles. A heater filament wire is enclosed in a protective tubular sheathing. The filament is looped within the sheathing with the positive and negative leads on one end. The looped wire halves are separated by a mineral insulation. The protective sheathing is durable and can be "coiled" around a cylindrical nozzle. These heaters provide excellent durability and long service life.

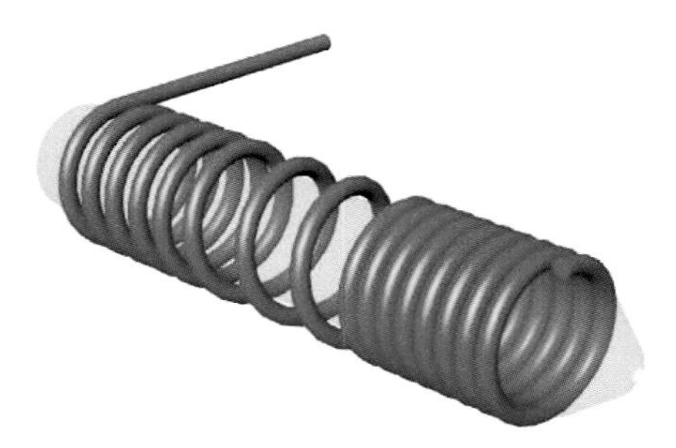

Figure 12.1 Profiled coiled heater for controlling heat distribution in a hot nozzle

The heaters have either a round or a somewhat rectangular shaped cross-section. The rectangular shape will improve contact when wrapped around a nozzle and can significantly improve the conductive heat transfer. The watt density of coil heaters can be as high as 100 W/in.^{2} and reach temperatures of 1000 °C. Thermal distribution along the nozzle can be easily profiled by adjusting the winding of the coil heater on the nozzle (see Fig. 12.1). The density of the windings can be increased in regions of high heat loss, such as tips and the base where the nozzle may directly contact the surrounding cooled mold.

Coil heaters can be wrapped externally to a nozzle, cast into it, or be housed in a band. The band is commonly made with a combination of copper and stainless steel. The copper improves the temperature distribution along the nozzle side, while the stainless steel helps insulate from heat loss to the surrounding mold.

12.1.2 Band Heaters

Band heaters are either mica- or mineral-insulated types. They are commonly used for heated sprues, nozzles, and tubular shaped manifolds (see Fig. 12.2). Band heaters include heater filaments that are wound within a flat cross-sectional housing that is formed into a tube. The filaments are separated by mica or mineral insulation.

Mica-insulated band heaters are low cost, operate at temperatures of up to 370 °C and are generally limited to watt densities of less than 45 W/in.^{2} to prolong their service life. It is important that the entire surface of the heater has excellent contact with the nozzle in order to avoid hot spots and premature failure of the heaters. These heaters are better for lower temperature applications and will require more frequent replacement than many of the other heater types.

Mineral-insulated band heaters extend the operating temperature to around 760 °C and watt densities can be increased to nearly 100 W/in.^{2}. These heaters are preferred for use with high temperature plastics and for extended service life compared to the mica-insulated band heaters.

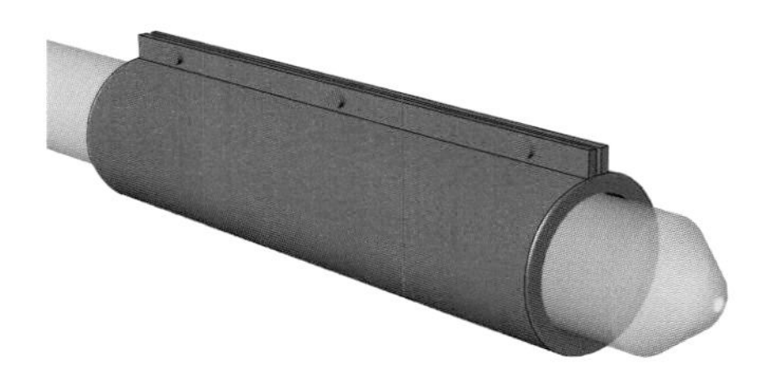

Figure 12.2 Band heater used on nozzle

12.1.3 Tubular Heaters

Tubular heaters are very common with externally heated manifolds. These heaters have a single filament inside a tube with its positive and negative wires on either end of the tube. This eliminates the need for insulation to separate the looped wire arrangement used in most other heaters. As a result, these heaters are extremely durable with a long expected service life. The tubular heaters can be formed to follow the contour of a shaped manifold.

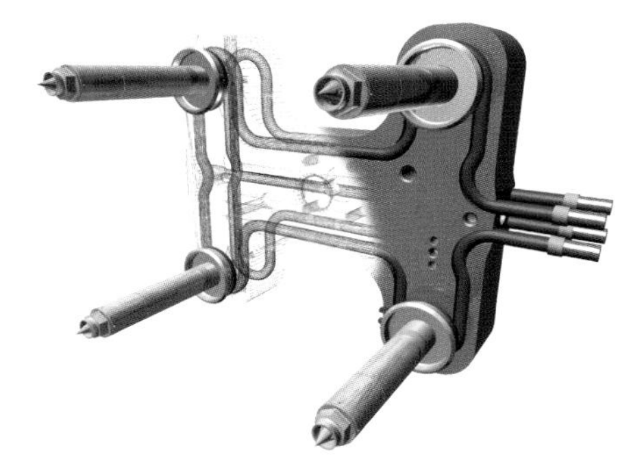

Figure 12.3 Tubular heater pressed into milled grove on hot manifold (Courtesy: Husky)

The tubular heater is commonly pressed into milled grooves along the surface of the manifold, which follow a path approximately parallel to that of the flow channel (see Fig. 12.3). The tube is somewhat malleable allowing it to be shaped to the contours of the milled groove. This shaping provides intimate contact between the heaters and the manifold resulting in excellent heat transfer and long life. The groove may be slightly undercut to aid in retention of the tubular heater. Some manufactures prefer to solder the tube in place with a copper-based solder. This improves contact of the tube with the manifold and can extend its service life. Upon failure, these heaters often must be milled out to install a replacement. This is best performed by the manufacturer of the manifold. In some cases, the tubular heaters are designed with a looser fit to the milled groove and are held in place with an aluminum plate. This provides for in-house replacement but at the cost of performance and service life of the heater.

Tubular heaters have excellent distribution of heat along their length. A single heater could be formed to heat most manifolds. However, to provide increased temperature control, most manifolds use multiple tubular heaters positioned in logical zones to provide control to various locations across the manifold.

12.1.4 Cartridge Heaters

Cartridge heaters consist of a resistance wire coiled within a rigid tubular sheathing (see Fig. 12.4). These heaters are provided as straight tubes that cannot be formed. They are generally fit within a drilled hole. Similar to a coil heater, the temperature distribution of the cartridge heaters can be profiled by controlling the density of the filament windings within the tubular sheathing. It is common for the density to be increased at either end of the heaters.

These heaters are used in internally heated drops and manifolds and in some externally heated manifolds. When used for internally heated drops and manifolds, the sheathing is designed to withstand direct contact with the plastic. When the heaters are to be placed in drilled holes, the holes must be well-matched to the diameter of the heater. Poor contact is a common problem and results in over-heating of the heater and a short service life. To minimize this problem, the heaters are commonly surrounded by a conductive paste.

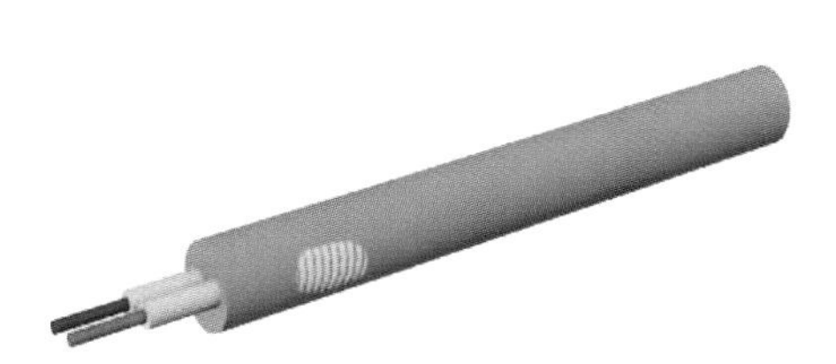

Figure 12.4 Cartridge heater

Cartridge heaters are offered in high and low watt densities. High watt density cartridge heaters are used when space is limited, but they have a relatively short service life. Therefore a low watt density cartridge is preferred whenever space allows or can be provided. High watt density heaters can have a watt density as high as 200 W/in.^{2} and can reach temperatures in excess of 650 °C.

12.1.5 Heat Pipe Technology

A heat pipe consists of a sealed container, usually copper or aluminum, whose inner surfaces contain a capillary wicking material. Inside this container is a liquid under its own pressure that enters the pores of the capillary material. Heat applied to any point on the surface of the heat pipe causes the liquid in the pipe at that point to boil and enter the vapor state. In the vapor state, latent heat of vaporization is present. This gas, which is at a higher pressure, moves inside the container to a colder location, where it condenses and gives up its latent heat of vaporization. This continues to move the heat from the input end of the heat pipe to the output end.

Heat pipes have an effective conductivity many times higher than the metal they are created from. The heat transfer or transport capacity of a heat pipe is specified by its Axial Power Rating (APR). This is the energy moving axially along the pipe. The larger the diameter of the pipe the greater is the APR. Similarly, the longer the heat pipe the smaller is the APR. Heat pipes can be built in many sizes and shapes.

Heat pipes provide excellent temperature uniformity. When used in hot runner systems, this property can help ensure consistency during molding. Also, there are no external forces, such as heater bands, needed to distribute the fixed wattage of the heat source to the variable wattage requirements of the hot runner. This helps to minimize the potential for downtime due to electrical equipment maintenance. In addition, systems containing heat pipe technology may provide more rapid start-ups, shorter recovery rates, and better temperature stability at the mold face. Synventive offers built-in heat pipe technology in the steel of their manifolds, drops, and nozzles rather than inserting a separate tube or pipe. The main advantage of heat pipe technology incorporated into hot runner systems is its ability to maintain a uniform heat distribution.

12.2 Heater Temperature Control

12.2.1 Thermocouples

Thermocouple placement has to be carefully considered to provide the desired control of the melt temperature. Type J and K thermocouples can be placed within most cartridge and band heaters. This is commonly done with cartridge heaters used in internally heated systems. When used for a hot drop, the thermocouple should be placed near the gate tip, where temperature control is most critical.

When used with externally heated systems it is important that the temperature of the steel near the flow channel, rather than that next to the heater, is measured. It is common for the thermocouple to be placed between the heater and the flow channel with a bias of being closer to the flow channel. It is important to keep in mind that the temperature of the melt in the flow channel, not the temperature of the heater is to be controlled. Placement in locations of high heat loss, such as near spacers or other regions of contact with the surrounding mold, should be avoided. These locations can give a low temperature reading causing the remainder of the manifold to be over-heated.

It should be expected that thermocouples will require replacement. Therefore, it is desirable that positioning and mounting of thermocouples provide for easy replacement. The thermocouple on the nozzle shown in Fig. 12.5 can easily be removed and replaced using a set screw. Note the thermocouple placement near the tip. The tip is the most critical region to control temperature.

Figure 12.5 Thermal couple designed for easy replacement (Courtesy: INCOE)

12.2.2 Temperature Controllers

Temperature controllers (see Fig. 12.6) range in sophistication from open-loop, to closed-loop, to PID controllers. Open-loop controllers are the most basic and provide the least amount of control. Here, the controller provides a constant current to the heater as set by the operator. The current is manually adjusted until the desired temperature is reached. If there are any changes in the environment or process that alter the required current the operator must identify this and make a manual change to compensate.

Closed-loop systems take into consideration that changes in process or environment will occur. These systems use thermocouples to monitor temperature change and automatically react to the change. How and when the closed-loop system reacts depends on its sophistication.

An integrated PID (Proportional Integral Derivative) provides the highest level of control. The first level of control is the proportional component. It monitors the difference between the process variable (PV) and the set point (SP). If there is a small difference between PV and SP, a small change is made in the energy output; if a large difference in these two values exists, a large change is made in the energy output. This proportional component alone will have the tendency to overcompensate or overshoot the set point temperature.

The second level of control is the integral component. The integral component takes into consideration the sum of errors over time. That is, it will enable the temperature controller to react to changes at a more appropriate time.

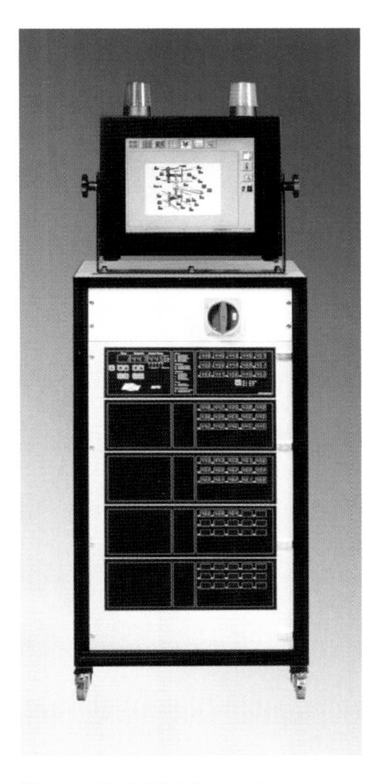

Figure 12.6 Multi-zone hot runner controller for controlling manifold and nozzle temperatures (Courtesy: INCOE)

The third level of control is the derivative component. This component takes into account the "dead time," or delay, between making a change in the output and seeing that change reflected in the PV. A classic example is getting an oven to the proper temperature. When the heat is first turned on, it takes the oven awhile to heat up. This is the dead time. If this initial temperature is then determined to be too hot, it is turned down, but it takes time to cool to this new SP. The derivative component compensates for dead time by holding some future changes back because the most recent changes in the energy output have not yet been reflected in the PV.

Nearly all heaters absorb moisture that can cause damage to the heaters during rapid temperature rise experienced during startup. This is particularly troublesome in humid weather. To help address this, the temperature controllers for a hot runner system should have a "soft start" sequence. During a soft start, only about 10% of heating power is applied. The temperature is slowly raised to about 90 °C and then "heat soaks" for 15 to 20 minutes, during which time the moisture is driven out of the heater. Following this, the heater can be raised to normal operating temperature.

12.3 Power Requirements

To ensure the proper operation of a hot runner system, the power requirement of the system must be known and adequately met. The following steps can be followed to calculate the power, or wattage, requirements of a given hot runner system [1].

Step 1: Calculate the power required to heat the manifold and the associated equipment in contact with the manifold. To determine the power necessary to heat the manifold, the following equation can be used.

$$\text{Power (kW)} = \frac{\text{weight of manifold material (kg)} \times \text{specific heat (J/kg °C)} \times \text{temperature difference (°C)}}{3.6 \times 10^{6}\ \text{J/kWhr} \times \text{time allowed for heat} - \text{up(hr)}} \quad (12.1)$$

Step 2: Calculate the power required to institute a phase change in the plastic during heat-up. The following equation can be used to find the power needed to melt the solid plastic present in the manifold at start-up.

$$\text{Power (kW)} = \frac{\text{weight of material (kg)} \times \text{heat of fusion (J/kg)}}{3.6 \times 10^{6}\ \text{J/kWhr} \times \text{time allowed for heat} - \text{up(hr)}} \quad (12.2)$$

Note: Heat of fusion is zero for amorphous materials.

Step 3: Calculate the power lost from surfaces. Use the following equation for conduction losses.

$$\text{Power (kW)} = \frac{\text{thermal conductivity (J/hr} \cdot \text{m} \cdot \text{°C)} \times \text{surface area (m}^2\text{)} \times \text{temperature difference (°C)}}{3.6 \times 10^{6}\ \text{J/kWhr} \times \text{thickness of material (m)}} \quad (12.3)$$

Step 4: Finally, it is necessary to determine whether more power (wattage) is required during start-up or during production. The greater need is the one that must be supplied to the hot runner system and thus determines the appropriate heater size. A safety factor of 15% is included in the following equations.

Power Required for Start-up:
Power (kW) = (Step 1 + Step 2 + 3/4 Step 3) × 1.15

Power Required for Sustained Operation:
Power (kW) = Step 3 × 1.15

12.4 Thermal Isolation of the Hot Runner

One of the objectives of designing a hot runner system is to contain the heat required to heat the plastic in the flow channels. This is desirable in order to maximize control and distribution of the heat of the plastic material and to minimize negative effects of transferring heat to the surrounding mold and molding machine. Figure 12.7 illustrates heat loss conditions in a hot runner design.

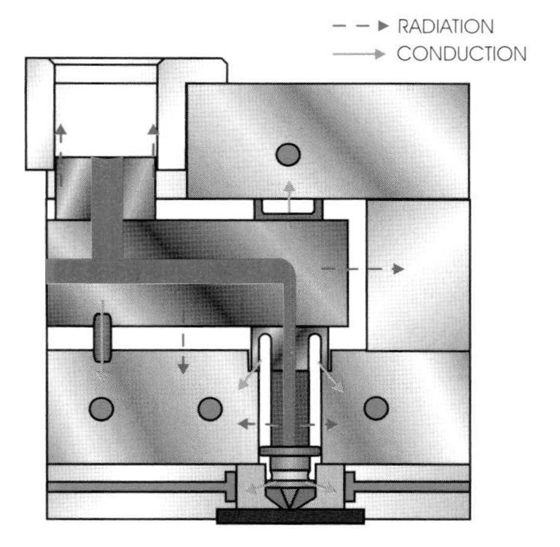

Figure 12.7 Typical radiant and conductive heat loss from a hot runner system

Heat is lost via conduction, radiation, and convection. The most effective means of heat transfer in a steel body is through conduction. Therefore, minimizing conduction from the hot runner to the surrounding mold is a primary focus in the design of the hot runner system. To minimize conduction, heated components are isolated by surrounding them with an insulating layer of air or plastic. In areas where direct contact with the surrounding mold is required, the area of contact is kept to a minimum. In addition, low thermal conductivity materials are used to minimize heat transfer. Titanium is commonly used owing to its excellent compressive strength and low thermal conductivity. Figure 12.8 contrasts the thermal conductivity of various materials. Note the difference between titanium and some of the high thermal conductivity materials that might be used to conduct heat in regions such as the gate tip.

The externally heated manifold is normally suspended, and supported, within a manifold housing using spacer pads. The spacer pads are designed to minimize surface contact. They are commonly made of low thermal conductivity materials including titanium and ceramics. Other points of contact to the mold include locating dowels, which control the position of the manifold relative to the hot drops. Other than the spacers, dowels, sprue, and nozzles, the manifold is surrounded by air.

The entire hot runner system should be enclosed to minimize any heat loss from convection. Heat transfer by radiation can be kept to a minimum by keeping heated and surrounding surfaces clean and polished. This will minimize both emissivity from the heated portions and radiation absorption from surrounding surfaces.

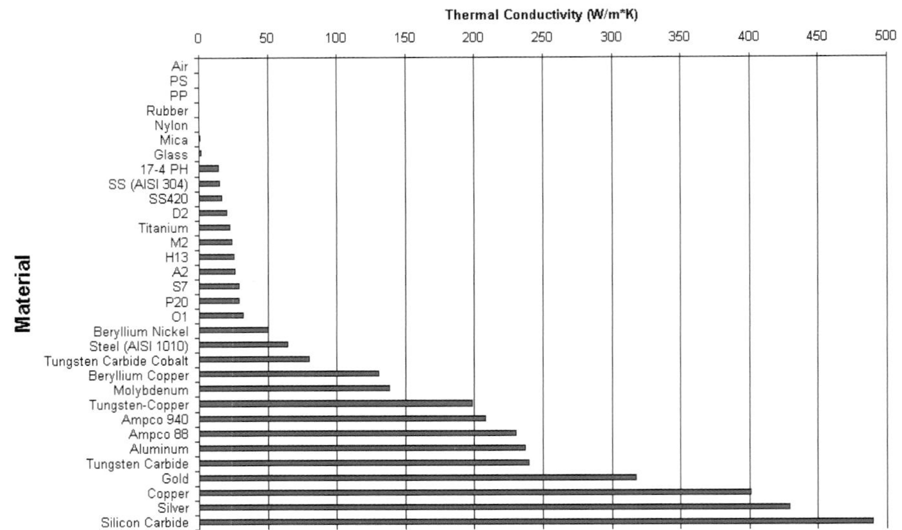

Material	k - Thermal Conductivity (w/m*K) @ 20 C	@ 200 C
Air	0.026	
PS	0.027	
PP	0.120	
Rubber	0.280	
Nylon	0.290	
Mica	0.523	
Glass	1.40	
17-4 PH	14.0	18.5
SS (AISI 304)	15.0	17.0
SS420	16.0	20.0
D2	20.0	21.0
Titanium	22.0	20.0
M2	24.0	28.0
H13	25.0	29.0
A2	26.0	27.0
S7	29.0	30.0
P20	29.0	30.0
O1	32.0	33.0
Beryllium Nickel	50.0	
Steel (AISI 1010)	64.0	59.0
Tungsten Carbide Cobalt	80.0	
Beryllium Copper	130.0	170.0
Molybdenum	138.0	134.0
Tungsten-Copper	198.0	
Ampco 940	208.0	
Ampco 88	230.0	
Aluminum	237.0	240.0
Tungsten Carbide Copper	240.0	
Gold	317.0	311.0
Copper	401.0	393.0
Silver	429.0	425.0
Silicon Carbide	490.0	

Figure 12.8 Thermal conductivity data

The common compression-fit nozzles have at least two critical points of contact with the mold: near the gate tip and at the base of the nozzle (see Fig. 12.9). The contact surface area in both of these locations is kept to a minimum to restrict heat transfer from the hot nozzle to the cold mold. The gate tip seal ring is commonly only about 0.4 mm wide, which is sufficient to provide a leak free seal while minimizing the contact area. The melt seal between the nozzle and the manifold requires the compression of the nozzle's flange between the manifold and the cavity support plate. Heat transfer between this heated flange and the cold mold is minimized again by controlling the land contact area. However, the contact area must be large enough to withstand the high contact stress required for the seal.

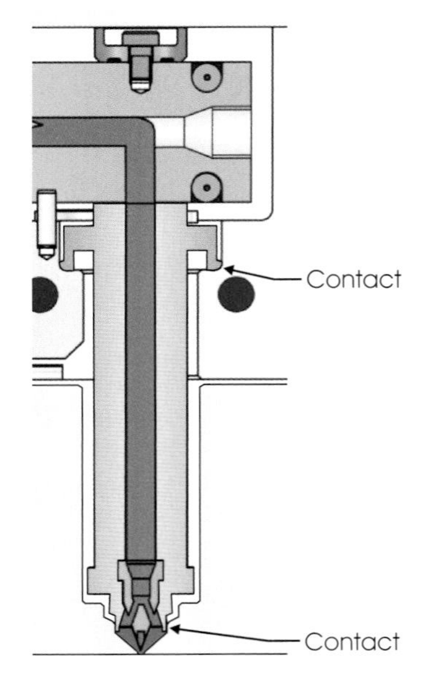

Figure 12.9 Critical contact locations of a hot nozzle to the surrounding cold mold

Nozzle heaters are normally designed to increase the watt density in the location of contact to the cold mold to compensate for the localized heat loss. Nozzles that are threaded into the manifold can avoid this additional contact surface.

Insulating heat from the machine platen is also an important consideration. This is often omitted as it is left to the molder. Cooling lines can be placed in the "A" half clamp plate. In addition, insulating sheets can be placed between the mold and the machine platen. Special high temperature insulating sheets are provided by a number of standard mold component suppliers. If heat is not properly insulated from the machine platens, tie bars and the clamping mechanism can be affected. Lubricating grease on the tie bars can begin to fail and the thermal expansion of the entire system can result in some presses becoming seized. This is particularly the case with toggle presses and some hydro-mechanical systems. Here, the thermal expansion can result in an increase in clamp tonnage beyond initial mold setup. Excessive thermal expansion may cause the clamping mechanism to lock-up, preventing the mold from opening until the system is cooled down and contracts.

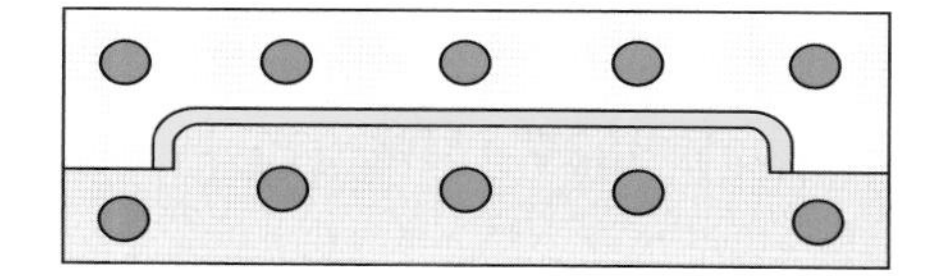

12.5 Gate Temperature Control

Temperature control of the gate region is critical to the success of a hot runner system. Too hot a gate region can increase cycle time, part stress, drooling, and stringing. Too cold a gate can cause gate freeze off.

The gate is the point at which the hot runner system meets the molded part. On one side of the gate a molten material must be maintained, while on the other side the material is supposed to freeze. The distance between these locations is a fraction of a millimeter. Complicating this issue is the problem of maintaining uniform controlled cooling of the part in the vicinity of the hot drop and gate.

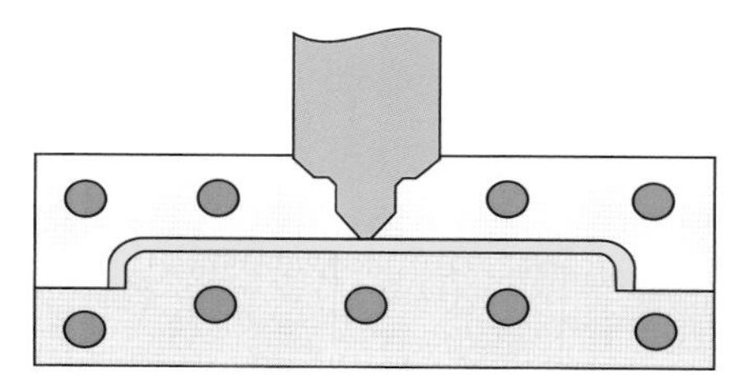

Figure 12.10 Non-uniformity of cooling created by the location of a hot nozzle directly gating into a cavity

As the hot drop approaches a part, it not only introduces localized heat to the part but also restricts the location of cooling (see Fig. 12.10). This cannot only affect cycle time in some applications but can also create non-uniform cooling that will increase residual stresses and potential for warpage. The negative influence of the non-uniform cooling is dependent on part geometry and material. A flat part molded of a high-shrink semi-crystalline material would pose a bigger problem than a center-gated cup-shaped part molded of a low-shrink amorphous material (see Fig. 12.11). The semi-crystalline material will experience more side-to-side shrinkage and variations in crystallization. In addition, the shape in Figure 12.10 has little structure to it and will more easily be influenced by variations in shrinkage. The structure of the center-gated cup is highly resistant to warpage regardless of the material.

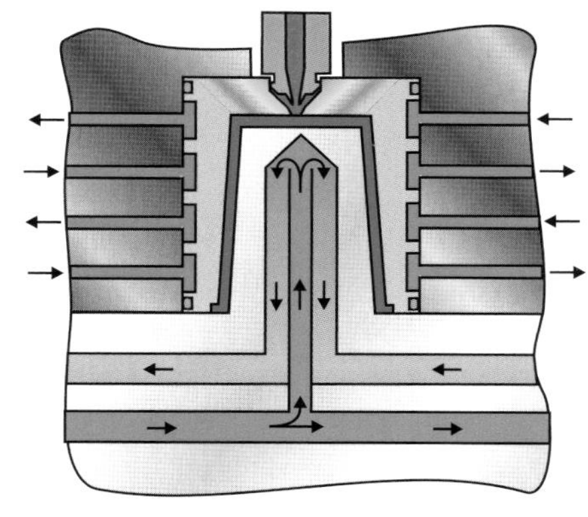

Figure 12.11 Cup-shaped parts are less sensitive to warpage from the non-uniform cooling resulting from the positioning of a hot drop directly gating a part

Thermal exchange between the hot drop/nozzle tip and the cooled cavity wall is generally controlled by three means.

1. A layer of frozen plastic material will develop in the gate region of all hot runners, which helps to reduce the thermal exchange between the hot drop and cold cavity wall (see Figure 12.12). This region is commonly referred to as the "cull."
2. Use of an insulting air gap provided by a gate insert (see Figure 12.12).
3. Use of direct cooling through use of a gate insert that includes a cooling channel. This channel may be machined or cast into the gate insert (see Figure 12.13).

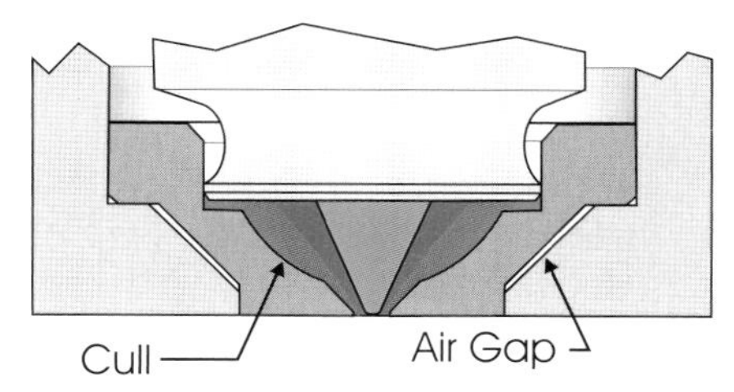

Figure 12.12 Air gap utilized to help insulate the plastic part from the heat of the hot drop

12.5.1 Gate Heating

It is sometimes desirable not to cool the gate. This may be the case with semi-crystalline materials, which are more prone to gates clogging from frozen material.

Gate inserts can also be designed to keep heat in the gate orifice. The insert shown in Fig. 12.12 includes an air barrier that helps insulate the gate orifice from the cold mold.

12.5.2 Gate Cooling

Depending on the demands of the plastic part being formed and the material being molded, it may be desirable to provide liberal cooling to the gate region. In addition to the issues discussed earlier, gate cooling may be desirable to reduce stringing (as discussed in Section 11.3.2) encountered with many amorphous materials. The increased cooling is intended to sufficiently freeze the material so it will break away from the gate opening rather than pull the material from the gate resulting in "stringing". Excellent gate cooling is also required for operation of thermal shut-off gates as discussed in Section 11.3.5. Gate cooling can be machined directly into the cavity steel (see Fig. 12.13A) or into an insert surrounding the hot drop. The gate insert in Fig. 12.13B is a design that includes a cooling channel to increase cooling in the gate region. Though this could be used with both semi-crystalline and amorphous materials, it is more likely to be seen with amorphous materials. The coolant in the gate insert could be fed from the same temperature controller as the entire mold or from a separate temperature controller, which would provide more discrete control of the orifice temperature. The use of an insert can improve the distribution of the cooling, compared to drilled channels, and allow closer proximity to the gate tip.

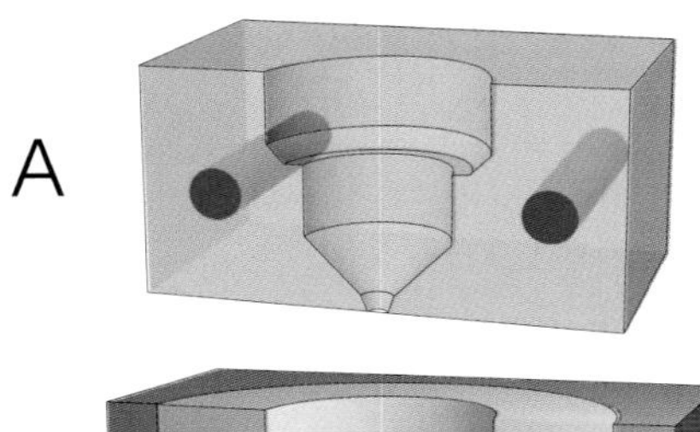

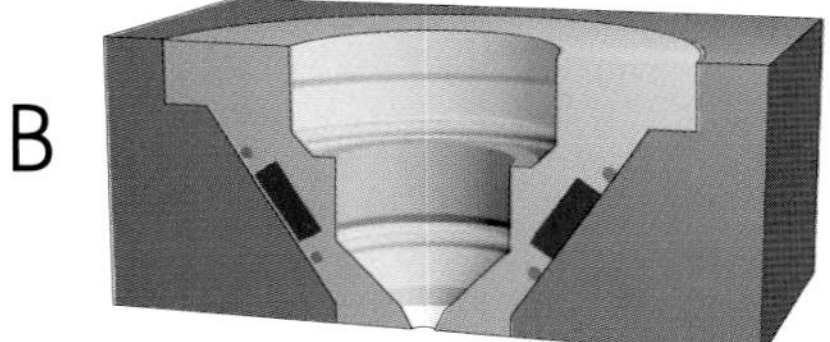

Figure 12.13 Use of a water cooled insert to insulate the plastic part from the heat of the hot drop

References

[1] Klees, J., "Hot Runner Molds" seminar manual. John Klees Enterprise, Inc., www.johnklees.com

13 The Mechanics and Operation of Hot Runners

13.1 Assembly and Leakage Issues

Hot runner assemblies range from relatively simple insulated systems to the more sophisticated externally heated systems. The focus of this chapter is on the dominating externally heated designs. Satisfactory operation of these systems requires precise engineering design, machining, assembly, and proper operating procedure. One of the most critical issues concerning assembly is the potential of system leakage. Leakage of a hot runner system can lead to catastrophic failure that can put a mold out of operation for weeks.

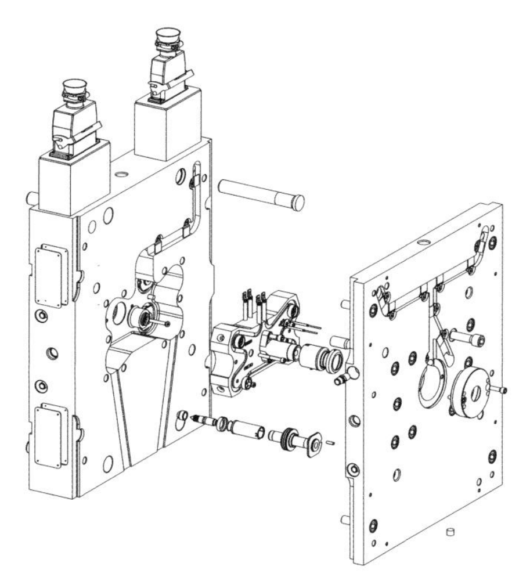

Figure 13.1 Exploded view of hot runner system (Courtesy: Husky)

Figure 13.1 is an exploded view of a four-cavity hot half of a hot runner system. Over 50 unique components are shown here with many having up to four duplicates. Though not all of these components are critical to resisting leakage, the figure illustrates the complexity of these systems.

Sealing methods include the use of direct or indirectly threaded fittings and the use of thermal expansion between components to create compression fittings. Virtually all externally heated systems include some critical seal region that is dependent on thermal expansion. This dependence on thermal expansion to create a seal emphasizes the importance of proper temperature settings and related process procedures to minimize the opportunity for leakage.

Proper seal-off of assembled components is dependent on five critical issues:

- Engineering and design
- Machining and assembly of the hot runner components
- Machining of the mold's hot runner interfacing components
- Operation of the hot runner system
- Interface of the mold to the molding machine

Leakage of the hot runner system in a mold can be highly destructive and potentially put a mold out of operation for an extended period of time. When a hot runner begins to leak, it can often go undetected until it results in catastrophic failure of the hot runner system. The leaking plastic can fill all of the air gaps around the manifold and drops, entombing the system in plastic. Figure 13.2 is a photo of the plastic removed from the manifold housing that resulted from a leaking nozzle. Note the heater and thermocouple wires entombed in the frozen plastic. In some cases, the

Figure 13.2 Results of a leaking nozzle entombing the heater wires

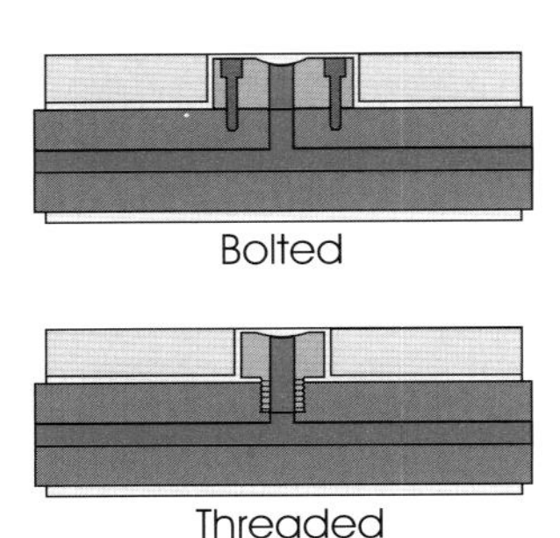

Figure 13.3 Bolted and threaded sprue connection to the manifold

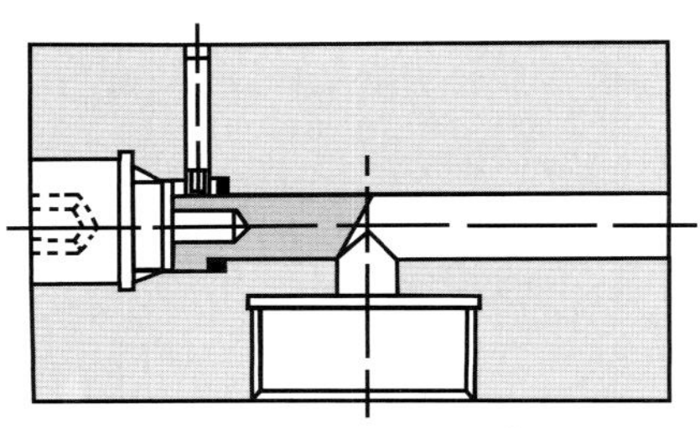

A) Cylindrical plug with collar

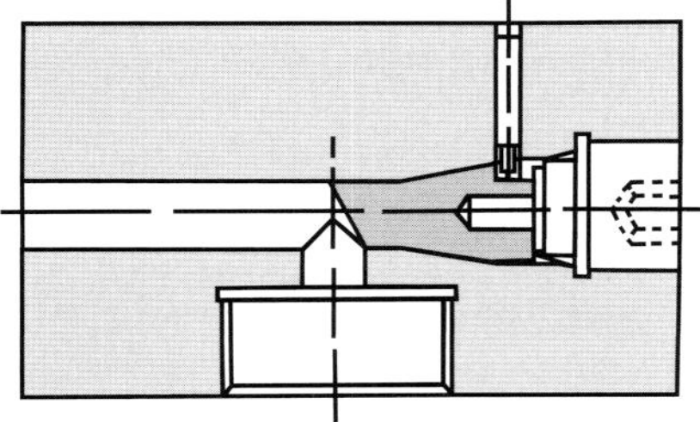

B) Plug with tapered sealing surface

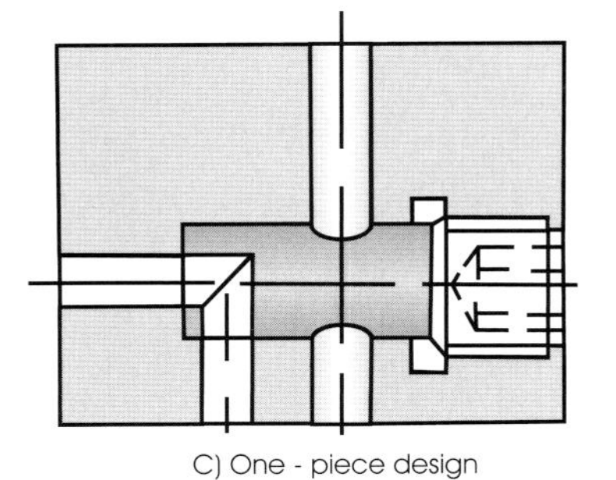

C) One - piece design

Figure 13.4 Variations of end plug designs to resist leaking in hot runner manifolds

system may be able to be repaired; however portions of it will likely have to be replaced. The removal of the plastic can result in destruction of all wires, heaters, and thermocouples, resulting in a downtime of two to four weeks waiting for the repair.

13.1.1 System Design

As indicated earlier, sealing of the hot runner components against leakage is based on either direct or indirect threaded fittings or differences in thermal expansion. The four critical assembly points are the assembly of the sprue to the manifold, the sealing and corner plugs for drilled flow channels, the seal of the nozzles to the manifold, and the seal between the nozzles and the cavity plate or gate insert.

Let's look at some critical assembly locations in the hot runner system progressing from the sprue to the drop tip:

1. The sprue bushings are most commonly either bolted to the manifold or threaded directly to it (see Fig. 13.3). The seal between these components may be augmented by use of a stainless steel "O" ring and by the pressure exerted by the injection barrel carriage as it presses the injection nozzle against the sprue bushing. This seal is generally the least troublesome.

2. End plugs for drilled flow channels in the manifold generally are threaded in place. They may also include "O" rings and dowel pins to seal and secure the plug. Figure 13.4 shows some variations of plug designs. Though the plugs could be welded in place, they are normally designed to allow removal for cleaning, often done by first drilling-out the plastic from the flow channels. It is critical that the manifold housing be designed such that if the plug were to blow out at high injection pressures, it be physically blocked from exiting the mold. A blown plug, at the high injections pressures of 20,000 to 30,000 psi, will act as a projectile that could have fatal consequences if it is allowed to exit the mold. Today, all properly designed hot runners are designed such that the manifold is totally enclosed.

   Split insert designs are used when a special geometry is required at corners and runner intersections. This approach is commonly used when shear- and thermally-sensitive materials like RPVC are to be molded. The desired fully radiused corners cannot be achieved with standard cross-drilling of the channels. A fully radiused flow channel can be located in an insert by either casting in the flow channel or milling two half round channels with the desired radius and then assembling them (see Fig. 10.6). The assembled halves may be brazed or press-fit together to reduce the risk of leakage. In addition, the insert may be isolated within the manifold by a sealing plug to prevent leaking.

3. The most troublesome seal regions of the hot runner system are the transition from the manifold to the nozzle and from the nozzle tip to the mold. The challenge is to provide a seal between the nozzle, whose tip position is relatively fixed, to the cold mold cavity it is feeding, as well as providing a seal between the same nozzle and the manifold that must grow and contract when heated and cooled. It must be realized that both of these seals are affected by the thermal expansion of the manifold in both its length and thickness as well as the nozzle in its length and diameter.

4. Three methods of attaching the nozzle to the manifold include:

   - *Direct Threads:* Directly screwing the nozzle to the manifold (see Fig. 13.5): This method is the most secure with regards to leaking but has limitations and creates challenges resulting from the thermal expansion of the manifold. Excessive expansion of the manifold, combined with a short nozzle can create a significant stress on the nozzle and corrupt the gate tip seal region (see Fig. 13.6). This method requires careful consideration of the deflection of the nozzle resulting from the expanding manifold. The nozzle steel must be able to withstand the expected flexing without compromising its seal at either end. To achieve this, the nozzle has to be long enough that it does not experience excessive strain when the manifold expands and contracts. INCOE's DF, or Direct Flo, system utilizes this method, allowing for a nozzle length which can be ½ of the distance between the center of the manifold and the nozzle. An additional benefit of this method is the minimized thermal transfer between the hot system and the cold mold. The only point of direct contact with the cold mold will be the nozzle tip seal ring.

   - *Indirect Thread:* Bolting the nozzle to the manifold through the use of a base flange (see Fig. 13.7). This method is most often used when close spacing of nozzles does not allow for the direct threaded method used above. Though not as positive as direct threading, this approach has been used with good success with small hot runner systems in which thermal expansion is minimum. Figure 13.8A shows a particularly compact nozzle design by Heitec. This design allows center-to-center spacing of only 7 mm. Note a similar design has been adapted for valve gates (Fig. 13.8B). On larger molds, thermal expansion of the manifold could cause excessive thermal expansion, stressing the bolts, and causing leakage at its base.

   - *Compression Fit:* Compression fitting based on thermal expansion is one of the most common methods used for fitting the nozzle to the manifold. Here, the nozzle is pressed between the cold cavity half and the hot manifold. There is no interlocking feature between these two components (see Fig. 13.9). The primary benefit of this method is that the manifold is free to expand along its length without bending or stressing the nozzle. Here, the lateral movement

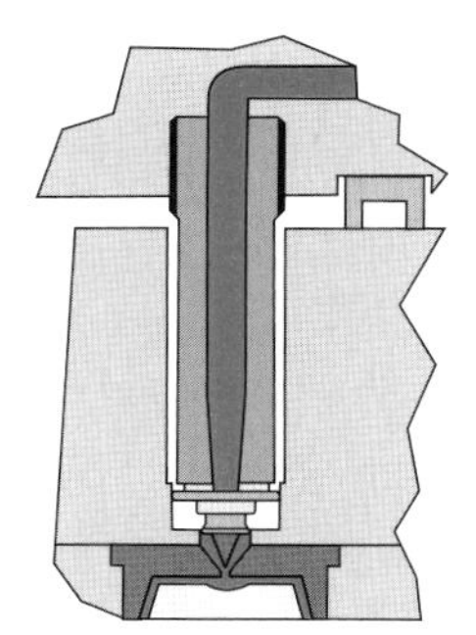

Figure 13.5 Heated nozzle screwed into manifold

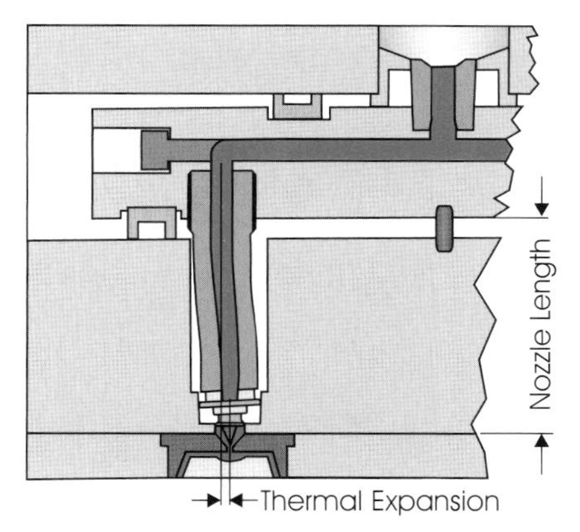

Figure 13.6 Potential excessive deflection and stress in a positively positioned threaded nozzle resulting from thermal expansion of the manifold during heating

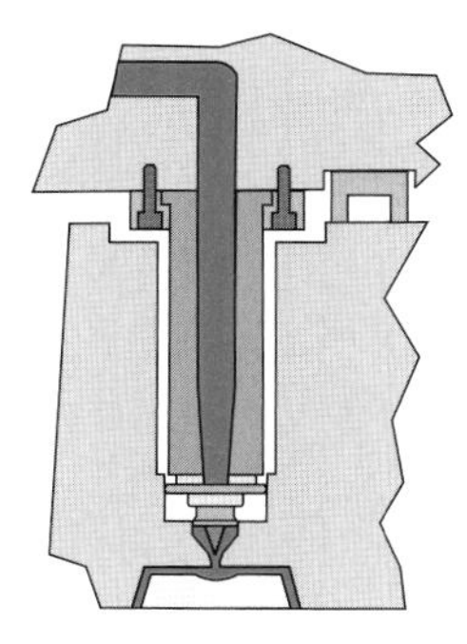

Figure 13.7 Nozzle assembled to the manifold with external bolts

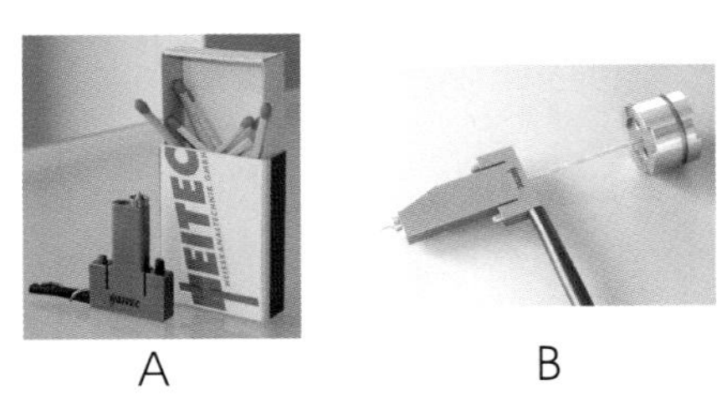

Figure 13.8 Miniature externally heated drop and valve gated hot drop used for specialty application. Drops are bolted to the manifold for assembly (Courtesy: Heitec)

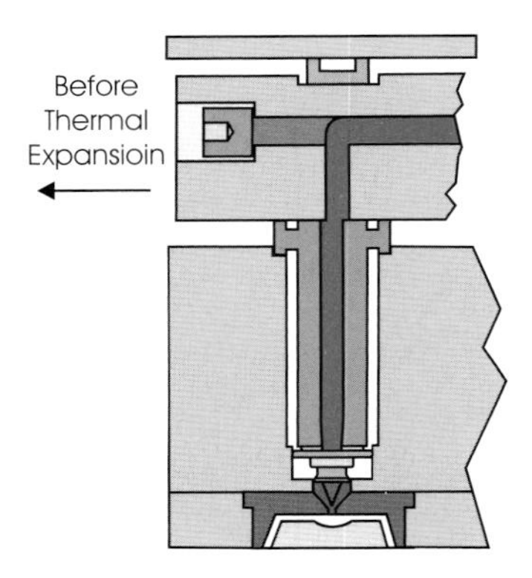

Figure 13.9 Misalignment of manifold and nozzle prior to heating and thermal expansion of the manifold

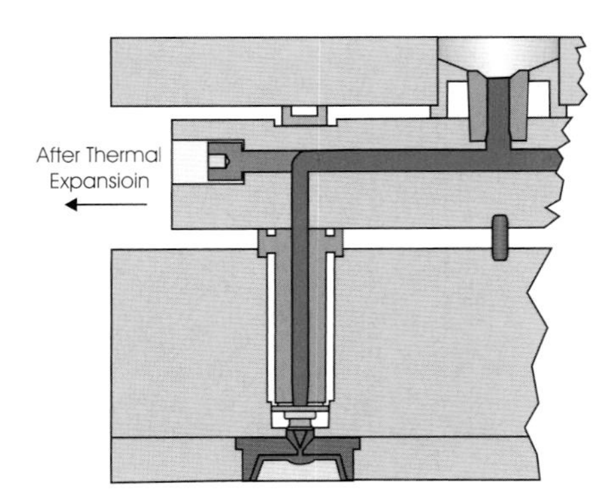

Figure 13.10 Alignment of the manifold and nozzle after heating and thermal expansion of the manifold (contrast to Fig. 13.9)

of the nozzle is fixed to the cavity plate (see Fig. 13.10). As the manifold and nozzle are heated, the manifold will expand laterally while the nozzle, fixed to the cavity plate, slides along the manifold's surface. Knowing the operating temperature of the plastic to be run, the outlet of the manifold flow channel will be machined such that when heated it will expand laterally so that it aligns with the flow channel inlet of the nozzle (see Fig. 13.10). The negative aspects of this method include:

- The system must be designed based on a specific temperature. Changes in temperature will affect:
  - The alignment of the flow channels between the manifold and the nozzle. Misalignment will create dead flow areas.
  - The thermal expansion-dependent compression seal between the nozzle and the manifold, potentially causing leakage (see Section 13.1.2).
- The method of sealing is much more dependent on operator and system compliance. The system must be brought up to operation temperature and allowed to heat-soak before operation. Either failure by the operator to adhere to this procedure or undetected failure of a heater or thermocouple resulting in low temperatures could result in leakage and damage to the system.

A discussion of the nozzle tip seal and operation can be found in Section 13.1.2 and in Figs. 13.13 and 13.14.

13.1.2 Hot Runner System Machining and Assembly

Most hot runner systems and components are purchased from a company specializing in the design, manufacture, and sale of these systems. These companies will also provide design specifications for the interfacing mold components. Though less common, there are still numerous companies that have developed their own internal expertise and design, and who build entire systems for their own use.

In an externally heated system, the design must minimize contact between the heated hot runner system and the surrounding mold. The manifold is generally suspended in the manifold housing through use of spacer buttons. These buttons are generally designed to minimize contact area between the assembled components. They are made of high strength, low thermal conductivity materials such as titanium.

This section will focus on the critical issues required for the successful installation of a hot runner system to a mold. We will assume that the system is built by a reputable hot runner provider and that they have provided a quality design and built the system to specification.

Many of the delays in startup and operation problems are the result of poor coordination between the hot runner supplier and mold builder and lack of experience by the mold builder and molder. If the hot runner system is purchased as individual components, it is the responsibility of the mold builder to correctly prepare the mold for assembly, assemble all components per instruction, and complete the wiring. Wire hook-ups are to be mounted externally to the mold.

It is becoming more popular these days to purchase a fully or partially assembled system from the hot runner provider. This may include purchasing of the entire hot half. Here, the hot runner system is assembled within a housing, which includes the "A" half clamp plate and all wiring completed. This significantly reduces the possibility for error because the hot half is simply bolted to the mold's "A" half (see Fig. 13.11). A variation of this, offered by INOCE, is the Unitized® System. This is a fully assembled and wired system that fits into a housing, which can be built by the mold builder. The advantage of this system is that it is easily accessed both in the press and on the bench. Quick-latch systems allow the hot half to be opened while in the press, fully exposing all components for easy inspection, service, removal, or replacement (see Fig. 13.12)

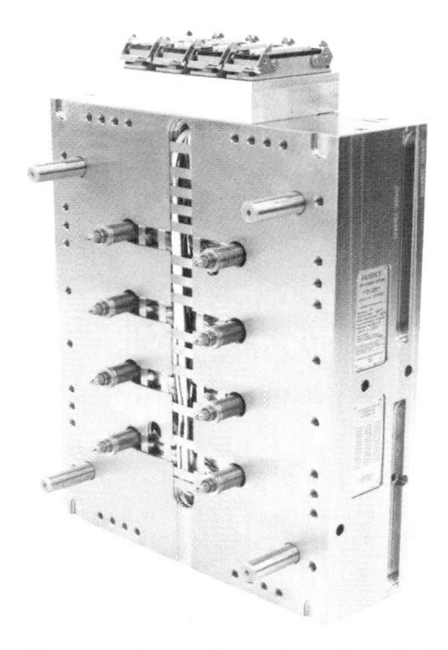

Figure 13.11 Complete "hot half" containing all hot runner components and housing – completely assembled and wired and ready to be attached to the mold (Courtesy: Husky)

Independent of which approach is taken, it is still the responsibility of the mold builder to provide the proper machined interface to the hot runner per the hot runner provider's instruction. Of particular issue are the seal-off regions between the nozzle to the manifold and between nozzle tip and the mold.

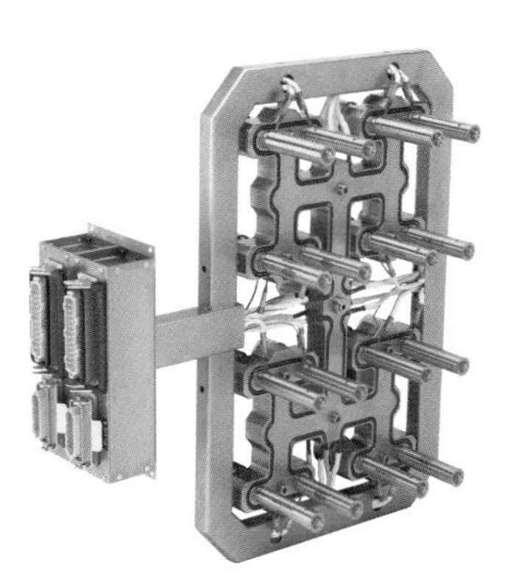

Figure 13.12 INCOE's Unitized System®, pre-wired and assembled and ready for attachment to the mold

In almost all heated system designs, the nozzle tip seal-off is provided by an interference fit between the OD of a nozzle tip sealing ring and the ID of a matching orifice in the mold's "A" half (see Fig. 13.13) The critical mold ID may be machined directly into the mold, the mold's cavity insert, or may be in a separate gate insert designed to interface with the nozzle and its particular tip design.

The OD of the nozzle tip sealing ring is generally jig-ground by the hot runner provider to a tolerance of + 0/(– 0.0005 to 0.0007 in.). It is important that the ID of the mating mold surface should be comprised of a hardened surface and should also be jig-ground to a tolerance of (+ 0.0005 to 0.0007)/– 0 in. It is a common mistake for the mold builder not to fully-harden the surface and try to save expense by electrical discharge machining (EDM) this surface rather than jig-grinding it. The tight interference between the nozzle seal ring's OD and the mold's ID can result in scraping of the surfaces in the direction of draw during assembly and disassemblies. This is the direction in which leaking plastic would follow to the detriment of the system. The damage to the seal can be increased by the more rapid deterioratetion of an EDMed surface. The hardened surface improves abrasion resistance. The jig-ground surface eliminates the pitted surface left by the EDM process, which is more rapidly eroded than the continuous flush or slightly ringed surface left by jig-grinding.

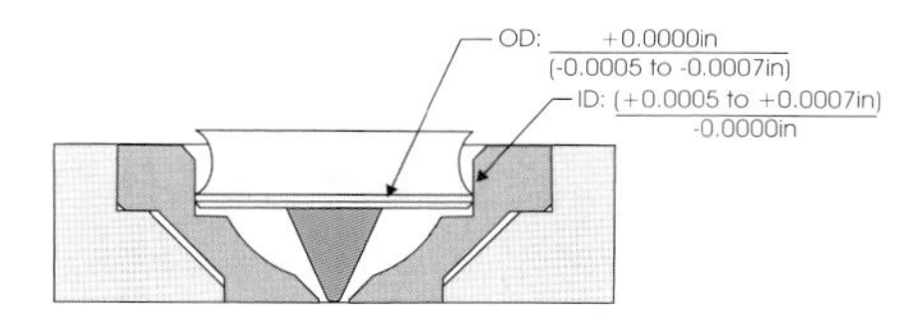

Figure 13.13 Critical dimensions of seal ring and mold to prevent leaking

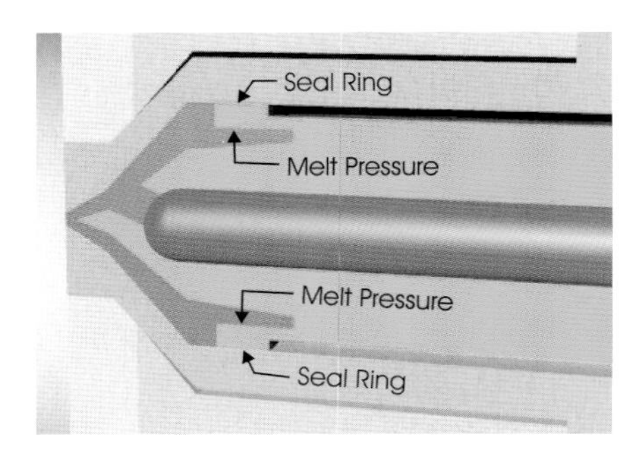

Figure 13.14 Sealing of seal ring in the hot nozzle is augmented by melt pressure creating a hoop stress acting to expand the ring into the mold

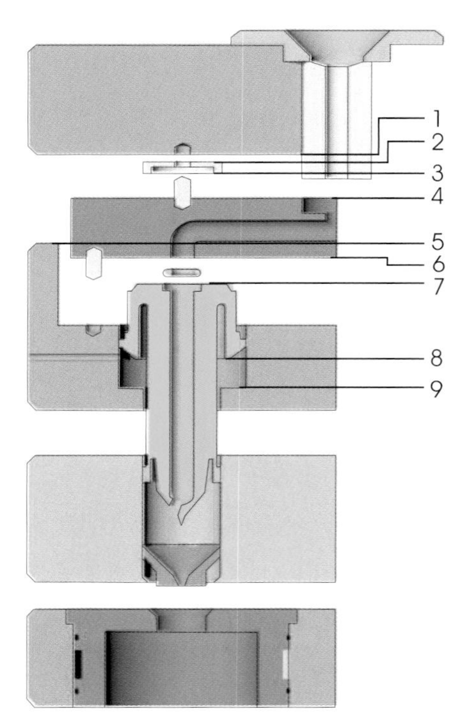

Figure 13.15 10 critical surfaces which must have a combined stacked tolerance of 0.0002 in. to help ensure a leak-free compression seal in many hot runner system designs

The OD and ID tolerances presented above are dependent on the system purchased. However, regardless of the system one should realize the critical nature of the seal. It is strongly recommended that the mold builder closely check and document the dimensions of the seal ring OD and the related mold ID. Considerer the gap that could potentially result, if both components were machined to the extremes of their tolerance. In this example, a gap of 0.0014 in. would result. During heating, this gap should close somewhat as the tip diameter expands relative to the cooler surrounding mold. The larger the diameter of the ring, the better is its ability to provide a seal. However, though the resulting gap should be quite small, at the extreme melt pressure that could develop some plastic materials may potential leak through this opening. When extreme melt pressures and low viscosity melts are to be used, such as in thin-walled molding, a mold builder may want to be particularly diligent in maintaining even tighter tolerances.

To further enhance this seal, the ring is often positioned on a surface designed to deflect during injection from the applied melt pressure. The higher the pressure, the more deflection there is and the better the seal (see Fig. 13.14).

Checks: To minimize potential of leaking at the nozzle tip:

- Check that the outer diameter of the nozzle seal ring is to specification
- Check that the inner diameter of the nozzle seal ring's mating mold surface is to specification
- Check that the mold seal ring's mating surface is of hardened steel and jig-ground

If the nozzle requires a compression seal, the tolerance on the stack height of the manifold assembly and the surrounding frame work is critical. Figure 13.15 shows all the surfaces that must stack up with a total tolerance of 0.005 mm (0.0002 in.). Each of the nine surfaces should be ground. Excessive stack height in a given nozzle will result in excessive stress in the local components and contribute to leakage in surrounding nozzles as it will decrease the compressive stress in these surronding nozzles. An under-toleranced nozzle will result in leaking. To accomplish the required fit of the assembled manifold components to the enclosing manifold housing, the mold builder must first measure the height of each of the stacked-assembled nozzle and manifold components. They are compared to the height of the space between the clamp plate and the cavity plate (the housing, which will contain and compress the manifold and the stacked nozzle and manifold components). This is the distance between the S1 and S9 surfaces in Fig. 13.15 when the mold is assembled. Each of the manifold spacer buttons is then ground so that when reassembled, the tolerance between each of the stacked nozzle and manifold components and the encasing manifold housing is within the specified 0.0002 in. All measurements, machining, and assembly should be conducted at approximately the same room temperature. It is important that the customized spacer buttons remain in their correct position during mold maintenance (disassembly and reassembly). Incorrect positioning could result in leaking.

When assembled at room temperature, there may be very little compression between the assembled components to prevent leakage. The fit of the assembled housing is designed so that at the operating temperature of the manifold the hot components will expand within the colder surrounding frame to create the required leak seal between the nozzle and the manifold (surfaces S6 and S7 in Fig. 13.15). If the components are stressed during assembly at room temperature, they may experience excessive stresses during expansion, which could cause damage.

To help with the seal a specially designed, hollow stainless steel, "O" ring is sometimes used (see Figure 13.16). The "O" ring is fitted in a slot so that it is not fully crushed. A hole in the "O"ring allows plastic, under the high injection pressures, to fill the hollow ring. During injection, the same high pressure trying to drive the melt past the "O" ring also pressurizes it from the inside. This internal pressure helps counter the external pressure and improves its resistance to leaking, as long as the material inside the "O" ring remains fluid.

Some hot runner providers have developed spring-loaded assemblies that are more forgiving of variation in the assembled components. Husky's patented compressive seal is shown in Fig. 13.17.

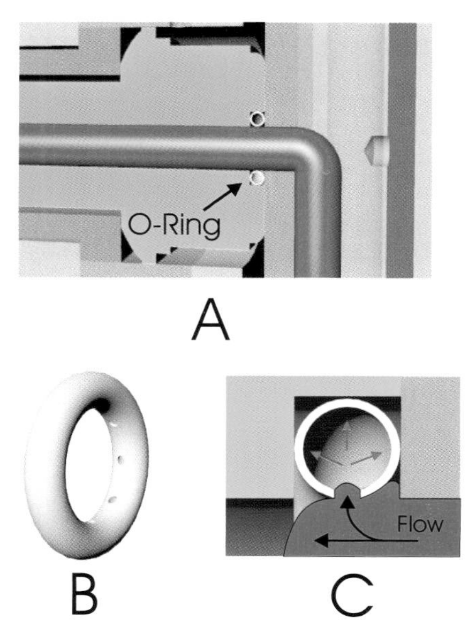

Figure 13.16 Manifold-to-drop hollow steel O-ring. Design has holes that allow plastic to fill the ring and maintain its seal from internal melt pressure developed during injection and hold phases

13.2 Mold and Machine Distortions

Although the mold and the hot runner system appear to be a static assembly, it is, in fact, subjected to high dynamic forces during the injection molding cycle. Cyclical bending forces, developed by the clamp and injection melt pressures, on the mold plates are partially transmitted into the hot runner system. In addition, the heating and cooling of the mold and hot runner system results in high forces during heat expansion.

These forces must be considered in the mechanical layout and design of the mold and hot runner system. Insufficient rigidity in the mold and the surrounding mold platens can result in melt leakage in the hot runner system and flashing in the cavities (see Fig. 13.18).

The number of mold plate screws must be sufficient to handle the internal forces caused by thermal expansion and melt pressure. Preload on the assembly screws prevents plate separation. Support pillars can be added around the contours of the manifold to resist plate deflection and provide additional fastening.

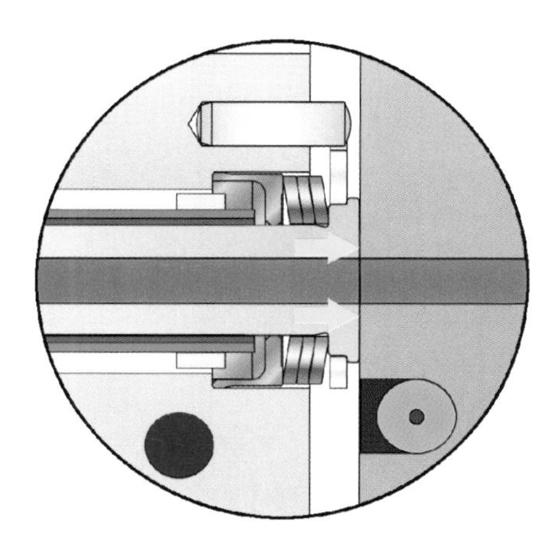

Figure 13.17 Patented spring for improving compression seal (Courtesy: Husky)

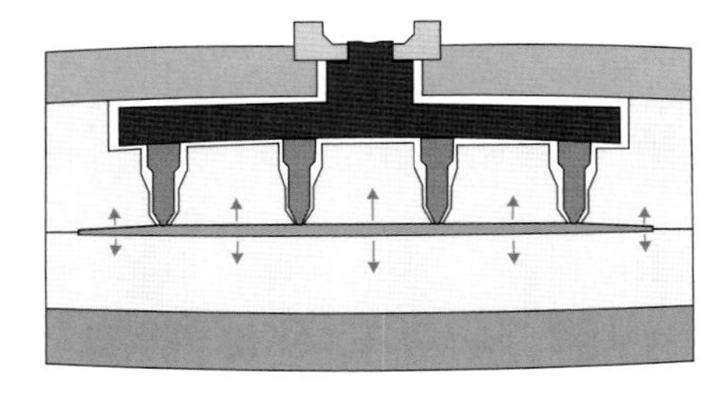

Figure 13.18 Deflection of mold, manifold, and nozzles if not properly designed to resist pressures and forces developed during the injection molding process

In addition to the mold, the molding machine's platens can distort. The distortion is dependent on the rigidity of the platens, the design of the clamping mechanism, and the relative size of the mold compared to the platen. The smaller the mold relative to the platen, the more the platen will bend around the mold. Both hydraulic and toggle clamping mechanisms rely on the stretching of the tie bars to provide the desired clamping force (see Fig. 13.19) In both cases, the force applied to the stationary platen is applied at the four corners by the tie bars. However, the means of transferring force to a mold can be quite different on the movable half. The hydraulic clamp system will provide a central force, whereas the toggle's clamp force is generally outboard on the platen, much like the stationary half.

Though these distortions seem to be minimal, one must consider the extreme effort that has been applied in the design of the mold and the hot runner system to achieve the desired compression seal. Small distortions of the platen, transferred through a non-rigid mold, can contribute to system leakage.

Heat generated from the hot runner system will cause the entire mold to grow and can also transfer into the molding machine's platens. These characteristics must be considered when running a hot runner system.

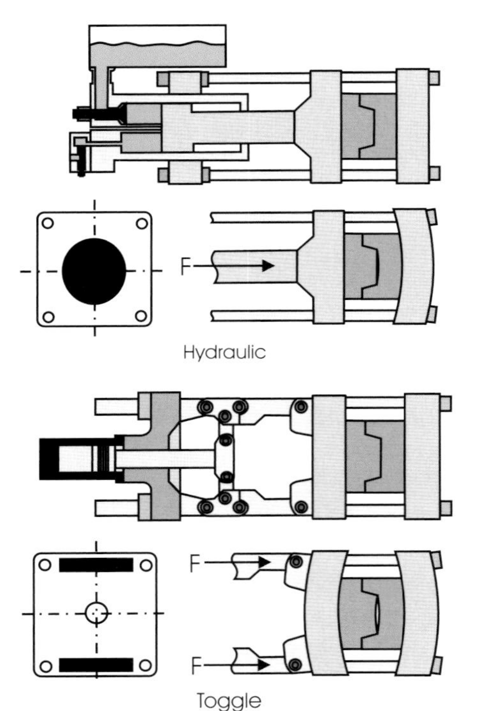

Figure 13.19 Distribution of potential deflection developed in an injection molding machine resulting from hydraulic and toggle clamp mechanisms

Thermal expansion of the mold can cause the clamp in a toggle or hybrid mechanical hydraulic system to lock up. Normally, the clamp tonnage on a molding machine is set after the mold is first installed and is still at room temperature. Most clamping mechanisms utilize a mechanical locking means (vs. hydraulic) to maintain their clamp tonnage by toggling past a threshold position during clamp-up. During the operation of the hot runner mold, the mold will expand. As it expands, it will increase tonnage by increasing the mold size between the fixed position platens. If this effect becomes excessive, the clamp may toggle through its threshold position to its locking position, but may not have the force to return back through the threshold point. This problem can be exaggerated if the heat is allowed to transfer from the hot runner system into the mold platens causing them to expand also.

Further problems can result as the platen is heated. Lubricants between the platen and the tie bars will become heated, diminishing their performance. Additionally, with time, heat can migrate into the tie bars and the clamp mechanism.

To minimize this problem, the "A" half clamp plate should be cooled. It is also common that an insulating plate be used between the "A" side clamp plate and the platen.

13.3 Startup Procedures

There are two critical issues that must be accommodated during the startup of a hot runner system.

- The cooling to the mold should be turned on before any of the heaters. This will prevent damage to any coolant "O" rings and ensure that the mold's thermal expansion is stable once the heaters are turned on.
- Soft start: All heaters will absorb moisture from the humidity in the air. If they heat up too quickly, they can be damaged by this trapped moisture. To prevent this, the heaters must go through a "soft start" which means that the heater's temperatures are initially raised slowly to approximately 90 °C and allowed to heat-soak for at least 15 minutes. This allows for the moisture to be driven out of the heater. Following this procedure, the heaters can be raised to their normal operating temperature. This soft start procedure is commonly programmed into the heater controllers.
- Heat soak: Once the hot runner system has reached operating temperature it must have an additional heat-soak at that temperature for at least 15 to 30 minutes. This assures that all components have expanded and thereby created the required seal to prevent leaking during injection.

13.4 Color and Material Changes

The major challenge during color change is to remove the material accumulated in low or stagnant flow regions. The more obvious regions that create problems during a color change are the low flow and stagnant flow regions that exist in corners or on the downstream side of internal-to-flow features such as heaters, gate tips, and valve pins. Less obvious is the flow channel wall. The nature of pressure-driven flow in a flow channel is that flow on the channel wall is considered to be zero, which means that an original colored material would always remain on the wall.

Recognizing that material in some locations within the hot runner flow channel will never be removed, it is recommended that the first material injected through the system be a natural or light-colored material, followed by producing parts made from the required colored material. This method will coat the low flow and stagnant flow regions of the flow channel with a light color that will be less obvious if it is washed into the molded part after a color change.

During actual color change it is generally recommended that the temperature of the melt and the hot runner system be increased, combined with a fast flush rate and short cycles. For best flush, inject at high injection rates while the mold is open (The "A" half of the mold must be securely fastened to the stationary platen). This allows a full shot to be developed in the barrel and purged through the system, thereby providing a more continuous flush. The fast flow rate creates a more plug like flow condition across the flow channel. This results in higher flow rates and higher shear rates near the channel wall. The higher flow rate and continuous flush will more thoroughly and quickly erode the previous colored material that is in the low flow regions. Returning to normal operation conditions, the new material should coat the remaining previous colored material.

The injection rate during purging should be at least the same as during normal injection. A faster injection rate will improve flush. Flushing through the system with the mold open requires that the "A" half of the mold be well secured to the platen. The force of the injection nozzle against the platen can reach several tons. The above method will erode at least the outermost of the original colored material (material closest to the flowing plastic). A small flow channel size will provide the best flush and is highly recommended for this purpose.

To further enhance color changes, a purging agent can be used. As these are plasticized in the injection barrel, they will reduce color change problems created both in the injection unit and the hot runner system. These are recommended for some of the more stubborn color changes.

13.5 Gates

The gates in a hot runner can be the source of numerous operation and maintenance issues. Some companies claim to provide over 50 variations in gate designs to help address any critical issue. Therefore, the mold and gates should be designed to provide for easy maintenance.

13.5.1 Vestige

A major objective of a hot gate is to minimize gate vestige and avoid stringing, while remaining open for each shot. To minimize gate vestige, a gate's diameter is kept to a minimum. However, the small gate diameter potentially results in freeze-off of the gate, partial freeze-off, potentially excessive shear, and high pressure loss during injection.

Therefore, the gate may have to be reworked as the user reacts to process problems. It is common that the molder will undersize a gate initially. If the small gate creates molding problems, it can easily be enlarged. If one gate is changed, it is critical that all other gates be changed to the same design and size.

13.5.2 Clog

Owing to their small size, gates are sensitive to clogging. Unlike a restrictive gate tip in a cold runner mold, if a contaminate clogs the gate, the clog is not ejected between cycles. The clog will generally remain until physically removed. A molder can try to release the clog by inserting a small brass pin through the clogged gate area. This is not recommended and requires extreme caution to assure that the system is not pressurized. Also, the clogging material is generally only pushed back by this method and returns to clog the gate during the next shot.

A filter nozzle is sometime used to try to catch debris that could clog the gate tip. However, these will increase pressure and can result in varying melt conditions as they progressively clog, thereby changing pressure drop through them. Regular monitoring of injection pressure is required to assure consistent molding.

The mold should be designed to provide reasonably easy access to the gate tips while the mold is in the press. It is recommended that the mold be designed such that the cavity retainer plates and cavity inserts can be separated from the hot tips while in the press. Many hot runner suppliers will provide design methods to allow this to easily occur or have designed systems that significantly simplify this type of maintenance.

13.5.3 Wear

All gates will eventually wear. Rapid gate wear will result from common aggressive fillers such as glass and carbon fibers. This wear will increase the outside diameter of the gate. The increased diameter will increase gate vestige and will need to be addressed. To prevent excessive wear, a gate tip insert should be used (see Fig. 12.12 and 12.13). A design providing for easy replacement of the insert is preferred.

13.6 Maintenance

Of primary concern is the leakage of the manifold. Therefore, the manifold should be periodically inspected. At a minimum this should be done after every run. In addition to the gate issues presented earlier, it is to be expected for thermal and mechanical parts to fail. Hydraulic valve-gated systems have moving parts that will wear and fail. Valve pins travel into the melt channel and therefore have a potential of causing melt leakage. Common sources of failure are heaters and thermocouples. Operators must be trained to quickly recognize that a failure has occurred. A simple diagnostic tool is the use of pressure transducers in the cavities. Ideally, the pressure transducer should feed a monitoring system through which an automated limit system can be activated to detect changes in process that

Figure 13.20 INCOE's integrated system exposed in the molding machine for easy inspection and service (Courtesy: INCOE)

could result from a failed component. The most common components requiring service are the gate tips, heaters, and thermocouples. Recognizing this, a mold and runner system should be used that allows these components to be easily serviced.

Figure 13.20 shows the INCOE Integrated System™, which is essentially their Unitized System™ assembled to the top plate. The photo shows an eight drop system opened in the molding machine. This system allows easy inspection and maintenance of the system while it is still in the machine.

14 Process of Designing and Selecting a Runner System (Gate and Runner) – A Summary

Following is a summary of the key factors that should be considered when designing a runner system. Most of these issues have been discussed in more detail throughout this text.

14.1 Number of Gates

A minimum number of gates is generally desirable for the part. However, manufacturing often requires that multiple gates are used. The fewer the gates, the fewer product problems resulting from gate vestige, blush, jetting, high residual gate stress, and welds from multiple flow fronts. However, a single gate can result in processing problems such as excessive fill pressure and clamp tonnage concerns. In some cases it may be desirable to reduce material shear rates and shear stresses by distributing flow during mold filling between multiple gates.

14.2 Gating Position on a Part

There are a number of factors to be considered when selecting a gating position on a part. Some of these are obvious, whereas others require a more in-depth understanding of the plastic part-formation process.

14.2.1 Cosmetic

Gates should not be placed in a cosmetic portion of a part. Not only is there a concern with gate vestige (a defect remaining when the gate is removed from the part), but it also raises concerns about other defects common in the gate region. These can include gate blush, sinks, flow marks, and high stress, which can lead to crazing and cracking of the part. If a part must be gated in a cosmetic region, try to gate the part on the back side of the cosmetic wall. Also, valve gates from hot runners can be used. However, in both cases it should be realized that some gate witness may still be present.

14.2.2 Effect on Shrinkage, Warp, and Residual Stress

Gating position can have significant effects on the shrinkage, warpage, and residual stresses in a molded part. This is due to variations in shrinkage that are established by orientation developed from plastic flow and by pressure distribution within a cavity dependent on relative position to the gate.

14.2.2.1 Orientation

As plastic material flows during the mold filling and compensation (packing) phases, shear and extensional stresses are developed that act to orient fillers and reinforcements within the material and to orient the material itself. Variations in orientation can result in variations in shrinkage, which in turn leads to residual stress that can warp a part and cause premature failure. A gating position that will result in a simple linear flow pattern is generally considered to minimize the potential for residual stress and warp. An exception to this rule is a cylindrical part for which a center gate resulting in a radial flow pattern is most desirable.

Anisotropic shrinkages are a particular concern with fiber-filled materials and with parts with a non-structural shape that will distort easily with stress. Fibers and other fillers/reinforcements, which have a high aspect ratio, can result in significant anisotropic shrinkages.

14.2.2.2 Volumetric Shrinkage (Regional)

Plastic materials will go through considerable shrinkage from their molten to their solid state. It is not unusual for this shrinkage to be 25% or more. As residual stress and warpage are a result of variation in shrinkage, it is desirable to minimize this volumetric shrinkage variation within a molded part.

Near Gate vs. Away from Gate Shrinkage

A part with constant wall thickness will almost always shrink more away from the gate than near the gate. The longer the distance from the gate to the end of fill, the greater is the potential for shrinkage variations between these regions. The exception to this normal shrinkage is when there is excessive heat in the gate region and when the pack time is too short. A short pack time can result in back flow through a gate and zero pack pressures near the gate after far regions have frozen under normal pack pressure. The normal expected shrinkage variation can sometimes be addressed by profiling packing pressure as the part freezes.

Shrinkage from Variation in Wall Thickness

If a part has variations in wall thickness, it should almost always be gated in the thicker region. Gating in thin regions will result in uncontrolled shrinkage in the thicker regions of the part. This is caused by the thin region freezing during the pack stage of the molding process earlier than the thick region. As this happens, compensating plastic material will not be able to feed the shrinking thicker regions. The exception to this guideline of gating into thicker regions is localized part features such as bosses and thick flanges. Gating into these features will often result in non-desirable filling patterns creating further molding problems (see Section 4.2.3).

Effect of Hydrostatic/Over-Packed Regions on Shrinkage

Unbalanced mold filling will result in variations in pressure across the part during the filling and packing phases of the molding process. Regions that fill early will become somewhat hydrostatic at a higher pressure than regions that fill later. Dead flow regions that fill early will build to a high pressure while pressure to a growing flow front ranges from zero at the flow front to maximum at the gate. The nearer the dead flow region is to the gate, the more pressure will build up in that region.

14.2.2.3 Unbalanced Filling

Unbalanced filling results in the poor pressure distribution described above and in transient flow conditions. Regions that fill early will result in flow being redirected to non-filled regions. This will result in variations in flow-induced orientation through the thickness of the part. This in turn will result in variations in shrinkage through the wall thickness and across the part. Both phenomena will lead to residual stresses and warpage (Section 4.2.7).

14.2.3 Structural Issues

A part's structure can be significantly affected by the gate location. The primary concerns are gate stress and orientation.

14.2.3.1 Gate Stress

Shear rates and shear stresses on the plastic material are extremely high at the point where the material enters the part from the gate. In addition, shear in this region is continuous throughout mold filling and packing. Relatively high shear rates and stress are continued throughout packing, because all regions of the part are fed through this region. This location may experience the highest pressures during mold filling but may actually be a region that is poorly packed-out. The poor packing can be caused by a number of issues: Eliminating pack before the gate freezes will result in back flow through the gate. The high sustained shear rates in the gate will locally heat up the mold steel. The gate freezing, or pack pressure being eliminated before this hot region is cool will cause local high shrinkage. All of these can result in the immediate gate regions shrinking significantly different from regions surrounding it. This will create a high stress in the gate area, which will often be the first place for a part to crack. Therefore, one should avoid putting the gate in any region of the part that will experience stress during application. In addition, a gate freeze study should be performed to determine the pack time required to avoid back flow.

14.2.3.2 Flow Orientation

Molecular and filler orientation can establish anisotropic properties in a molded part. Generally, the part will be stronger, in a tensile loading application, along the direction of orientation and stronger transverse to orientation under flexural or impact loading.

14.2.4 Gating into Restricted, or otherwise Difficult to Reach Locations

It is sometime desirable to locate a gate in a position that is relatively difficult to get to. This includes gating into the inside diameter of a tubular part or a location along the side of a part with vertical walls.

Cooling requirements and ejection requirements, including side action, lifters, etc., can interfere with desirable runner and gate placement. Use of side action and complex parting lines may be a particular problem with runner and gate locations in a two plate cold runner mold.

14.3 Cavity Positioning

Similar to Section 14.2.4, the physical size of a hot drop, and the required insulating air space around a hot drop, limit how close multiple hot drops can be placed next to each other. This potentially limits how close cavities producing small parts can be placed to each other. Hot drops with multiple gate tips may be considered for feeding small closely-spaced parts.

14.4 Material

Though both cold and hot runner molds can be run with virtually any type of commercial thermoplastic materials, the cold runner is considered easier to work with in many cases. Thermally sensitive materials such as rigid polyvinyl chloride are particularly troublesome in hot runners and are generally avoided. Use of recycled materials, with their increased potential of contamination, increases the potential for blockage of a restrictive hot gate tip. A blocked tip will require the molding process to be interrupted, the cavities separated from the drops, and the gates manually cleared. A molder producing parts which do not allow use of regrind will often prefer a hot runner.

14.5 Jetting

Jetting results from the inertial effects of material exiting at a very high velocity from a restrictive gate. To avoid jetting, the melt must immediately impinge on an opposing wall or some kind of mold feature that prevents the material from jetting across the cavity. This generally implies that a gate cannot be placed along a parting line parallel to the wall it is feeding.

14.6 Thick vs. Thin Regions of the Part

Though it is ideal for a molded plastic part to have a constant wall thickness, it is more common for some variations in wall thickness to exist. These variations can cause a number of mold filling problems that need to be considered during gate location selection.

14.6.1 Volumetric Shrinkage

As discussed earlier, a part should always be gated in the thicker region of the cavity. Material should fill from thick to thin.

14.6.2 Hesitation

Hesitation occurs when plastic filling a mold encounters a thin wall section, while there are other thicker walls yet to be filled. The plastic entering the thin region will slow down as a result of the increased flow resistance. As soon as it slows down, the benefit of non-Newtonian shear thinning is reduced. In addition, the slow fill speed will reduce frictional heating and the plastic will lose temperature to the relatively cool mold. The combined effect can significantly increase the material's viscosity relative to the material still filling thicker regions of the part. This compounds the resistance from the thin wall causing the material to quickly slow down and potentially freeze off. To minimize hesitation it is desirable to gate as far away as possible from thin features.

14.7 Number of Cavities

The higher the number of cavities the more critical the runner design becomes. Balancing filling and melt conditions to a cavity is a particular challenge. Any branching in a runner will cause an asymmetric condition in the melt stream. This will affect runner balance and melt conditions within a cavity. Balanced melt conditions in multi-cavity molds require careful consideration of runner layout and use of melt management technologies within the runner (Chapter 6 and 7).

14.8 Production Volume

Injection molding is less economical compared to some alternative processes if production volumes are less than 10,000 parts per year. Hot runner molds are more difficult to justify unless production volumes are quite high. Hot runner molds are more expensive, require longer mold commissioning times, and longer job start up times.

14.9 Precision Molding (Precision Size, Shape, Weight, Mechanical Properties and Consistency)

High precision molding requires particular attention to runner and gate design and gating location. The more critical the part, the higher the demand for specific attention. The runner must provide balanced melt conditions to each cavity (not just a geometrically balanced system or balanced filling). Both the gate and runner must have sufficient cross-sectional area to provide for filling and packing. As many higher-precision parts are made of more expensive engineering grade resins, runner sizing must be controlled in cold runner molds to minimize regrind, which may not be reusable in the same application.

14.10 Color Changes

A cold runner mold is completely purged of any prior material during every cycle. This provides for very simple color changes, dependent only on purging the screw of the injection molding machine. This is not the case with hot runners. Internally and externally heated hot runners require that a new colored material be repeatedly flushed through the system to remove any old material. Low flow and dead flow regions in the flow channel require specific procedures to allow for a color change. Such regions may be any location along the manifold, drop, and gate tip. In many cases, the original material shot into the hot runner may never be removed unless the hot runner is completely disassembled and the flow channels reamed out. Despite this, successful color changes are regularily achieved when the proper procedure is followed (see section 10.4.3). Insulated hot runners provide for thorough color change in as little as 15 minutes by opening the plates, containing the runner while in the press, and removing the runner. However, these systems are very limited in terms of the variety of materials that can be used and in terms of tolerance requirements that can be fulfilled in these type of runners.

14.11 Material Change

Cold runners are normally preferred if material changes are expected because they are completely purged between every shot. In addition, a cold runner is much easier to modify if the new material has any special requirements. This is particularly beneficial if the material change was not anticipated. It is not unusual for a mold to be rushed into production without a clear knowledge of the plastic material to be molded. If necessary, the entire runner system can be remachined at a minimal cost and time by remachining or replacing a runner bar. Hot runner systems are designed to a specific material and generally have little tolerance for material changes. Channel diameters are fixed, temperature changes between materials will affect the standard compression sealing of the system, the alignment of components, the positioning of gate tips, heaters may be improperly sized, gate tip and region may be of the wrong type for the new material, channel geometry may be inappropriate, and so forth.

14.12 Regrind of Runners

Cold runners require handling of the runner produced during every molding cycle. This generally requires runner handling/sorting equipment, granulators, and operator attendance. Depending on the product requirements and material characteristics, all, a portion, or none of the reground runner may be reused in the product it was originally used to produce. This dictates how the runner is to be handled. Some cases allow for runners to be reground at press side and fed directly back into production. In the other extreme, the runners are reground and sold. For each mold, the cost of runner handling must be evaluated and compared to the cost and issues of using a hot runner.

14.13 Part Thickness

Part thickness can affect the decision process of selecting a runner. Factors such as fill pressure and packing can be affected.

14.13.1 Thin Part

A thin part normally implies high pressures to fill and fast cooling times. If the cavity fill pressure is expected to be quite high, it is logical that the pressure required for the melt delivery system should be low. In a cold runner channel, the larger channel may determine cycle time and result in a disproportionate amount of runner material to be reground. The hot runner can eliminate this concern. However, be cautious of the increased residence time and low flush that may result from the large diameter hot runner when used with a small thin-walled part. With many materials and applications these concerns are minimal and the hot runner is the obvious choice.

14.13.2 Thick Part

For proper packing, it is required that the runner and gate do not freeze prior to the part being adequately packed out during the compensation phase of mold filling. In order to pack out a thick part with a cold runner system, a relatively large diameter runner and gate are often required. This will result in a fair amount of runner regrind and possible gate vestige issues. A hot runner does not have this problem as it is always molten. A relatively small diameter runner channel and gate can be used. If there is concern with the freeze-off of the hot gate or gate vestige from the hot gate, a valve gate could be used.

14.14 Part Size

Part size and part thickness are closely linked when considering a runner system. A very small part may prefer a hot runner because the hot runner will reduce the disproportionate amount of material in the cold runner that may otherwise result. A large part may be expected to result in a high pressure to fill or use of multiple gates. As with a thin part, the hot runner provides a better opportunity to reduce pressure drop in the runner without negatively effecting cooling time. The concern of long material residence time with hot runners is less of an issue with large parts vs. small thin-walled parts. The increased volume of the part may be sufficient to flush the runner during every shot. If multiple gates are required to fill the part and cosmetics and strength are a concern at the resulting welds, a hot runner utilizing sequential valve gating should be utilized.

14.15 Labor Skill Level

The use of hot runners requires a commitment to a higher level of training for all personnel involved with production. This includes designers, process engineers, process technicians, operators, and tool room maintenance personnel. The greater commitment a company makes to the use of hot runners and training, the more successful they will be in their operation. Companies who infrequently use hot runners will experience more problems with their hot runner than companies who are experienced. Custom molders will often have the biggest problem as they may inherit a mold with a hot runner system that they are not familiar with.

14.16 Post Mold Handling

With some parts it is desirable for a cold runner to remain attached to the parts. The runner may maintain positioning of parts for secondary operations such as decoration. Sixteen parts on a cold runner can easily be positioned for printing and the runner can be removed afterwards. This would minimize the complexity of robotics or handling required if the parts were degated during ejection. This same principle can help with automated assembly of parts from a family mold. The runner can maintain order and presentation of the parts. This is also the case with parts such as toys having many varied parts, which may require assembly by the consumer.

14.17 Part/Gate Stress Issues

Some parts require a minimum of gate stress, or the use of a gate that improves the melt flow pattern in the part. A hot runner is limited to fairly restrictive gates. Hot valve gates can be less restrictive but are still somewhat limited in how they can deliver melt to a part. Use of cold gates provides an almost limitless array of gate design opportunities to suit almost any situation. These include fan, film, ring and diaphragm type gates. Though these gates are normally associated with cold runners, they can be added to a hot runner system. The combination of hot and cold runners and gates is a fairly common practice.

14.18 Hot and Cold Runner Combinations

A hot-to-cold runner and gating arrangement may be desirable in many cases. This not only provides the advantage of the cold gates presented in Section 14.17, but it can also improve the economics of a hot runner mold. A 16-cavity mold can be addressed with a four drop hot runner, with each hot drop feeding a small "X"- or "H"-patterned cold runner, which feeds four parts. This approach can also be beneficial when cavities are positioned too close together to accommodate the hot drops, or when it is desirable to put some distance between the part and the heat of the hot drop.

14.19 Two-Phase Injection Processes

Most two-phase injection processes will not work with hot runner systems. This includes structural foam and MuCell®. Also, gas assist injection processes that introduce the gas through the molding machine nozzle cannot use hot runners. Specialized manifold systems have been developed for co-injection molding.

15 Troubleshooting

This chapter provides three different approaches to troubleshooting injection molding and includes contributions from John Bozzelli and Brad Johnson. Section 15.1 presents the *5 Step Process*™ developed for troubleshooting new and existing molds. This method is particularly useful in diagnosing cavity-to-variations in multi-cavity molds. It provides a means to isolate and quantify the contribution of shear-induced melt variations from variations in the mold's steel which may include sizing of runners, gates, cavities, etc. Section 15.2 presents *Injection Molding Troubleshooting Guidelines for Scientific Injection Molding* by John Bozzelli. Here, the reader will find broad guidelines for recognizing and correcting many of the problems found in injection molded plastic parts. Finally, Section 15.3 by Brad Johnson presents directions on establishing a *Two-Stage Molding Set-Up*, which is a proven methodology for optimizing the injection molding process. This method, and variations of it, have been referred to as "scientific molding" by John Bozzelli and "fully decoupled molding℠" by RJG Enterprises. Software has been developed by both BTI (*5 Step Process*) and John Bozzelli (*Scientific Molding*) to simplify the diagnostics and set-up presented below. Though the software simplifies the processes, it is not required to conduct the procedures.

15.1 The 5 Step Process™

The *5 Step Process* is a fast, easy method for isolating and quantify mold filling imbalances. It allows isolating and quantifying molding variations resulting from shear-induced runner effects from steel variation in the mold. The *5 Step Process* does not require any special instrumentation and achieves its purpose by comparing the weights of short shots of the multiple "flow groups" within the mold. To effectively utilize the *5 Step Process*, it is important to understand the factors that contribute to mold imbalances and how to group the numerous flows within a multi-cavity mold so the variations can be isolated and quantified individually.

The main factors that cause imbalances can be broken down into the following categories:

15.1.1 Shear-Induced Flow Imbalance Developed in a Geometrically Balanced Runner

As discussed in Chapter 6, despite their geometric balance, which has traditionally been referred to as a "naturally balanced" runner system, these designs can introduce significant variation in melt conditions (i.e., temperature, pressure, and material properties) delivered to the individual

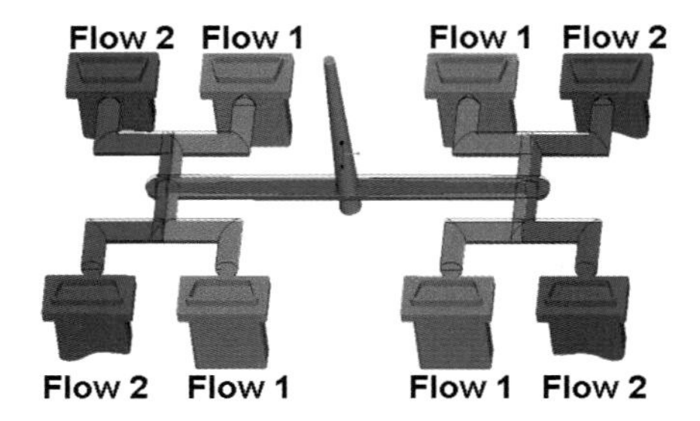

Figure 15.1 Development of two flows in a traditional 8-cavity mold

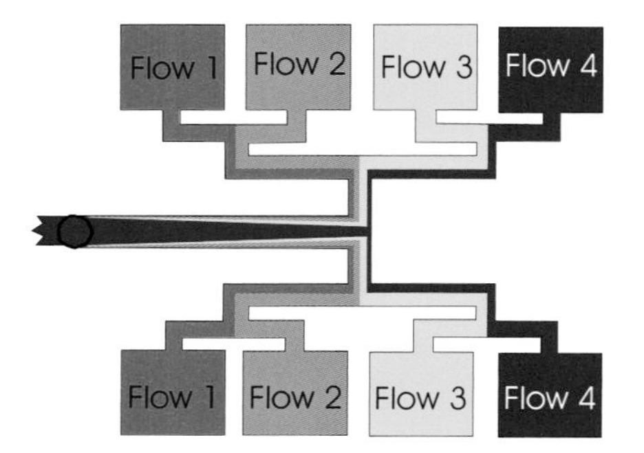

Figure 15.2 Development of four flows in a traditional 16-cavity mold

cavities within a mold. What must be recognized is that conventional geometrically balanced runners actually create multiple flows, much like the old "tree" or "fishbone" branching runners. For instance, there are normally two flow groups in an eight-cavity mold, four flow groups in a sixteen-cavity mold, eight flow groups in a thirty-two cavity mold, and so forth. These flow groups in turn produce multiple families of parts in the mold, which create variations from cavity to cavity such as dimensions, warp, flash, sink, short shots, and so forth. Figures 15.1 and 15.2 illustrate various runner layouts and the different flows they create. Figure 15.1 shows the development of two flows while Fig. 15.2 illustrates the development of four flows in a 16-cavity mold. This concept of flow development was introduced earlier in Chapter 5 and 6. Figure 5.13 shows variations of flow groups being developed in 4-, 8- and 16-cavity molds.

Note that the flow led by the outer laminates of the primary runner is typically the dominant flow and is referred to as Flow #1. In a mold with two flows, as illustrated in Fig. 15.1, the outer branching flow, fed by the center laminates of the primary runner, is called Flow #2. If there are more than two flows, as in a sixteen- or thirty-two cavity mold, only Flow #1 is obvious. The progression of fill of the remaining flows become less obvious. *Therefore the numbering of these flows simply follows a logical sequence relative to Flow #1 and will not necessarily dictate the actual filling pattern.* The remaining flow numbering will progressively increase from Flow #1, as illustrated in Fig. 15.2. There will always be at least one cavity per flow in each quadrant of the mold.

15.1.2 Steel Variations in the Mold

Since parts within a *given flow* and *given shot* should be identical in material properties, any measurable differences between parts can only result from variation in the physical makeup of the mold. These part variations can be caused by the runner layout, differences in the size of the cavities, gates sizes, gate land geometry, runner lengths, runner diameters, venting, and so forth. Cold slugs may also be considered a steel variation because they are easily resolved by mold design modifications. Basically any variation that is non-shear-related would be categorized as a steel variation.

15.1.3 Cooling Affects

Variation between parts molded in different cavities within a *given full shot* can also be caused by dissimilarity in cooling between the different cavities. Cooling differences between cavities have the largest impact during the packing and cooling phases of the molding cycle (i.e., surface finish, shrink, warp, and sink) and very little on mold filling. Therefore, variation in mold temperature would have a minimum impact on the weight of samples molded from the *5 Step Process* as there is no packing stage.

Because of this, only the shear-induced flow imbalance or the dimensional variation in the mold steel can cause the variation in part weight within a cold runner mold.

15.1.4 Hot Runner Systems

Hot runner manifold systems exhibit the same flow imbalances typically seen in cold runner systems. However, hot runner molds have many additional variables that are difficult to isolate and will normally be grouped under "steel imbalances." Hot runner variations can include temperature fluctuation, temperature-influenced variation in part cooling, thermal expansion differences, heater band and thermocouple placement, and the melt delivery geometry itself.

15.1.5 Summary of Test Data

Figure 15.3 shows a summary plot of a study performed at Penn State University using the *5 Step Process* (though data exists for molds with higher cavitation, for clarity only data on eight-cavity molds is presented). The plot compares the average steel imbalances within Flow #1 and Flow #2, and the average shear-induced imbalance between Flow #1 and Flow #2. The figure clearly shows that the shear-induced imbalances are the largest contributing factor to variations in multi-cavity molds.

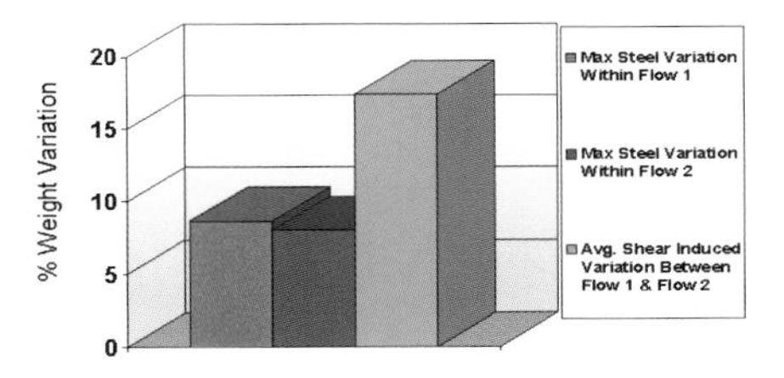

Figure 15.3 Steel- vs. runner-induced imbalance influence on variations within eight cavity molds. Note shear-induced imbalance creates nearly twice the variations as developed by variations in the tool steel dimensioning.

15.1.6 The 5 Step Process: Method of Application

Step 1 Identify Flow Groups: Identify each flow group for the given layout and number of cavities. Identify each cavity by combining the flow group number and a cavity-identifying letter. There will normally be one cavity per flow group per quadrant. For example: 1A through 1D for flow group 1 that contains four individual cavities (Fig. 15.4).

Step 2 Mold Short Shot Samples: Once a reasonable process for the mold is established, set the hold-pressure and hold-time to the minimum value permitted by the process controller (zero values are not recommended for some machines). Reduce the screw feed (if necessary) until the best-filling cavity in the mold is about *80%* full to avoid masking any imbalances due to unvented air, thin regions, or other hesitation effects. The original injection rate should remain constant. To minimize external variations, such as cold slugs and machine variations, a minimum of five shots is recommended.

Step 3 Weigh Parts: Collect all the molded parts from each individual shot and weigh them separately, being sure to label each part from each shot as indicated in Step 1. This can be done immediately because the samples do not need to be conditioned.

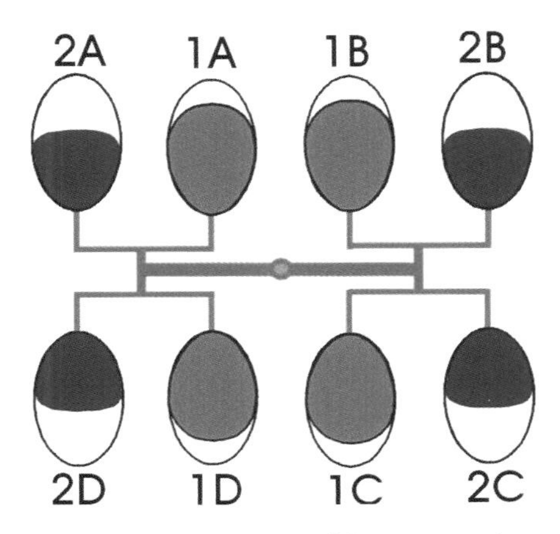

Figure 15.4 Illustration of the suggested new numbering method for 8-cavity molds and the short shot pattern expected (when the best-filling cavity is 80% full). The dominant Flow #1 is indicated in red (1A through 1D)

Step 4 Determine Steel Imbalances: Identify and separate the part weights molded from Flow #1. Compare the part weights within Flow #1 to each other using a graph, and calculate the percent imbalance. Repeat the graphs and calculations for each flow identified in Step 1. The resultant percent imbalance within each flow group is related to steel imbalances. Causes of steel imbalances are non-shear related and generally are found to be discrepancies in the mold steel.

Step 5 Determine Shear-Induced Imbalances: Determine the average weight of the parts molded in each flow group and use this to calculate the percent imbalance between each of the various flow groups. The runner layouts will dictate which flow groups to compare. For example, based on Fig. 15.4, comparing Flow #2 to Flow #2 will give the maximum shear-induced imbalance for that layout. For a better understanding of shear-induced imbalances, care should be taken to compare inside versus outside flow groups because they represent the shear history of the material as it flows through the runner system. The shear-induced imbalances are a result of material property variations created by the shearing action of the polymer throughout the melt delivery system. Shear-induced imbalances are independent of steel imbalances and *cannot* be properly corrected by modifying the size of cavities, runners, or gates in the mold.

Sample Analysis – Conventional Imbalance vs. *5 Step Process* Calculations

Step 1: Flow groups were identified and the cavities were labeled according to Step 1 of the *5 Step Process*.

Step 2: Short shot samples at 80% full were molded.

Step 3: The eight short-shot parts were collected, properly labeled and weighed. The weights are listed below and plotted in Figs. 15.5 and 15.6. The weights are compared in graphs using conventional cavity numbering versus the *5 Step Process* numbering system respectively (weights are in grams)*.

Cav. 1A (2)-5.19; Cav. 1B (3)-5.33; Cav. 1C (6)-4.62; Cav. 1D (7)-5.26

Cav. 2A (1)-4.21; Cav. 2B (4)-4.29; Cav. 2C (5)-4.30 Cav. 2D (8)-4.11

Step 4 Flow #1 Steel Imbalances: A quick review of Fig. 15.6 reveals that cavity 1C is significantly different from cavities 1A, 1B and 1D. Efforts can be immediately focused on finding the steel imbalances creating this variation. The following formula was used to quantify the maximum steel imbalance within Flow #1 by comparing cavity 1B and Cavity 1C:

$$100\times\left(1-\frac{\text{weight of cavity 1C}}{\text{weight of cavity 1B}}\right)=100\times\left(1-\frac{4.62\text{ g}}{5.33\text{ g}}\right)=13.3\% \qquad (15.1)$$

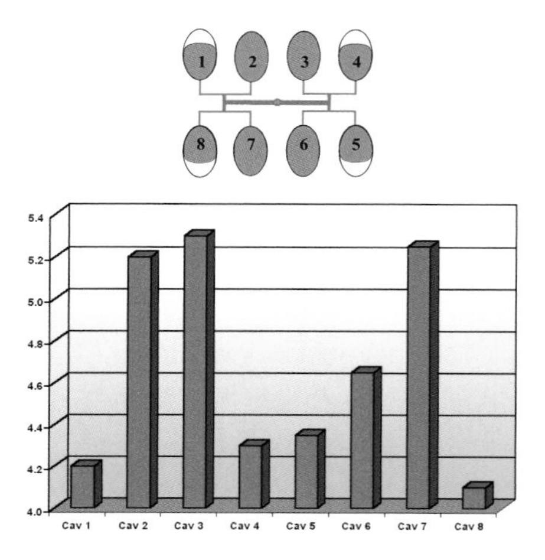

Figure 15.5 Imbalance graphs contrasting part weight from 8-cavity mold using traditional numbering

Flow #2 steel imbalances: Repeat the above for Flow #2. Upon reviewing the data in Fig. 15.6, Flow #2 cavities show only a minimal steel imbalance. The maximum steel imbalance was found to be only 4.42%.

Step 5 Shear-induced imbalances: The average weight of the parts from Flow #1 was calculated to be 5.10 grams. The average weight of part molded from Flow #2 was 4.23 grams. The percent imbalance from the shear-induced imbalance was calculated to be 17% using the following formula:

$$100\times\left(1-\frac{\text{avg. weight of Flow \#2}}{\text{avg. weight of Flow \#1}}\right)=100\times\left(1-\frac{4.23\text{ g}}{5.10\text{ g}}\right)=17.1\% \qquad (15.2)$$

**NOTE:* When evaluating a mold's imbalance, it is imperative to separate out steel imbalances versus shear-induced imbalances by identifying the flow groups as shown in this example. Based on conventional imbalance calculations (heaviest vs. lightest cavity), a 22.9% imbalance is calculated through cavity 2D and cavity 1B. It must be noted that the overall 'heavy vs. light' approach only provides one imbalance value, and does not identify the source of the variations (steel imbalances vs. shear-induced imbalances).

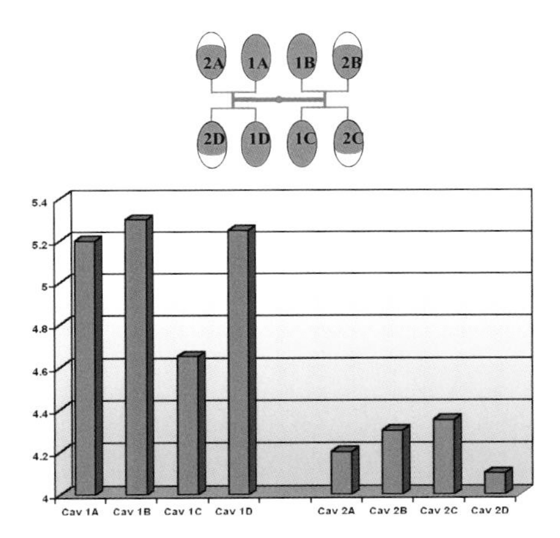

Figure 15.6 Imbalance graphs contrasting part weight from eight cavity mold using the *5 Step Process*

15.2 Injection Molding Troubleshooting Guidelines for Scientific Injection Molding

JOHN BOZZELLI

The troubleshooting guide presented in the following is designed for injection molders and processors who comprehend and use velocity controlled first stage.

Velocity control means the hydraulic or plastic pressure limit allowed on first stage is higher than the hydraulic or plastic pressure during first stage, usually at the position of transfer from first to second stage. Time and hydraulic pressure modes of transfer are not acceptable. If you are filling and packing the part on first stage, you should seek other resources. If you are not cutting off before the cavity is full or you begin to pack the part on first stage, this strategy for troubleshooting may cause you to damage your tool, do NOT use this troubleshooting strategy. An explanation and procedure for processing under velocity control conditions can be found in Section 15.13.

Please also note that all pressures are in melt or plastic pressure. Hydraulic pressures cannot be used because intensification ratios vary on hydraulic machines and electric machines are, by design, calibrated in plastic pressure. This troubleshooting guide will work on either hydraulic or electric injection molding machines. For an explanation of intensification ratio see Section 15.4.

Further, this troubleshooting guide and strategy assumes the press is doing what you are asking it to do. If machine problems are encountered, the strategy must be to fix these problems before further troubleshooting. If you are assuming the machine is working fine with no data to substantiate this assumption do not use this troubleshooting guide, rather seek other resources as this may be a waste of your time.

It is critical that when a root cause is found that the problem be fixed, not a "work around" be implemented.

Safety Note:

You are responsible for your own and others safety. Think about what you are doing. Injection molding machines and molds operate under high temperatures and pressures. Plastic pressure can be near 50,000 psi. Never bypass safeties or circumvent guards. When processing resins, be careful never to mix, blend, or purge any of the following combinations of resins. Any of these resins will catalytically decompose to gas at high pressures within the injection screw, barrel, hopper, or nozzle tip. These pressures can be so high that barrels, nozzles, hoppers, etc. have shattered into projectiles and other bomb like fragments.

Severe Danger

Never mix, purge or process after one another:

- **Acetal (POM) with PVC or CPVC**
- **Santoprene® (Exxon Mobile) with PVC or CPVC**
- **Santoprene® (Exxon Mobile) with Acetal (POM)**

If the same machine barrel and screw must process both these resins, disassemble and clean properly before processing. Small amounts of these resins mixed will catalytically decompose to gas at extreme pressures.

Air bubbles

See also "Bubbles"

Ball Check Valve

While not the same as a non-return valve its function is the same, see also "Non-Return Valve"

Balance

Do all cavities fill at an even progression? This is important for preventing shorts, flash, and sinks. See Section 15.1 for the correct 5-step procedure to determine mold balance. Differentiate between poor filling balance due to melt-flow vs. tool or steel variations.

Black Streaks

See also "Black Specks" and "Color Mixing"

Possible causes	Possible remedies
Wrong style nozzle tip	Do NOT use a general purpose or standard nozzle tip, it has dead spots. Use either a full flow or "ABS" type or a nylon type nozzle tip.
Sharp angle or corner at the gate	Polish gate area, radius sharp corners, change gate type, or enlarge gate
Hot runners	Check for hang-up areas or dead spots in the flow path. Check for proper temperature control; burnt-out heating elements, open thermocouples and correct placement of thermocouples. Check depth of probe relative to gate surface, account for thermal expansion.
Poor nozzle temperature control	Nozzle temperature is notoriously poorly controlled. Placement of the thermocouple should be 1/3 the distance of the nozzle from its tip, if the nozzle is over 3 in. (75 mm) long. For 3 in. (75 mm) lengths and shorter it is OK to have the thermocouple imbedded into the hex on the nozzle body. For longer nozzle bodies, the thermocouple should be underneath the heater band. Use a butterfly type thermocouple and wrap it with glass tape. The nozzle should be PID temperature-controlled, use of a % or variac is unacceptable. Do not allow the thermocouple to be attached via the screws on the clamp of the heater band. Clean out the nozzle, inspect for blockages.
Sharp angle on screw tip or broken non-return valve	Check entire flow path of plastic looking for burrs, sharp corners, grooves, etc. in any of the plasticating components: screw, tip, check ring, flights, etc.
Screw gouged or flight chipped	Check screw to make sure there are no gouges or nicks on the flights.
Wet resin	Dry resin to the resin producer's specification

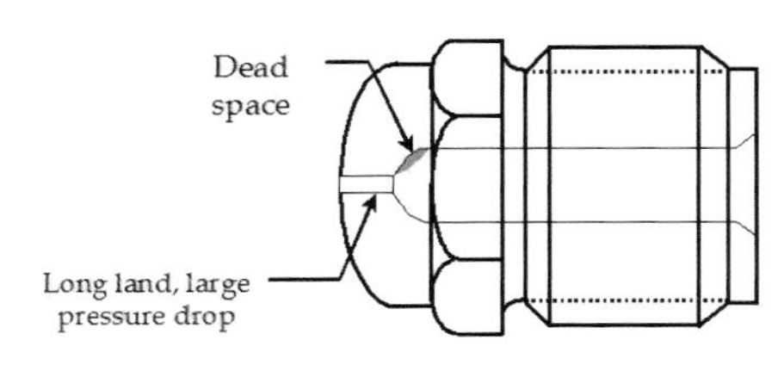

Figure 15.7 Standard nozzle with dead space

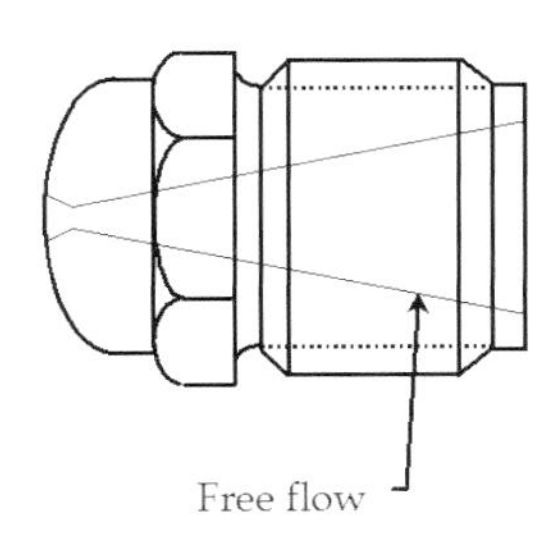

Figure 15.8 Free flow nozzle, recommended

Black Specks or Flakes in the Part

See also "Color Mixing"

Carbon or other contamination that affects cosmetics and part performance, can be any shape or size from chunks to fine particles or flakes. Can be caused by resin degradation or by contamination.

Figure 15.9 Severe black specks

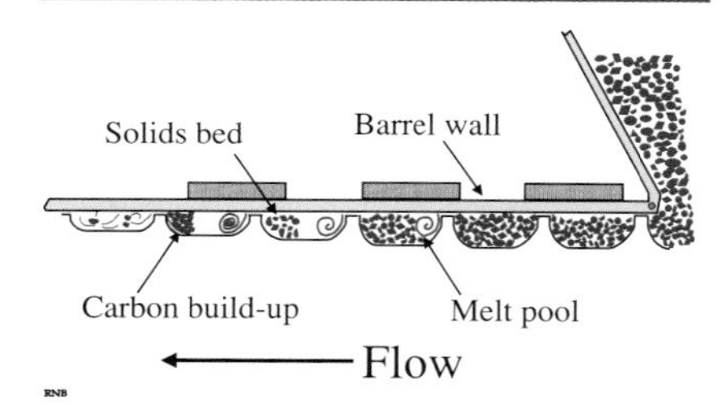

Figure 15.10 Poor screw flight design, dead spots lead to black specks

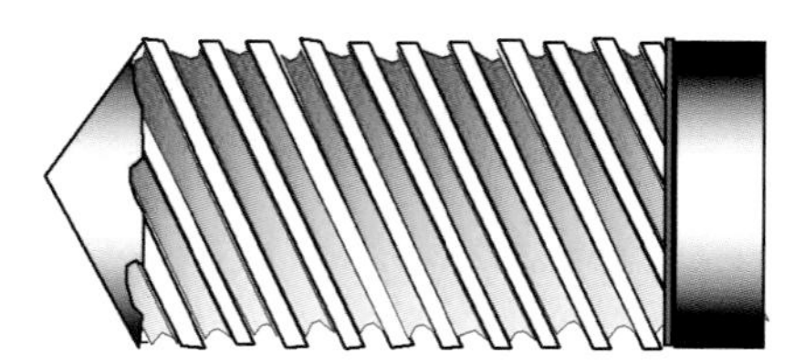

Figure 15.11 High shear dispersive mixing element

Possible causes	Possible remedies
Poor screw or nozzle design	Pull and clean the screw barrel. Check for carbon buildup behind the flights in the metering and transition sections of the screw. Most general purpose screws have this problem due to dead space behind the flights. Molding of acrylic, polycarbonate (PC), styrene acrylonitrile (SAN) for lens application with a general purpose screw is unwise. Specify a correctly designed screw with no dead spots and melt uniformity. Check non-return valve, nozzle end-cap, and nozzle tip for abrupt changes in flow path. See "Screw Design". See Figs. 15.7, 15.8.
Excessive nozzle length	Make sure the nozzle length is as short as possible and it is PID temperature-controlled.
Faulty or poorly designed hot runner flow channels	Check for open or incorrect thermocouple readings. Repair and replace any that are working on % or variacs. Redesign hot runner. Check watt density and location of heaters. Check location of thermocouples.
High melt temperature, hot spots in screw or barrel	Check melt temperature via the hot probe technique or an appropriate IR sensor. Few IR sensors work correctly. Adjust temperatures only if necessary. Check screw and barrel for hot spots. Are any zones overriding their temperature setting? If so, raise that zone or the one before it. Lower back pressure. Slow screw speed RPM.
Contamination in virgin resin	Check virgin resin using a white pan; spread 1.5 kg of resin over a 1,500 cm^2 area and inspect for 5 minutes under appropriate lighting. If there are black specks, find out if they are on the surface or imbedded in the granules. Inform resin supplier.
Contamination in reprocessed or reground resin	Check screw design, material grinding and handling procedures. Discard or resell the material
Excessive high temperatures or long residence times at melt temperatures	Use 25–65% of the barrel capacity. Using lower than 25% provides long residence time for the resin or additive to degrade. Go to smaller barrel if possible. Check heater band function, nozzle should be PID-controlled, check screw for high shear dispersion mixing elements. Repair.
Excessive fines	Remove fines before processing, a must for lens applications.

Possible causes	Possible remedies
Special note:	It is rare for a purging compound to solve a black specking problem if the cause is screw design. 80% of the time it will be wiser to pull the screw and clean it. Screws should have a high polish.
Vented barrel	On vented barrels, often the vent is poorly designed with a dead space or hang-up area. This is a hard to get to area for mechanical cleaning, but it must be cleaned.
Wet resin	Dry resin to the resin producer's specifications

Blisters on the Part's Surface

See also "Delamination"

Thin film of plastic that bubbles up from the surface.

Possible causes	Possible remedies
Gas traveling across surface during fill or pack	Check for moisture, trapped air or excessive volatiles. Also, check for excessive decompression or suck back
Trapped air due to inadequate venting	Check number of vents, check vent depth and compare to manufacturers recommendations. Clean all parting line and core vents. Vents can be checked with pressure sensitive paper. Pull a vacuum on the mold.
Trapped air due to flow pattern	Perform a short shot sequence changing transfer position or shot size to make various sized short shots from 10 to 95% full parts. Observe flow path for any back flow or trapped air in blind ribs. Jetting also traps air. Changing first stage pressure cannot remedy this correctly!
Trapped air due to excessive clamp tonnage	Reduce clamp tonnage, especially for small molds: mold should take up 70% of the distance between the tie bars.
Trapped air due to decompression or suck back	Minimize decompression or screw suck back, especially with hot runners or hot sprues. Be careful to maintain proper non-return valve function.
Trapped air due to low L/D screw	Especially for general purpose screws with 18:1 L/D or lower. Raise backpressure to 1,000 to 1,500 psi melt pressure.
Degraded resin or additive package	Check melt temperature for proper range as recommended by resin supplier. Minimize residence time by shortening cycle by reducing cooling time first. Try virgin only, a new lot.

Figure 15.12 Blisters and delamination

Blooming

See also "Mold Build-up", "Blush" and "Surface Finish"

This can be a solid powder, wax, or liquid buildup on the surface of a part or mold, usually an additive migrating to the surface. The build up can happen in one shot (immediately) or over a period of time. Difficult to eliminate during processing, usually requires a different additive or formulation.

Possible causes	Possible remedies
Additive migrating to the surface of the part	Try different lots or grade. Try slower injection rate. Seek a different additive or formulation.
Inadequate venting	Clean vents. Check vents by pressure sensitive paper, bluing, or plastigauge. Add more vents.
White powder on part or mold surface	If the resin is acetal it may be formaldehyde depositing on the mold or part. Dry the acetal before processing.
Melt temperature too high or low	Try higher and lower melt temperatures. Usually not much of a benefit here, wasted time.

Blush

See also "Gate Blush," "Surface Finish," and "Gloss"

A haze or dullness on the surface, often seen as rings or half circles near the gate; usually worse with impact-modified resins such as HIPS or ABS. One of the more difficult cosmetic issues to resolve. The problem is a different morphology of the impact modifier. This develops as shear rate changes at the flow front during fill.

Possible causes	Possible remedies
Improper temperature control of nozzle	PID temperature-control nozzle and tip, make sure temperature is uniform over length of the nozzle. Minimize length of nozzle. % or variac heating of the nozzle is not acceptable. See "Black Specks" for correct thermocouple placement. Try higher and lower nozzle temperatures to see if blush changes.
Sharp corners at the gate	Break the corners, provide a minimal radius to reduce shear at the sharp corner
Improper gate type	Change gate type for less shear or sharp corners
Improper gate location	Change gate location
Size of the gate	Decrease land, increase area of the gate, this will change gate seal time!

Possible causes	Possible remedies
Fast injections speeds	Change injection velocity. May need to profile injection; use no more than 2–3 velocities. Slow-fill as plastic enters gate area then increase velocity. If possible use one slow velocity to allow for repeating in other machines.
Too high or low melt temperature	Target middle of melt range specified by resin producer, try 25 °F (12 °C) higher, then same amount lower.
Too low or high a mold temperature	Try high end and low end of recommended mold temperature range suggested by resin supplier.
Trapped gas	If blush is near an area of non-fill or short, check venting. Pull a vacuum on the mold cavity during fill.
If just in a small area	Look for excessive ejector speed, part sticking remedies, or hot spot on the mold surface. See "Sticking".

Bridging in Feed Throat or Hopper

Usually this is caused by plastic pellets that have agglomerated in the feed throat blocking the normal gravity feed of pellets into the feed section of the screw. If it happens in the drying hopper, most likely it occurs as one desiccant bed is transferred from regeneration to process air and there is a temperature spike.

Possible causes	Possible remedies
In feed throat	
Excessive feed throat temperature	Check cooling line at feed throat for proper water flow. Adjust temperature to 90–130 °F or 30–55 °C. Check water flow and temperature in and out of the feed throat cooling lines.
Too large granule size	Make sure regrind is properly ground in size.
Excessive screw pull back	Sometimes when the screw is manually pulled back melted or partially melted plastic gets pushed into the feed throat. Empty the screw before screw retraction.

Possible causes	Possible remedies
In hopper	
Excessive drying temperature	Check set point of drying hopper for correct temperature. Measure temperature at inlet and compare to set temperature. If all this checks out, you will have to put the dryer into regeneration and measure the temperature of the process air as beds change for regeneration. Often a temperature spike occurs. This can be done with a peak-pick pyrometer.

Brittleness

Parts crack easily or upon cooling; causes can be from over- or under packing and often from molecular weight degradation or contamination.

Possible causes	Possible remedies
Defining the brittle areas	First find out if the brittleness is throughout the part or localized in one area. If localized, it may be a packing issue. If throughout the entire part, look for molecular weight degradation via melt flow rate testing of the material before and after molding. If brittle at lower temperatures, check material for proper resin selection.
Over- or under-packing the gate	If brittleness is near the gate, do a gate seal analysis. Check if brittleness changes with or without gate seal. Running without gate seal allows plastic to discharge and may cause under-packing or excessive residual tensile stress. Running with gate seal may over pack the gate and lead to brittleness due to over packing, too many molecular chains packed too closely together and they do not have space to get closer with cooling. This leads to retained compressive stress.
Molecular weight degradation due to hydrolysis	Check to see if this polymer can undergo hydrolysis. This is where minute amounts of water react with the polymer in the barrel of the machine. Water can act as a pair of scissors cutting the long polymer chain into short segments. This leads to lower melt viscosity. Therefore check moisture level. Check the melt pressure at transfer from 1st to 2nd stage and see if it is lower than normal or “good” runs.

Possible causes	Possible remedies
Molecular weight degradation due to excessive temperatures	Check melt temperature, via the hot probe method or appropriate IR temperature sensor. Adjust only if necessary. Check barrel heats and duty cycles. Check for temperature zone override. Check for proper screw and barrel wear and condition. Check for cracked, chipped, or restrictive non-return valve or check ring. Check residence time in the barrel. Add some colored granules to the empty feed throat and count how many shots it takes to see the color come out. Multiply the number of shots times the cycle time for the true residence time. Note also if it comes out in streaks or blended. If streaks, there are dead spots in the barrel. See "Screw Design"
Too much degraded regrind or contamination	Check amount and quality of regrind. Check for contamination. Run 100% virgin and check properties.
Part not conditioned properly	Certain resins, especially nylons, need to regain moisture to achieve full physical properties. Have the parts been conditioned according resin manufactures specifications.
Too much retained orientation or stress	Check gate location and for orientation effects. Check part design for sharp corners, appropriate nominal wall and its uniformity.

Bubbles

Bubbles are either trapped gas or vacuum voids. Before proceeding you must determine which one you have. TEST: On a *freshly* molded part, *gently and slowly* heat the area containing the bubble. A hot air gun is best, a small lighter next, and a torch if you know what you are doing. As the resin softens, the bubble will expand or contract. If it expands it is trapped gas; if it contracts or pulls the side wall in; it is a void.

Possible causes	Possible remedies
Trapped gas	Progressively short shoot the mold using shot size or position transfer to make 10, 20, 40, 60, 70, 80, 95% full parts. Note, if any show the plastic flow front coming around on itself. Is there a racetracking effect? This includes jetting. Note, if ribs are covered before completely filling. This test cannot be done correctly if you reduce 1st stage pressure or velocity. The process must be velocity-controlled for 1st stage. The concept is to find where the gas is coming from

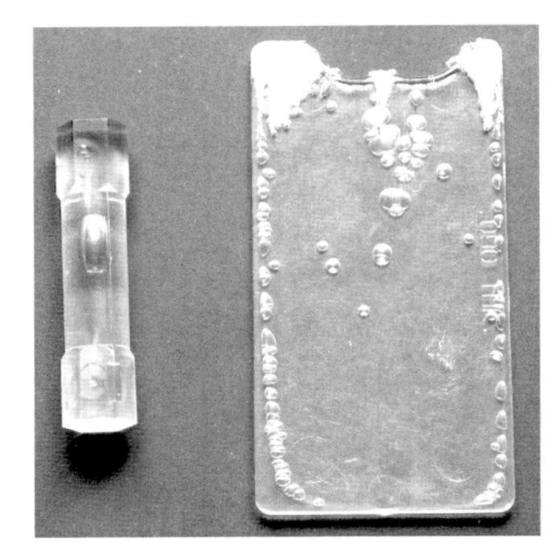

Figure 15.13 Bubbles

Possible causes	Possible remedies
	and eliminate its source. Vent the tool properly or use porous steel to eliminate gas traps. Jetting can cause gas or air to be trapped. Check for moisture, steam is a gas. Change gate location. Pull a vacuum on the mold cavity during fill.
Vacuum voids	Thin up the nominal wall in this area by adding steel to core-out thick sections. Change gate location, fill thick to thin. These occur upon cooling in thick sections. You need to pack more plastic into the cavity with: 1) consistent cushion, make sure you are not bottoming out the screw, 2) higher 2nd stage pressures, 3) longer 2nd stage time, 4) very slow fill rates, 5) use counter pressure, 6) open the gate for longer gate seal times to allow more packing during 2nd stage, or 7) increase the runner diameter. Also, raise the mold temperature significantly and/or eject the part sooner. These will allow the outside walls to collapse more upon cooling. Reduce melt temperature.

Burns

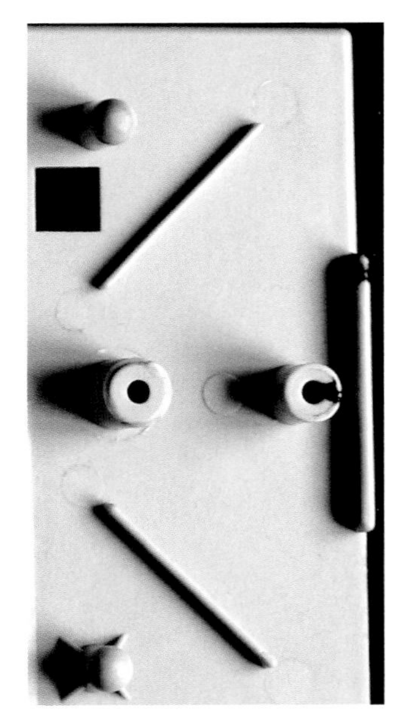

Figure 15.14 Burn marks

See also "Black Streaks"

Black carbon soot, or white residue, glossy areas on a part. Burns are a result of dieseling, which can lead to pitting on the mold surface in that area. Due to this mold damaging effect, do not continue molding until a remedy has been found, severe tool damage may occur.

Possible causes	Possible remedies
Trapped air or volatiles dieseling	Air trapped in a mold can be pressurized to thousands of pounds of pressure. Air with a little hydrocarbon as fuel will diesel when pressurized. Check for proper venting in this area. If there is a question regarding venting: rub the suspect area with a hydrocarbon solvent and shoot a part. If carbon forms, venting is not adequate. Glossy spots on parts are often an indication of trapped gas.
Inadequate venting, plugged vents, hobbed over vents, undersized or not enough vents.	Normally the recommendation would be to slow down the injection rate to allow time for air to escape. This is unacceptable in Scientific Molding, as this will add seconds to the cycle time, significantly decreasing profits. 30% of the perimeter should be vented, this is to include runners. Porous steel may

Possible causes	Possible remedies
	be used to vent blind pockets. Use an ejector pin as a vent. Vacuum may also be applied to the mold to avoid trapping air and dieseling. Are the vents cut to the resin manufacture's specifications? Solve the venting issue, do not work around it. Dieseling will pit and erode the mold steel. The extra time coupled with increased mold maintenance will continually erode profits.
Core pins not vented	Core pins must be vented because air is trapped as plastic flows around the core pin.
Excess volatiles	Resin may have a large percentage of mold release agents, lubes, or other additives that volatilize during injection filling. Try a lower melt temperature, if allowed by the resin supplier. Try a different grade of resin with reduced additive concentration.
Excessive decompression	Decompression pulls air into the nozzle because nozzle contact force is off. Since nozzle contact force is "on" before injection, air is trapped in the nozzle and is pushed into the sprue, runner, etc., possibly causing splay or burns. This is especially important in hot sprues and runners. Reduce amount of decompression.

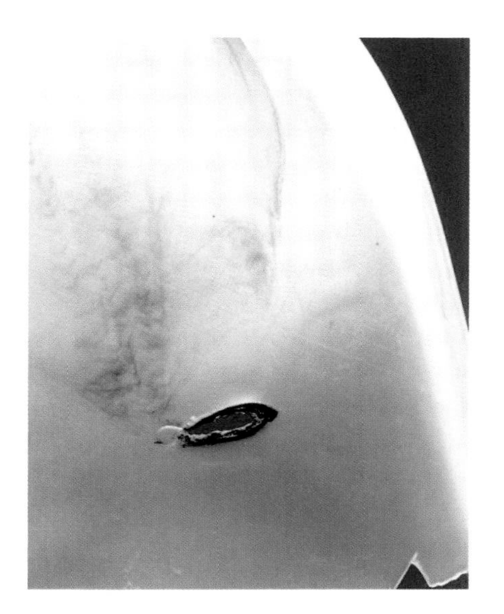

Figure 15.15 Burns, trapped air

Burns at the Gate

See also "Burns" and "Black Streaks"

Cloudiness or Haze in Clear Parts

Possible causes	Possible remedies
Excess additive or non-compatible additive package	Try a different lot, change material supplier.
Contamination	Clean barrel and screw thoroughly. Disassemble screw, barrel, and non-return valve. If there is carbon on the back of the screw flights, see screw design. Check resin for contamination in virgin lot, or regrind, or excessive fines.
Non-uniform melt	Check melt quality by adding a small amount of compatible color granules. If they come out as streaks rather than a uniform tint or pastel, the screw is not adequately mixing or providing melt uniformity.

Possible causes	Possible remedies
Incorrect mold temperature	Raise mold temperature. Lower mold temperature for amorphous PETG.
Moisture in resin	Check moisture content, function of dryer, residence time, dew point, etc.
Worn or improper mold texture or surface	Check mold surface for deposits, plate-out. Check for appropriate polish.
Crazing	Check for stress in the area; for example stress whitening near ejector pins or places where the part sticks. Is there any contact with chemicals such as solvents or mold spray that attacks this type of plastic?

Check Ring

See also "Non-Return Valve"

Color Mixing

Color streaks, marbling, dark streaks, and unmelted solids all fall into this category. In almost every case, the root cause is poor screw design or, less common, a worn screw and/or barrel. The goal is to distributively mix the color concentrate (master batch) or liquid coloring agent *uniformly* into the base resin. Do not try to achieve dispersive mixing with injection molding screws. For a full description of screws and their method of melting plastic see Spirex's "Plasticating Components" (free from Spirex 330-728-1166; www.spirex.com) or Westland's Cylinder & Screw Handbook, (free 800-247-1144). Others are available but these two are the ones I am most familiar with. It is not recommended to use powder colorant. It is cheaper, but the fine dust is detrimental to your lungs, machinery components, and good housekeeping.

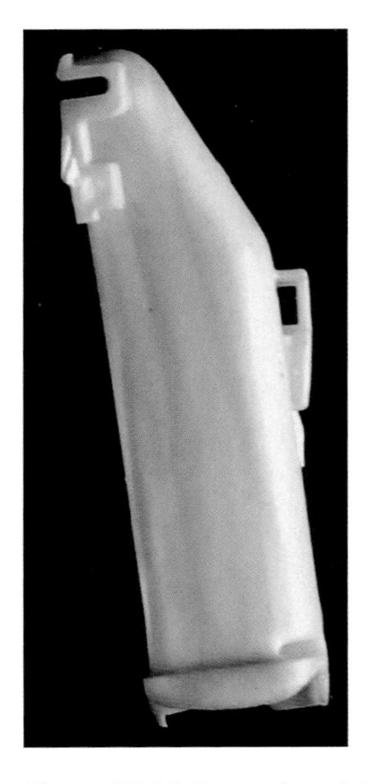

Figure 15.16 Poor color mixing

Possible causes	Possible remedies
Back pressure incorrect	Increase back pressure to 1,000–1,500 psi. by adjusting the backpressure set point on the controller or adjusting the back pressure valve. Inspect after 3–4 shots. Maximum back pressure to try is ~ 5,000 psi. Be careful using high backpressures for resins that tend to degrade such as acetal or PVC. High backpressure cannot be used on vented barrels due to vent flooding. Rarely will you have problems of color dispersion or melt uniformity with vented barrels and screws.

Possible causes	Possible remedies
Melt temperature	Make sure the melt temperature is within the resin supplier's recommendation. Use the hot probe technique or appropriate IR sensor. Adjust if necessary.
Color concentrate carrier incompatible with base resin	Check the color concentrate carrier by calling the supplier and asking. It should be of the same resin family, usually only higher melt flow, as the part. There is no such thing as a universal color carrier.
Incorrect colorant let down ratio	Weigh-blend a few pounds to make sure there is the correct ratio of colorant to base resin. Too low or too high a concentration can cause problems. Do not use more than recommended as the cost of using extra concentrate will hurt profits and often part properties are compromised with too much colorant.
Melt uniformity	If possible, run natural resin and add a few colored granules into the feed throat. Wait until the color appears in parts. If the color comes out in streaks, you do not have melt uniformity. If the color comes out as a uniform pastel (lighter shade), then you have melt uniformity. If you have a melt uniformity issue, specify a new screw; see "Screw design" below. Worn barrels and screws provide better mixing due to some back flow over screw flights.
Low L/D screw	Try higher back pressure as above. Most likely, this will help but not alleviate the mixing problem. Check L/D, if below 18:1, forget processing fixes and use precolored resin or get a melt uniformity screw. The screw should not have mixing pins or a dispersive mixing element in it. Not recommended but a possible solution is to replace the existing nozzle with one fitted with mixing elements similar to a reversing helix. With mixing nozzles, be careful of high back pressures. Also, some are difficult to purge. Strongly discouraged are pineapple mixers or dispersion disks; they are cheap but detrimental to the resin.
Fast screw RPM's	Slow screw rotation speeds provide the best mixing. Use all but 2–3 seconds of the cooling or mold closed time to get the screw back ready for the next shot.

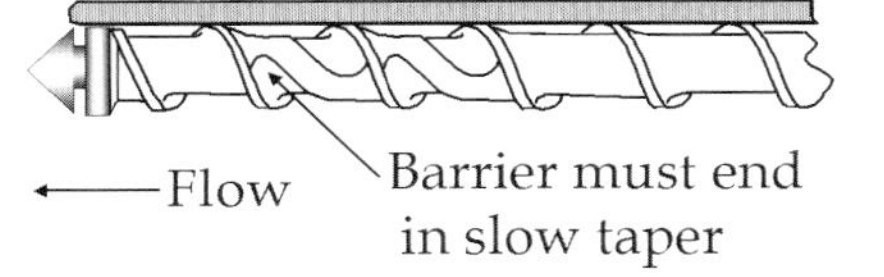

Figure 15.17 Screw designed for melt uniformity or color mixing

Possible causes	Possible remedies
Screw design	Standard general-purpose screws are known to provide unmelted solids within the melt. Do the melt uniformity test above. Melt uniformity screws are strongly recommended. These have a gentle mixing section near or within the metering section as well as filled radiused rear flight sections. Minimum recommended L/D is 20:1 for normal injection molding situations.
Miscellaneous	Use the maximum amount of regrind allowed. Regrind acts like additional coloring agent.

Color Variation, Instability, or Color Shift

Possible causes	Possible remedies
Melt temperature too high	Make sure melt temperature is within resin supplier's recommended range via hot probe or appropriate IR sensor. Adjust if necessary.
Colorant thermally unstable	Run the colorant and base resin at the minimum temperature allowed by the resin supplier. Then run the colorant at the maximum temperature allowed by the resin supplier. View identical parts in a light booth. If no light booth available: look at the identical parts at the same angle under florescent light, then sun light, and then incandescent light. Thermally stable colors will look the same.
Incorrect coloring agent	Double-check the coloring agent for correct color, type, and base resin. Call the supplier if necessary.
Long residence time	Use 25–65% of the barrel capacity. Minimize cycle time to decrease residence time.
Virgin resin color shift or instability	Check color of different virgin lots, should be similar. Master batch or coloring agent can not hold color tolerance if base resin shifts in color. Is natural resin thermally stable color wise?
Gloss differential between parts	Often in an assembly two mating parts look like different colors due to gloss differences. Check gloss level for each part.
Surface finish of the mold	Often in an assembly two mating parts look like different colors due to surface finish differences. Check mold cavity for identical surface finish.

Possible causes	Possible remedies
View angle different	Often in an assembly two mating part will look different in color because the angle of viewing each part is different because of the assembly. Disassemble the components and look at them at the same angle.
Colorant let down ratio not constant	Weigh-blend a few pounds at the correct blend ratio and test for color accuracy. If this matches the color target, recalibrate feeder and make sure auger only turns during screw rotation. Auger feeders can get out of sync with screw rotation and it has been known to happen that operators incorrectly adjust feed rate.
% regrind not constant, or discolored	Check virgin only samples for color accuracy. Check level and quality or regrind.

Core Pin Bending or Core Shift

Cores and core pins move due to non-uniform pressure on the core during packing.

Possible causes (if throughout the part)	Possible remedies
Non-uniform pressure on the core or core pin during 1st stage or filling	Take 2nd stage pressure to minimum to make a 99% full shot. If minimum is above 300-psi plastic pressure, take 2nd stage time to 0 seconds. Do not reduce 1st stage pressure to do this test, You must be in velocity control for 1st stage. Look at the part for evidence of core or core pin shifting. If the pin has shifted, try faster fill rates. Find the tool maker and see if there is a way to lock or support the core during fill. Would a gate location change help? If a 99% full short shot does not result in a core shift, the problem occurs during 2nd stage. Contrary to popular opinion, core shift most often occurs during pack and hold.
Non-uniform pressure on the core or core pin during 2nd stage or pack, hold and cooling	Do a gate seal analysis. Inspect parts for core shift with and without gate seal. Set 2nd stage time from the best results of the gate seal study. Next determine the best 2nd stage pressure by starting out with 5 shots with the lowest 2nd stage pressure that makes a full part. Inspect parts for core shift. Then 5 shots

Possible causes (if throughout the part)	Possible remedies
	with 1,000 psi higher melt pressure and check parts for more or less core shift. Repeat until 2nd stage pressure begins to flash the part or mold damage would result with that 2nd stage pressure. Choose pressure that provides least core shift.
Core or core pin shift due to mold misalignment	Check mold for proper alignment upon closing. Wax or silly putty will work. Check for thermal expansion mismatch. Correct as needed.
Unmelted solids	Check for contamination in resin. Check melt uniformity as in "Color mixing" above.
Cores not interlocked	Interlock all cores/stripper bushings with a nesting taper on the open end of the cavities. Cores to float and align on clamp-up. This is an expensive tool build.

Cracking

First determine if the cracking is localized on one area of the part or if it is occurring throughout the part; see also "Brittleness" and "Sticking".

Possible causes (if throughout the part)	Possible remedies
Molecular weight degradation	Check part and resin for proper melt flow rate ASTM 1260; check for contamination, try a different lot, check regrind level and quality.
Improper resin	Check grade and type of resin
Improper type or amount of colorant	Check let down ratio and type of colorant carrier

Possible causes (if localized)	Possible remedies
Solvent, surfactant, or chemical attack	Inspect mold and part handling for possible contamination of oils, solvents, fingerprints, mold sprays, soaps, cleaners, etc. Remove if source is found.
Contamination	Check localized area for off-color or foreign material. Check regrind for trace contaminates.

Possible causes (if localized)	Possible remedies
Exposure to radiation	Check the resin for UV, sunlight, or gamma radiation stability

Crazing

Fine lines or small cracks usually confined to a small area

Possible causes	Possible remedies
Solvent, surfactant, or chemical attack	Inspect mold, part handling or assembly for possible contamination of oils, solvents, fingerprints, mold sprays, soaps, cleaners, etc. Remove if source is found.
Exposure to radiation	Check the resin for UV, sunlight, or gamma radiation stability
Part distorting upon ejection	Note location of craze, if near or at an ejector pin, the part may be sticking in the mold; see Sticking

Cycle Time too Long

Possible causes	Possible remedies
Thick or non-uniform nominal wall	Use minimum nominal wall and keep it uniform within the guidelines for good piece part design. Maximum variation for amorphous is 20–25%; for semicrystalline it is 10–15%. Thick is not stronger in plastic parts. Thinner with ribs will provide better performance, save plastic and cycle time. Teach these principles to your designer and clients.
Slow filling	Increase injection rate, make sure 1st stage is velocity-controlled filling 95–99% of the cavity.
Improper cooling	Note temperature in vs. out for each water line, maximum allowable difference is 4 °F or 2 °C. Make sure water flow is turbulent, Reynolds # > 5,000.
Robot movement too slow	Optimize or update robotics.
Mold movement too slow	Optimize mold opening speed and distance. Optimize mold closing and mold protection.
Long or excessive ejection strokes	Optimize speed of ejection, but not too fast to cause pin marks; if part sticks, fix the problem; do not compensate with extra ejection strokes.

Possible causes	Possible remedies
Screw recovery too long	Start with the rear zone at the lowest temperature setting recommended by the manufacturer. Note recovery time. Raise rear zone temperature 15 °F (7 °C) and note recovery time. Repeat until rear zone is at maximum recommended temperature setting of resin manufacturer. Plot data and pick best temperature for minimum recovery time. Keep backpressure and rpm constant for this test.
Not enough ejector pin area	Add or enlarge ejector pins. Larger surface area will allow ejection earlier.
Excessive 2nd stage time	Perform gate seal experiment to optimize 2nd stage time setting.

Dark Streaks

See also "Black Streaks," "Color Mixing" and "Black Specks"

Deformed or Pulled Parts

See also "Sticking" and "Warp"

Degradation

See also "Black Specks"

Delamination

See also "Blisters"

Can be on the surface or when a part breaks and you see layers at the break.

Possible causes	Possible remedies
Contamination	Delamination is most often caused by incompatible resin contamination. Run only virgin, inspect virgin, try a new lot, change resin supplier, check color carrier for compatibility to base resin. Run natural resin without colorant. Check regrind.
High molecular orientation at the surface	Try a hotter mold, slow injection rate down for longer fill times. Review gate location.
Hot spot in the mold	Sometimes seen opposite a gate due to plastic impinging on the metal. The incoming polymer remelted the skin layer. Check cooling lines for appropriate delta T (< 4 °F (2 °C)) Reynold's number, > 5,000. Try lower mold temperature and slower injection speed, see also Fig. 15.12.

Dimensional Variations

If this is the first trial of the mold, make sure processing is done via scientific molding before going through this troubleshooting list. That is, verify that the steel is dimensioned correctly. Develop a stable process and center the process as close to the center of the part specifications as possible. Part size variations can be caused by one or a combination of the following: pressure gradients, flow imbalance, improper measuring technique, different amounts of material in the mold, degree of crystallinity, cooling rate, amount of molecular chain orientation and/or fiber orientation.

Possible causes, parts too small	Possible remedies
First stage not consistent	Take 2nd stage pressure to minimum. If minimum is above 300-psi plastic pressure, take 2nd stage time to 0 seconds. Do not reduce 1st stage pressure to do this test, you must be in velocity-control for 1st stage. Make parts 99% full and run 20 shots. Are they all consistent in size? If not, fix the non-return valve. If consistent, go to next possibility.
Pressure gradient variations	Do a gate seal experiment. It is critical to know if you are using the correct amount of 2nd stage time. This time should not vary run to run. The test takes 10–20 minutes and should be done on every mold before release to production. Running with out gate seal allows plastic discharge from the gate, therefore parts would be smaller. Running with gate seal, (longer 2nd stage times) packs the most plastic into a part and holds it in. Thus, larger parts may be possible. Careful not to overpressurize the gate. Also check resin viscosity and percent of water if hygroscopic.
Too little 2nd stage pressure	Increase the amount of 2nd stage pressure; if flash occurs, do not run with flash. Make sure machine switchover from 1st to 2nd stage is less than 0.1 s.
Non-return valve malfunction or screw and barrel wear. Different amount of material in the mold	Check non-return valve function, check cushion repeatability. Try adding more decompression to help seat the check ring. Watch out for splay development with too much decompression. If non-return valve is leaking, replace with one that has a stepped angle for better sealing. If non-return valve checks out OK, then check barrel for roundness for the entire stroke distance. Repair if more than 0.003 in. clearance between barrel wall and flight. Check machines repeatability of 2nd stage pressure, it should be < +/– 5–10 psi.

Possible causes, parts too small	Possible remedies
Degree of crystallinity	If a semi-crystalline resin, check cooling rate by measuring mold temperature. A hot mold allows the plastic to cool slowly and yield higher crystallinity, which provides more shrinkage and a smaller part. Cooling must be identical run to run and throughout production. Check colorant for change in formulation, colors affect crystallinity and therefore size of parts. Try the resin with a nucleating agent.
Too much post mold shrinkage	Not advised but longer cooling time may hold the part to size. This is using the mold as a shrink fixture, which is costly. Not recommended but try cooling the part in water after ejection. Rapid cooling prevents some shrinkage.
Low melt temperature	Increase melt temperature, this may lower density and allow for more shrink.
Gate too small, gate seal occurs too quickly	If the gate is too small and gate seal occurs before the part is fully packed out, the part may be too small. Larger gate may be indicated. Obtain proper approval before modifying the tool. Adding a filler such as glass fibers can also reduce shrinkage. Again, obtain proper approval first.
Improper measurements	Check for identical conditioning of parts before measurements are taken. Have the parts been cooled the same way and for the identical times after molding. Even laying down the part with the same side up consistently is important. Check measuring device for calibration

Possible causes, parts too large	Possible remedies
Pressure gradient variations	Do a gate seal experiment. It is critical to know if you are using the correct amount of 2nd stage time. This time should not vary run to run. The test takes 10–20 minutes and should be done on every mold before release to production. Running with gate seal, longer 2nd stage times allows for keeping all the plastic in the mold and a more robust process. Try decreasing 2nd stage time to allow for gate unseal. Also, check resin viscosity and percent of water if hygroscopic.

Possible causes, parts too large	Possible remedies
Too much 2nd stage pressure	Reduce 2nd stage pressure
Degree of crystallinity	If a semi-crystalline resin, check cooling rate by measuring mold temperature. A colder mold will prevent crystalline growth, thus making the part larger. Cooling must be identical run to run and throughout production. Check colorant for change in formulation, colors affect crystallinity and therefore size of parts. If possible, try a resin with no nucleating agent to increase crystallinity for a smaller part.
Not enough post mold shrinkage	Reduce cooling time to eject the part hotter to allow for more shrinkage.
Improper measurements	Check for identical conditioning of parts before measurements are taken. Have the parts been cooled the same way and for the identical times after molding. Even laying down the part with the same side up consistently is important. Check measuring device for calibration.

Discoloration

See also “Color Instability”, “Color Mixing” and “Black Specks”

Ejector Pin or Push Marks

See also “Sticking”

These are blemishes or part deformation near ejector pins, often stress-whitening is seen.

Possible causes	Possible remedies
Part sticking to the ejector side	See sticking in mold, especially gate seal analysis, mold issues, and cooling or mold temperature
Ejector velocity is to fast	Slow velocity of ejection
Not enough ejection pin surface area	Add ejector pins or use larger diameter ejector pins
Ejector plate cocking	Check the length of the knock-out bars, they should be identical in length, maximum difference is 0.003 in.
Ejector pins not all the same length	Check ejector pins for proper length

Figure 15.18 Stress whitening

Flash

In separating 1st stage (fill) from 2nd stage (pack and hold) it is critical that 1st stage fills the part to 95–99% full. This troubleshooting procedure mandates that 1st stage ends before the cavity is full. Flash may be caused by a combination of the following causes.

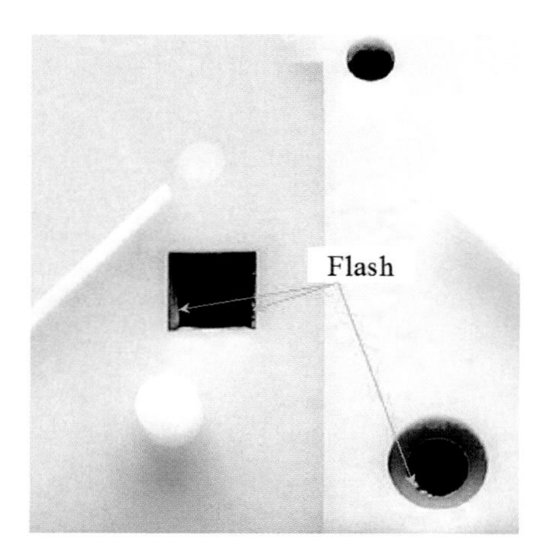

Figure 15.19 Example of flash

Possible causes	Possible remedies
Parting line mismatch or mold damage	Take 2nd stage pressure to minimum. If minimum is above 300-psi plastic pressure, take 2nd stage time to 0 seconds. Do not reduce 1st stage pressure to do this test, you must be in velocity control for 1st stage. If part is short, inspect part for flash. If flash is present, first clean mold surfaces and carefully inspect for any material on the surface or in the cavity that would prevent proper clamping at the parting line. Ensure there is no flash behind slides. Mold another shot under identical conditions. If flash is still present, there is a parting line mismatch, tool damage that must be repaired, or clamp problems. Use pressure sensitive paper to check pressure on parting line in flash area and non flash area while the tool is in the press. Close clamp and note depth of color on paper. Repeat if necessary. If this is a thin-walled part and flash is present on short shots see the following possible causes.
Too much plastic delivered to cavity during first stage	Take 2nd stage pressure to minimum. Do not reduce 1st stage pressure to do this test, you must be under velocity control for 1st stage. If part is full, increase the position cut-off or reduce shot size. Also check machine response on switchover from 1st to 2nd stage. If there is a sharp rise in pressure on switchover, the machine needs repair. Switchover should occur with no rise or dip in hydraulic pressure as it transfers from 1st to 2nd stage.
Clamp tonnage too low; cavity and runner pressure exceeds clamp force	If flash occurs during 2nd stage or pack and hold, reduce 2nd stage pressure. If this makes an unacceptable part. Check machine level, level if necessary. Check tie bar stretch for each tie bar, all should be within 0.002 in. maximum of each other. Check platen parallelism. Try mold in a larger clamp tonnage press.
Clamp pressure too high	If flash is concentrated in the center of the tool, lower the clamp tonnage. Small molds in large platens tend to cause platen wrap. The mold should take up 70% of the distance between the tie bars.

Possible causes	Possible remedies
Melt temperature too high	Take melt temperature via hot probe technique or appropriate IR sensor. If not within the resin manufacture's specified range, adjust accordingly. Reduce residence time if degradation is possible. Checking melt flow rate (MFR) before and after molding may tell you if molecular weight changed. A 30% change in MFR is acceptable for unfilled resins; 40% is unacceptable. Higher for filled resins.
Viscosity too low	If resin is hygroscopic, check moisture content. Check for correct resin. Try different lot. Check for degradation as above. As a very last resort, you can lower injection rate which will increase viscosity. This is NOT recommended.
Mold improperly supported	Check number, placement, and length of support pillars in mold.
Sprue bushing too long	For non-sprue gated parts, check length of sprue bushing under production conditions. The sprue bushing can be too long due to a) thermal expansion and b) nozzle contact force, 5–15 tons, can bow the center of the mold (stationary side) into the moving side.
Check balance if a multi-cavity tool	See text for 5-Step Process for determining mold balance.
Clamp misaligned or not parallel	As above, check level of machine, stretch of tie bars. If they check out OK, get a laser check on platen parallelism.

Flow Lines

See also "Weld Lines"

Gate Blush

See also "Blush"

Possible causes	Possible remedies
Gate design	Break any sharp corners on the gate. Change gate type.
Gate location	Change gate location.
Angle of impingement from gate	Change the angle at which the plastic enters the cavity.
Injection velocity too fast	Slow injection rate to avoid blush.

Possible causes	Possible remedies
Wrong mold temperature	Raise or lower the mold temperature, significantly.
Wrong melt temperature	Check nozzle for proper size and temperature control. Nozzle and tip should be as short as possible. Temperature control should be PID, NOT a % or variac.

Gloss

See also "Blush"

Too high or low a gloss level on the surface of a part. Gloss is how shiny a surface is. Usually measured as a Gardner Gloss at 60%. Angle is important!

Possible causes	Possible remedies
Mold surface finish	Clean and inspect mold surface finish under good lighting conditions. Make sure there is the correct finish and that it is clean of any mold buildup. See "Mold buildup"
Incorrect mold temperature	Generally, a hotter mold will provide higher gloss. Raise mold temperature to increase gloss, lower mold temperature to decrease gloss.
Pressure in the cavity	Check 2nd stage pressure. Is the part packed well enough to replicate the steel surface? Generally, higher 2nd stage pressures provide higher gloss. Lower 2nd stage pressures provide lower gloss.
Injection velocity	Increase injection velocity for higher gloss, decrease injection velocity for lower gloss.
Resin type	Certain ABS and PC/ABS resins are made specifically for high- or low-gloss applications. For high gloss you want emulsion type ABS. For low gloss use mass type ABS.

Hot Tip Stringing

Possible causes	Possible remedies
Residual melt pressure	Plastic is compressible. During screw rotate there may be 1,000 to 5,000 psi of back pressure compressing the melt. Use screw decompress to relieve this pressure; be careful not to pull air into the nozzle.

Possible causes	Possible remedies
Volatiles	Moisture, low molecular weight additives, or degraded resin can turn to gas when heated in the barrel or hot manifold. This gas formation can provide enough pressure to push molten plastic out the tip. Trapped air in the hot manifold or runner system will do the same. Check moisture content of the resin. Check for resin degradation, or excessive residence time.
Poor hot tip design or geometry	Check tip design and clearance with gate surface. Check thermocouple placement and heater energy distribution. If incorrect replace.
Poor temperature control of the hot tip/high temperature	Check thermocouple placement in hot tip. If poor design, redesign for proper temperature control and cooling.

Jetting

This appears as worm tracks or squiggly lines on a part

Possible causes	Possible remedies
Gate location	The gate should not be in the center of the nominal wall. Plastic must have something to stick to or impinge on as it enters the cavity. Extend runner to go beyond the gate. This allows a small amount of plastic to enter the gate providing a normal flow front before full injection flow is started into the cavity.
Gate type	Break any sharp corners of the gate. Change gate type. Extend the runner to go beyond the gate. Increase the gate depth.
No flow front formation	Provide an impingement pin 40 to 70% of the nominal wall thickness just opposite the gate. This pin may be withdrawn during 2nd stage or pack and hold by a spring or hydraulic mechanism.
Fast injection rate	Slowing the injection rate is usually the first approach to solve jetting, it is the most expensive due to the increase in cycle time. Consistency and overall part quality often suffer.

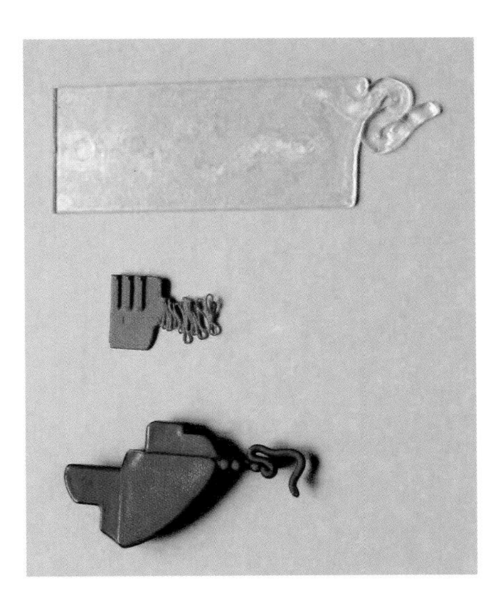

Figure 15.20 Three examples of jetting

Knit Lines

See also "Weld Lines"

Living Hinge Failures, Tears, Cracks or Breaks

Living hinges that can withstand millions of flexes are possible. If the hinge tears easily, cracks or breaks after a few flexes something is wrong. Plastic parts made with living hinges do NOT have to be flexed immediately after ejection if the choice of resin, hinge design, gate location, and cooling rate are done correctly.

Possible causes	Possible remedies
Cooling	For a living hinge to perform properly, the polymer molecules must be oriented like strands of string across the hinge and be in the amorphous state. Therefore, the living hinge area must have cooling lines near and parallel to the length of the hinge. The hinge needs to be cooled faster than the rest of the part. These cooling lines must have a separate cooling circuit. To get the best elongation or hinge performance the resin at the hinge area must be quench-cooled.
Hinge design	Check with the resin producer for their recommended hinge design. "Thin is in."
Resin selection	Semi-crystalline resins are preferred for living hinges; heterophasic copolymer polypropylene is a common choice.
Gate location	Gate location should ensure the flow front flows *evenly* across the gate. The flow front must not get to the hinge area and stop, while the rest of the part is filling and then blow across the hinge area. Similarly, if there are two or more gates in the part, the flows must not meet at or near the hinge area.
Too slow a filling rate	In order to get the desired molecular orientation across the hinge area, fast injection rates are best.

Mold Build-up or Deposits

See also "Surface Finish"

Mold build-up can be a solid (dark or white), liquid, or wax accumulation on the mold or part. Usually the cause is an additive in the plastic resin. Ideally, take a cotton swab and accumulate as much of the material as possible on the tip and send to an appropriate analytical lab. Make sure the lab is capable of interpreting the results, not just do the analysis.

Possible causes	Possible remedies
Additive package: i.e., mold release, stabilizers, colorant(s), antioxidants, anti-static agents, and flame retardant packages	Try natural resin with no colorant. Request a different grade with lower additive content, try a different supplier's resin.
Additive package temperature stability	Reduce melt temperature to the lower end of the acceptable range as stated by resin supplier. Minimize residence time in the barrel by shorter cycle times or by using a small barrel. Shot size should be 25–65% of the barrel capacity.
Inadequate venting	Check vents for proper depth and number, add or repair as needed. Provide a vacuum to the cavity during fill.
Moisture	Check moisture level in resin. If high humidity environment, run a trial with dehumidifier blowing over mold surface.
Mold temperature	Check for high mold or steel temperatures, is the area of build up a hot or cold spot. Remedy any cooling issues.
Contamination	Check for contamination.
Infrequent mold cleaning	Some residue build-up is to be expected over time. Molds must be cleaned on a routine basis.
Injection rate too high	Fast injection rates can provide shear separation of low viscosity additives, which bloom to the surface. Slow injection rate.
Mold steel – resin incompatibility	Certain metals and resins are incompatible, for example some flame retardant packages and PVC can form acid gases that will corrode mild steel, like P-20.
White powder on surface of mold or part	Sometimes a flame retardant additive package blooms. If on acetal parts, try drying the resin before processing.

Non-Fill

See also "Shorts"

Non-Return Valve

See also "Check Ring"

A mechanical device that acts like a valve in a car engine. When closed, it becomes a piston pushing the plastic forward through the nozzle. It often leaks or does not seal properly. It is critical for this valve to hold a consistent cushion. Note that it is possible for the ball check or non-return valve to be fine but the barrel is worn or egg-shaped.

Stepped Angle Non-Return Valve

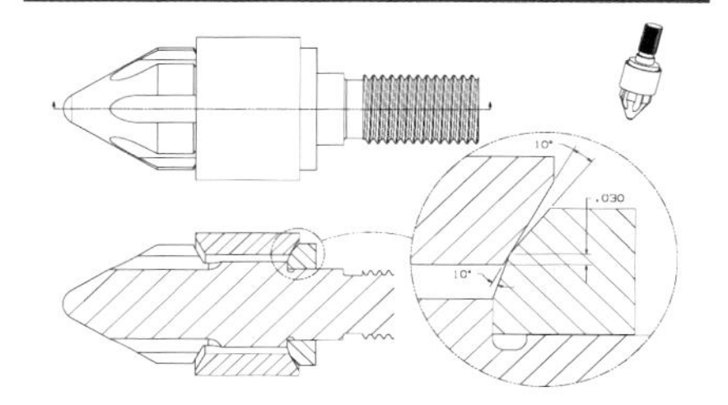

Figure 15.21 Correct non-return valve design

Possible causes	Possible remedies
Performance test	Turn 1st stage pressure limit down to the normal 2nd stage pack or hold pressure for this mold. Bring the screw back to 90% of the barrel capacity and inject for 10 seconds with the previous part still in the mold, runner and sprue. Note if the screw drifts forward. It should not. Repeat at the 50% and again at the 10% shot position. Any screw forward drift is an indication that the barrel or non-return valve is worn. Replace or repair; *both barrel* and valve must be inspected. There is no process technique to work around a broken or worn non-return valve.
Proper design	Commonly the seat and the sliding ring have mating angles. This is not correct. Like a valve in a car it works better to have a step angle to assure a more positive seat. *Few valves are built correctly.* Buy the stepped angle valves. It is also best not to have it rotate with the screw. Avoid ball checks.

Nozzle Drool or Stringing

See also "Hot Tip Stringing"

Possible causes	Possible remedies
Volatiles	Moisture and low molecular weight additives turn to gas when heated in the barrel. This gas formation can provide enough pressure to push molten plastic out the nozzle. Trapped air in the screw flights will do the same. Check moisture content of the resin. Increase back pressure. Make sure screw L/D is greater than 18:1.
No decompression after screw rotate	Plastic is compressible. Plastic is compressible. During screw rotate there may be 1,000 to 5,000 psi of back pressure compressing the melt. On sprue break or mold open plastic will be pushed out of the nozzle. Use screw decompress to relieve this pressure; be careful not to pull air into the nozzle.

Possible causes	Possible remedies
Incorrect or poor temperature control on the nozzle	Nozzle temperature is notoriously poorly controlled. Placement of the thermocouple should be 1/3 the distance of the nozzle from its tip if the nozzle is over 3 in. (75 mm) long. For 3 in.(75 mm) lengths and shorter it is OK to have the thermocouple imbedded into the hex on the nozzle body. For longer nozzle bodies, thermocouple should be underneath the heater band via a butterfly type thermocouple. The nozzle should be PID temperature-controlled, use of a % or variac is unacceptable. Do not allow the thermocouple to be attached via the screws on the clamp for the heater band.
Wrong nozzle length	Try a shorter nozzle.
Wrong nozzle tip	Avoid standard tips, try a reverse taper, nylon, or a full taper, ABS tip.

Nozzle Freeze or Cold Slug

Possible causes	Possible remedies
Nozzle tip too cold	The sprue bushing is a heat sink for the nozzle tip and draws heat from the tip while in contact. Use sprue break if possible. Insulate the tip from the sprue bushing with aramid fiber woven cloth.
Poor temperature control of nozzle	Use PID temperature control. Ensure thermocouple is 1/3 the distance of the nozzle from its tip if the nozzle is over 3 in. (75 mm) long. For 3 in. (75 mm) lengths and shorter it is OK to have the thermocouple imbedded into the hex on the nozzle body. For longer nozzle bodies, thermocouple should be underneath the heater band via a butterfly type thermocouple. Do not allow the thermocouple to be attached via the screws on the clamp for the heater band.

Nozzle Stringing

See also "Nozzle Drool"

Odor

See also "Black Specks"

Any smell or odor that is unusual. Odor due to off-gases of plastics may have a potential to be a health and safety issue. Get help, find the source, vent it, or remedy the cause but do not continue to "live" with it.

Possible causes	Possible remedies
Polymer degradation	Certain polymers when overheated or oversheared break down to acid gases. PVC, acetal (POM), and certain flame retardant resins are the most common. Proper processing does not provide these odors. If detected, check for excessive temperatures, high back pressure, worn or scored screw or barrel or a high-shear check ring or non-return valve.
Contamination	Check base resin for foreign material. Contaminates may cause the odor themselves or cause the plastic to degrade. Check regrind, how clean is your regrind? Try virgin resin to see if odor disappears. Omit coloring agent.

Orange Peel

See also "Surface Finish"

Surface cosmetic imperfections in gloss or smoothness, sometimes concentric lines called record grooves or recording.

Possible causes	Possible remedies
Mold build up or deposits	Check for residue or deposits on the mold/cavity surface. If there are mold deposits, see "Mold buildup."
Mold surface finish	Check surface of cavity for proper polish or finish and whether it is clean. Repair and clean.
Slow filling	Increase injection rate, this decreases resin viscosity and allows more pressure to be transferred to the cavity. If 1st to 2nd stage switchover is < 0.1 s, ensure velocity is not pressure-limited.
Low cavity pressure	Increase 2nd stage pressure. Increase 2nd stage time, and if possible remove the same amount of time from the cooling or mold closed timer to keep cycle time constant.
Mold temperature	Increase mold temperature. Decrease mold temperature.
Melt temperature	Check melt temperature, adjust to within the manufacture's guidelines if temperature is outside limits. Try higher end and lower end of resin supplier's guidelines.
Uneven filling of a single cavity	Balance flow path with flow leaders if possible. Increase injection rate.

Possible causes	Possible remedies
Unbalanced filling in multi-cavity molds	Adjust runner size to balance filling. Do not adjust gate size to balance filling; this will provide various gate seal times and vary part dimensions, weight, etc.

Pinking of the Part

Relatively rarely it happens that parts will turn pink while in storage. The cause is usually carbon monoxide gas reacting with components of the plastic.

Possible causes	Possible remedies
Carbon monoxide	Minor amounts for carbon monoxide are known to discolor certain resins. Remove parts to open area and see if discoloration disappears. Exposing the part(s) to sunlight can accelerate the disappearance of the discoloration. If discoloration reverses, remove all gas fueled lifts etc. from storage area. Improve storage area ventilation. Go to battery operated fork lifts.

Pitting

Possible causes	Possible remedies
Trapped gases dieseling	See "Burns." If it is due to dieseling, do not run the mold, further damage will result.
Corrosion or chemical attack by the resin or additive on the steel	Check for resin compatibility with the steel of the mold. If acid gases are possible, a more chemically resistant surface may be required. A different steel or coating of the existing surface should be specified.
Abrasive wear, erosion	Highly filled resins can pit and erode a mold's surface finish. Change gate location, coat cavity with a wear resistant finish. Rebuild tool with appropriate hardened steel.

Poor Color Mixing

See also "Color Mixing"

Race-Tracking, Framing, or Non-Uniform Flow Front

The flow front should be a continual half-circle fill from the gate.

Possible causes	Possible remedies
Non-uniform wall thickness	Thicker sections of part fill preferentially due to lower melt pressures required to fill. Plastic flow will accelerate in thicker sections and hesitate filling a thin section. This may allow the plastic to "race-track" around the perimeter or section of a part and trap air or volatiles. Try faster injection rates but it is unlikely this will solve the problem as you are fighting a law of physics. Round the edge or taper the junction between the nominal wall change. The correct fix is to redesign with a uniform nominal wall.
Gate location	Gate into the thick area and provide flow leaders to the thin areas to provide uniform filling.
Hot surface or section in the mold	Allow the mold to sit idle until mold is at uniform temperature. Make and save first shot for 99% full. If flow path is different than in later shots, it is a tool-steel temperature and cooling issue. Check mold for hot spots. Get uniform cooling.

Record Grooves, Ripples, Wave Marks

These are concentric grooves or lines usually at the leading edge of flow. The flow front is hesitating, building up pressure then moving a short distance and hesitating again. This is almost always related to lack of adequate pressure at the flow front or slowing of injection velocity.

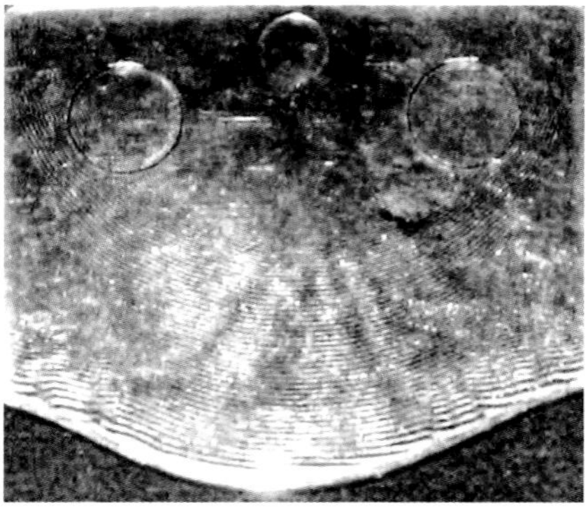

Possible causes	Possible remedies
Pressure limited 1st stage or lack of velocity control	Double check that the pressure during 1st stage is 200–400 hydraulic psi lower than the set first stage limit. Make sure there is enough pressure differential (delta P) between the highest pressure during 1st stage and the set pressure limit for 1st stage. First stage pressure limit should be higher than the pressure used during 1st stage.
Incorrect position transfer	Take 2nd stage pressure to 300-psi plastic pressure or if the machine does not allow this, take 2nd stage time to zero. The part should be 95–99% full. If this is a thin-walled part, the part should be full with only slight underpack near the gate. Adjust position transfer to provide appropriate fill volume.
Melt temperature too low	Check melt temperature via the hot probe technique or appropriate IR sensor. Make sure it is within the resin supplier's recommended range.

Figure 15.22 Record grooves or orange peel

Possible causes	Possible remedies
Poor 1st to 2nd stage switch-over response	Note response of hydraulic pressure at switch-over. It should rise to the transfer point, then drop rapid to the set 2nd stage pressure. If hydraulic pressure drops much below set 2nd stage pressure, the flow front may be hesitating and building a high viscosity. Repair machine.
Low pack rate or volume	Increase pack rate or volume of oil available for 2nd stage.
Low mold temperature	Increase mold temperature 20–30 °F. Decrease cycle time. This will raise steel temperature in the mold.

Screw Recovery, Slow Recovery, Screw Slips or Does not Feed

The metering section of the screw pumps plastic forward, which pushes the screw back.

Possible causes	Possible remedies
Feed throat temperatur	Run throat temperature at 110 to 140 °F for most resins. For high-end engineering resins you may want to go higher. Do not run feed throat at 60–80 °F. Feed throat should be PID temperature-controlled.
Feed problems	Check size of granules and flow through hopper and feed throat. Ensure that material gravity-feeds correctly when resin is being loaded into the hopper. Vacuum loading may interrupt normal gravity-feeding, especially with single shot loaders. If coloring at the press, check recovery without colorant. Certain color concentrate carriers can increase recovery times, too much wax or oil.
Heavily carbonized or blocked flights	Standard general purpose screws are notorious for dead spots behind flights. These can have large carbon or other deposits that block plastic flow. Check screw for clean polished flights.
Worn screw and/or barrel	Worn screws and barrels will provide better mixing but slow recovery rates as plastic back flushes over flights. Flights should be sharp, screw root should be highly polished, no nicks or scratches.
Moisture	Check moisture content of plastic, check feed throat for cracks leaking water.

Possible causes	Possible remedies
Granule size	Plastic granules should be uniform in size and shape. A wide range in granule size, fines and small granules along with large chunks of regrind will cause feeding problems. This includes large and small pellets in virgin.
High back pressure	Target 1,000 to 1,5000 psi melt back pressure. Try lower back pressure.
High RPM	Try lower screw rotate speeds; better melt uniformity and mixing are obtained with slow screw speeds. Use all but ~ 2 seconds of the cooling time for plasticating. Do not lengthen the cycle.
Incorrect barrel temperature settings	Start by setting front and center zones to the center of the resin suppliers recommended range. Set rear zone at the minimum of the range. Back pressure set at 1,000 melt psi. Average recovery time for 10 cycles. Repeat with rear zone 10 °F higher until you have reached the rear zone setting at the maximum recommended by the resin supplier. Pick the temperature that gives you the minimum recovery time.
Poor screw design	See "Color mixing" and "Screw design"

Screw Slip

See also "Screw Recovery"

Screw Design

Possible causes	Possible remedies
Standard general purpose screw	These are known to produce unmelt due to solids bed break up. They should be replaced with melt-uniformity screws: Minimum L/D is 20/1. This is an industry problem, most (99%) of general purpose screws do not provide uniformly melted polymer coming out of the nozzle.
Screw and barrel metals	Recommend: bimetallic or hardened barrels and soft screws like stainless steel. Chemically resistant screw material is especially critical for clear resins. Screw should be polished with sharp edges on flights with the back of the flights rounded with a large radius to prevent dead spots and carbon buildup. Screw root and flight channels should be highly polished with no nicks or scratches. A modified barrier should lead from the transition zone to the metering zone.

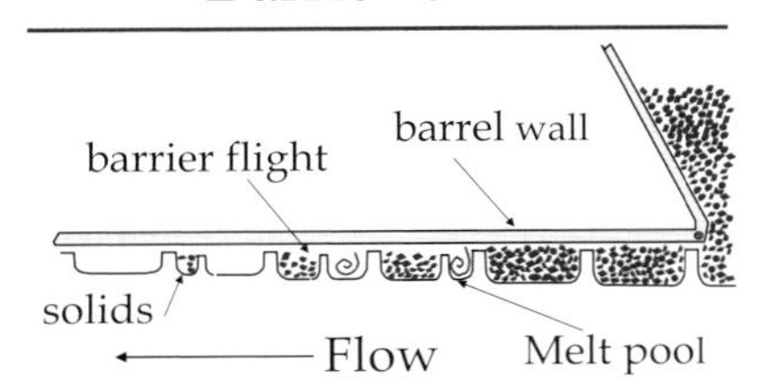

Figure 15.23 Barrier screw

Possible causes	Possible remedies
Barrier flights	Generally not recommended, unless short and at the end of the transition zone or beginning of the metering section. Often cause severe degradation and overheating.
Vented barrels	Vented barrels, though uncommon, do have their purpose. They provide excellent melt uniformity and process resins that are not subject to hydrolysis more uniformly. Unfortunately, their design is often poor. The two-stage screw must be designed with a continuous flight through the decompression section. The first stage should be cut such that it cannot overpump the second stage. Vented barrels require near zero backpressure to prevent vent flooding. This presents purging and residence time problems. See Figs. 15.10, 15.17, 15.23.

Shorts or Short Shots or Non-Fill

Part is short or some section of the part, such as a rib is not completely filled out.

Possible causes	Possible remedies
Consistent short shots	
Incorrect shot size	Take 2nd stage pressure to 300-psi plastic pressure or lower ; if the machine does not allow this, take 2nd stage time to zero. The part should be 95–99% full. If this is a thin walled part, then the part should be full with only slight underpack near the gate. Adjust position transfer to provide appropriate shot size or 1st stage volume. Do not use 1st stage pressure-limit to adjust the amount of plastic that enters the cavity on 1st stage.
Pressure-limited 1st stage or lack of velocity control	Double check that the pressure during fill or 1st stage is 200–400 hydraulic psi lower than the set first stage limit. Make sure this is enough pressure differential, delta P, across the flow control valve.
Injection rate	Increase injection rate to decrease viscosity.
Non-return valve or barrel worn or broken	Check non-return valve and barrel, are they in specification? If the non-return valve is OK, double check barrel for wear and ovality. Repair or replace as needed. Note: non-return valve should not have mating angle between seat and sliding ring. This should be a stepped angle for positive shut off.

Figure 15.24 Short shot series filling a part; the part is center-gated

Possible causes	Possible remedies
Large pressure drop	Perform a short shot analysis for pressure loss. Note pressure at transfer for shots making: 1) 99% full part, 2) Sprue, runner and gate, 3) sprue and runner only, 4) purge full shot through the nozzle into the air. This is best done using one velocity during first stage. Use intensification ratio to calculate pressure drop for a) nozzle, acceptable range 200–4,000 psi b) sprue and runner, acceptable range 200–5,000 psi c) gate, acceptable range 200–5,000 psi d) part, acceptable range 200–40,000 psi. Evaluate where largest pressure drop is and remedy. A restriction of flow will be discovered with this method.
Trapped gas or air	Progressively short shoot the mold using shot size or position transfer to make 10, 20, 40, 60, 70, 80, 95% full parts. Note if there is any sign of the plastic flow front coming around on itself. Is there a racetrack effect? This includes jetting. Note if ribs are covered before completely filling. This test cannot be done correctly if you reduce 1st stage pressure or velocity. You must be velocity-controlled for 1st stage. The concept is to find where the gas is coming from and eliminate its source. Vent the tool properly or use porous steel to eliminate gas traps. Jetting can cause gas and air to be trapped. Check for moisture, steam is a gas. Change gate location. Pull a vacuum on the mold cavity during fill.
Insufficient 2nd stage pressure	Make sure 1st stage is at the right shot volume; part should be 95–99.9% full at the end of 1st stage. If OK, raise 2nd stage pressure.
No cushion	Ensure adequate cushion to allow for transfer of packing or 2nd stage pressure.
Melt temperature	Verify that melt is within the resin supplier's recommended range.
Mold temperature	Try higher mold temperatures and or faster cycle times.
Resin viscosity	Change resin to a higher melt flow rate. Be careful as properties may decrease due to lower molecular weight. Parts must be fully tested in the application for correct performance.
Long flow length	For every mm of flow there is a pressure drop. Long flow lengths have large pressure drops. Add gates or flow leaders. Last resort increase nominal wall.

Possible causes	Possible remedies
Check balance if a multi-cavity tool	See text for 5-Step process for determining mold balance.
Thin nominal wall	Add flow leaders if possible. Increase nominal wall if all other avenues fail. Thicker nominal wall will reduce pressure loss.
Intermittent short shots	
Non-return valve or barrel worn or broken	Check non-return valve and barrel, are they in specification? If the non-return valve is OK, double check barrel for wear and ovality. Repair or replace as needed. Note: non-return valve should not have mating angle between seat and sliding ring. This should be a stepped angle for positive shut-off.
Cushion not holding	Note cushion repeatability, if varying by more than 0.200 in. or 5 mm, check non-return valve as above. Try a larger decompression stroke to help "set" the check valve. Be careful not to suck air into the nozzle and cause splay.
Contaminated material	Check gates and parts for foreign material. Check quality of regrind. See introduction to "Color mixing."
Check balance if a multi-cavity tool	See text for 5-Step process for determining mold balance.
Melt temperature	Verify that melt is within the resin suppliers recommended range.
Mold temperature	Try higher mold temperatures and or faster cycle times.
Unmelt	Look for unmelted granules in the part, color streaks, see "Screw design" and introduction to "Color mixing". Provide uniform melt to the gate.
Cold slug	Occasionally it is possible for a cold slug from the nozzle to go beyond the sucker pin and plug a gate. Check nozzle for cold slug formation. See "Nozzle drool."
Insufficient 2nd stage pressure	Raise 2nd stage pressure.
Trapped gas	See above, see also "Bubbles."

Shrinkage

See also "Short Shots" and "Flash"

An industry nightmare! At first, ASTM Test method 955 was not intended to be used to determine shrinkage for parts out of molds, yet it is what most people use. Its purpose was to compare "lot-to-lot" differences. Also, the test specimen is 0.125 in. (3.17 mm) thick. Then in 2000, ASTM modified the procedure to include use for mold shrinkage numbers. Shrink depends on: molecular orientation; nominal wall thickness (thicker shrinks more); cooling rate (higher mold temperature, the slower cooling rate the more the part shrinks); nucleation in semi-crystalline resins; and pressure gradient in the cavity.

Possible causes	Possible remedies
Too much shrinkage	
Insufficient plastic	Make sure cut-off position provides the right volume of plastic on first stage. Do a gate seal analysis. Add more 2nd stage time to achieve gate seal. Add more 2nd stage pressure. Check that you have adequate cushion.
Degree of crystallinity	If a semi-crystalline resin, lower mold temperature to increase cooling rate. This will decrease crystallinity and decrease shrink. Look at a nucleated resin. Watch for post mold warp over weeks of time. Do a thermal cycle test on parts to show amount of internal stress and warp. Check all properties.
Low 2nd stage pressure	Raise 2nd stage pressure, see "Insufficient plastic" above. See "Short shots."
Too little shrinkage	
Too much plastic	Make sure cut-off position provides the right volume of plastic on first stage. Do a gate seal analysis. Decrease 2nd stage time to achieve gate unseal; add the time taken off 2nd stage to cooling or mold-close time. Decrease 2nd stage pressure. Check that you have adequate cushion.
Degree of crystallinity	If a semi-crystalline resin, raise mold temperature to decrease cooling rate and increase crystallinity. This will make a smaller part. Check all properties.
High second stage pressure	Lower 2nd stage pressure. Lower 2nd stage time to allow for gate unseal. See "Flash"

Sinks

See also "Bubbles" in the "Void" section

Sinks are depressions on the surface of a part that do not mimic the mold steel surface. Sinks and voids are signs of internal stress and are warning signs that the part may not perform as required.

Possible causes	Possible remedies
Insufficient plastic	Sinks occur on cooling in thick sections. You need to pack more plastic into the cavity with; 1) consistent cushion, make sure you are not bottoming out the screw, 2) higher 2nd stage pressures, 3) longer 2nd stage time, 4) very slow fill rate or fast fill rate, 5) use counter pressure, 6) open the gate for longer gate seal times to allow more packing during 2nd stage, or 7) increase the runner diameter. First determine where the sink or sinks are; near the gate or farther down the flow path. If near the gate, check gate seal time. If farther down the flow path; increase injection speed to decrease viscosity and allow more pressure. Other possibilities (not first choices) include: lower the mold temperature significantly to freeze the outside wall of plastic, though this might produce voids internally. Reduce melt temperature. Thin-up the nominal wall in this area by adding steel to core-out thick sections. Change gate location, fill thick to thin.
Check balance if multi-cavity tool	See text for 5-Step process to determine mold balance.
Post mold cooling	Thick sections can remelt a part's outside surface once ejected, allowing the surface to collapse. Try cooling in water or between aluminum sheets rather than air.
Thick nominal walls	Core out to thin the nominal wall. Thicker is not stronger in plastic parts. Thick nominal walls should be redesigned with ribs if strength is needed. This will save plastic and cycle time as well as provide improved strength or stiffness.

Splay or Silvery Streaks

"Splay" (also called chicken tracks, tails, and hooks)

These are cosmetic blemishes on the part's surfaces. Splay is usually caused by a gas, usually water (steam), but can be caused by unmelt, dirt, chips or flakes from three-plate molds, cold slugs, excessive volatile additives, air, and/or decompression. Remember, if the resin is subject to hydrolysis (reacts with water), you may need to discard samples with splay rather than regrind them. Water permanently degrades molecular weight in some resins such as polycarbonate, nylon, and polyesters.

Possible causes	Possible remedies
Moisture in the plastic granules	Check moisture content with a moisture specific method (not weight loss): Karl Fisher or Computrac 3000. Weight loss does not tell you if it is moisture or volatiles. Check dryer for proper function.
Dryer not functioning properly	Check for the correct drying temperature set point. Remove the bottom 20–50 lbs. of material in the dead space of the cone of the hopper and put it on the top. Check for funnel flow, hopper must drain via mass flow. This can be achieved by having the angle of the cone at the bottom of the hopper 60° included. Check temperature of the incoming air, check temperature of the air exiting the hopper: should be within 20–40 °F (10–20 °C) of the incoming air. Check for fines clogging the filters. Check that the regeneration heaters are functioning. Check the temperature of the air going into the desiccants; it should be below 150 °F (65 °C). Check dew point of air before and after desiccants if possible. Dew point should be lower after desiccant beds.
Condensed moisture on pellets	If resin has been brought inside from a cold climate, condensation can occur like condensation on glass. Even though the resin is not hygroscopic it still must be "dried". For condensation moisture, drying at 150 °F (65 °C) for 1–2 hours should be fine.
Moisture due to high humidity	In un-air conditioned shops during high-humidity days (summer time) there are times when non-hygroscopic resins pick up moisture, or the air around them carries enough moisture that a dryer is needed.
Moisture in the feed throat	Check feed throat temperature and look for cracks. Run feed throat at 110–150 °F (40–65 °C) via PID temperature control.
Moisture from mold	Check mold, particularly cores, for condensation. Raise mold temperature if condensation is present. Check for water leaks: from fittings dripping into the mold and from cracks or leaks in the mold. Repair all leaks!

Possible causes	Possible remedies
Poor nozzle temperature control	Nozzle temperature is notoriously poorly controlled. Placement of the thermocouple should be 1/3 the distance of the nozzle from its tip if the nozzle is over 3 in. (75 mm) long. For 3 in. (75 mm) lengths and shorter it is OK to have the thermocouple imbedded into the hex on the nozzle body. For longer nozzle bodies, the thermocouple should be underneath the heater band via a butterfly-type thermocouple. The nozzle should be PID temperature-controlled, use of a % or variac is unacceptable. Do not allow the thermocouple to be attached via the screws on the clamp for the heater band.
Air trapped in melt from screw or decompression	Check L/D of screw; if less than 18:1, there is a good possibility that the screw is at fault. Raise backpressure, slow screw rpm and extend cooling time. Raise rear zone temperature. If any of these work, order a melt uniformity screw (see screw design). If you have decompression on, minimize or turn off; decompression, especially with hot runner or hot sprue molds, traps air in the nozzle. Too much decompression sucks air into the nozzle, which gets trapped in the flow stream because nozzle contact force is loaded before injection.
Chips or flakes from 3-plate molds	Open mold and carefully inspect for plastic chips, flakes or particles. These can upset the flow front and cause marks similar to splay.
Contaminants, dirt, or cold slugs,	Solid particles of any kind can upset the flow front and cause marks similar to splay. Inspect plastic granules for contamination or dirt. Check for degradation on the screw.
Unmelted granules and gels	Solid particles of any kind can upset the flow front and cause marks similar to splay. Unmelt can be a cause due to poor screw design or short L/D. Unmelt is to be expected from general-purpose screws. Gels are high molecular weight or crosslinked resin that does not melt. Gels are rare and can only be remedied by the resin supplier. Trial a different lot or grade of resin.
Volatiles due to degradation of polymer or additives	Due to high shear, hot spots in the barrel, flame retardant packages, degraded regrind, etc., a polymer can produce off-gases that will mimic moisture and provide splay. Check temperature set points vs. actual melt temperature as measured with a hot probe.

Sprue Sticking

Possible causes	Possible remedies
Nozzle tip radius mismatch with sprue bushing	Check if nozzle radius matches with sprue bushing by inserting a piece of cardboard and bring in nozzle contact. Look for perfect tip imprint in cardboard, any tears, cuts or mismatch – change tip. There are inexpensive gauges that have the ¾ in. and 1/2 in. radius.
Nozzle tip orifice too large	Check orifice of sprue and tip. Tip orifice should be at least 0.030 in. (0.75 mm) smaller in diameter.
Scratches or incorrect polish on sprue	Sprue should be draw polished to a #2 finish. Vapor honing is better for soft touch material. Circular polishing provides minute undercuts that get filled with plastic and "stick" the sprue. Look for scratches or gouges that form undercuts, even small ones will stick a sprue, too.
Sprue puller problems	Is the sprue puller large enough and of the right design? Add undercuts, Z-puller, or more reverse taper.
Sprue too soft, not frozen	If sprue is still soft at mold open, down-size sprue or cool the sprue bushing. Try lower nozzle temperature; then melt temperature. Last resort, add more cooling time.
Improper taper on sprue	Is taper 1/2 in. (13 mm) per foot (305 mm). Increase taper
Overpacked sprue	Do a gate seal study, if possible take off some 2nd stage time and add to cooling or mold closed timer.

Sticking in Mold

It is important to watch mold open and ejection; the use of a high speed camera will often show if it is a mold, part, or ejection issue.

Possible causes	Possible remedies
Improper polish or undercuts causing part to stick the "A" side of the mold upon mold opening	In manual mode, open the tool SLOWLY and note any noises or cocking of the part. Before ejection, inspect part for deformation. If deformed before ejection, the part is sticking to the opposite side of the mold and deforming on mold open. Draw polish or remove undercuts in areas where part sticks. Many soft touch polymers need a vapor honed surface to aid release. High polish forms a vacuum.

Possible causes	Possible remedies
Improper mold polish or undercuts causing part to stick to the "B" side of the mold on ejection	Slowly eject the part and note if a corner or section hangs-up or sticks or if there is cocking of the part at this stage. Draw polish or remove undercuts if present. Many soft touch polymers need a vapor honed surface to aid release.
Incorrect 2nd stage time or improper gate seal time	Perform a gate seal analysis. Determine if gate seal or unseal changes the problem.
Overpacking, not enough shrinkage	Reduce pack and hold or 2nd stage pressure. Make a few parts with shorter pack and hold time to achieve gate unseal. Add any time taken off 2nd stage to the cooling, cure, or mold closed time to keep cycle time the same. Also, try a reduced cycle time by taking time off of the cooling or mold closed time.
Under packing causing excessive shrinkage	If reducing pack and hold pressure makes it worse, the cause could be the part shrinking onto a mold detail or core. Increase pack and hold pressure, increase 2nd stage time and take the identical number of seconds off cooling to keep cycle time the same. Also try a longer cycle by adding time to the cooling or mold closed time.
Vacuum, high polish can cause a vacuum to form during molding, which holds the part to the steel	Provide a vacuum break before mold opening or ejection.
Check filling balance if a multi-cavity tool	See text; follow 5-Step process to determine mold balance. This can lead to over- or under-packing a part.
Excessive ejection speed	Slow ejection velocity.
Parts handling or robot removal	Check end of arm tooling and movement of the part on the robot arm.
Contamination	Check regrind for contamination or excessive fines.
Plate out on the mold surface	Inspect mold surface for buildup, clean appropriately; for example with toilet paper and semichrome for non optical parts, for optical finishes check with mold maker.
Incorrect mold finish	Too high a polish can cause a vacuum to hold the part to that side of the mold. Change polish or texture.

Possible causes	Possible remedies
Crazing	Check for stress in the area, for example stress-whitening near ejector pins or places where the part sticks. Is there any contact with chemicals such as solvents or mold spray that attacks this type of plastic?
Mold temperature too cold or too hot	Raise and or lower mold temperature significantly as long as you do not cause mold damage. Check cooling lines for water flow rate in gallons/minute (GPM). You need a Reynold's number of 5,000 or greater to achieve turbulent flow (optimum cooling). Make sure the difference in temperature between in and out of water lines is less than 4 °F or 2 °C.
Ejector plate cocking	Check for even length of knock out bars. Should be within 0.003 in. of each other.
Mold deflection	Take dial gauge measurements on top and bottom of both sides of the mold. Repeat for both halves. Note mold deflection, if severe can cause undercuts which in turn can cause sticking.
Vacuum	May need an air blow to release a vacuum formed on ejection.
Inadequate mold release in resin	Add mold release. Try same resin from different supplier. To solve the problem avoid using mold sprays except for start-up situations.

Streaks

See also "Color Mixing" and "Black Specks"

Surface Finish

See also "Orange Peel", "Gloss" and "Gate Blush"

Possible causes	Possible remedies
Mold build up or deposits	Check for residue or deposits on the mold/cavity surface. If there are mold deposits, see "Mold buildup."
Mold surface finish	Check that surface of cavity has proper polish or finish and that it is clean.
Poor venting	Check for vents, proper location, are they clean and open?
Slow filling	Increase injection rate, this decreases resin viscosity and allows more pressure to be transferred to the cavity if 1st to 2nd stage switch-over is < 0.1 s. Ensure velocity control is not pressure-limited.

Possible causes	Possible remedies
Low cavity pressure	Increase 2nd stage pressure. Increase 2nd stage time, and if possible, remove the same amount of time from the cooling or mold closed time to keep cycle time constant.
Mold temperature	Increase mold temperature. Decrease mold temperature.
Melt temperature	Check melt temperature, adjust to within the manufactures guidelines if temperature is outside limits. Try higher end and lower end of resin supplier's guidelines.
Uneven filling of a single or multi-cavities	Balance flow path with flow leaders if possible. Increase injection rate.
Unbalanced filling in multi-cavity molds	Adjust runner size to balance filling. Do not adjust gate size to balance filling; this will provide various gate seal times and vary part dimensions, weight, etc.
White powder build-up on acetal parts	Formaldehyde is condensing on the mold or bloom to the part surface. Dry the acetal before processing.

Unmelted Particles

See also "Screw Design" and "Color Mixing"

Vent Clogging

See also "Mold Build-up"

Voids

See also "Bubble" and "Sink"

Are you absolutely sure it is a void and not a bubble of gas? You cannot tell by looking, it must be tested. Test is found under "Bubbles."

Warp

Dimensional distortions in a part. Often one of the hardest problems to resolve through processing due to the most common root cause, that is poor part design, or non-uniform nominal wall sections, especially in semi-crystalline resins. Warp is a result of retained compressive, tensile, orientation, and/or crystalline stresses. The trick is to figure out, which stress is ruling. Always, after setting the production process, run a thermal cycle study to see if the part is stable. That is, slowly heat the part to the maximum possible application temperature for a few hours, then cool it slowly to the minimum temperature it might see in use or application, normally ~ – 20 to – 40 °F. Repeat this cycle for an appropriate number of times determined by the application. Does the part hold its shape and dimensions for proper performance in the application?

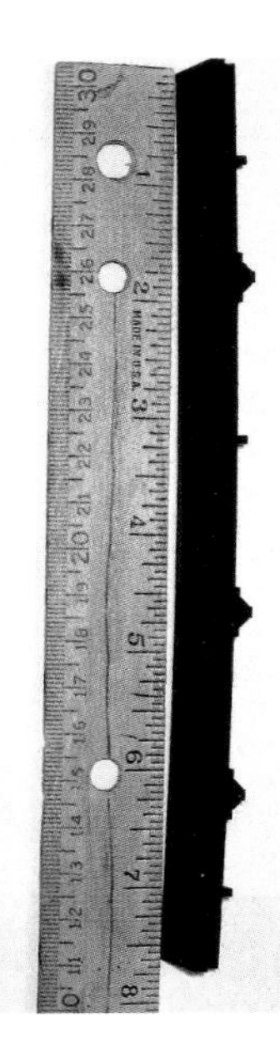

Figure 15.25 Warp

Possible causes	Possible remedies
Warp for amorphous resins, low shrink	
Molecular orientation	Take 2nd stage to less than 300-psi plastic pressure and mold a 99% full part. See "Short shots" for this procedure. Inspect the 99% full part for warp. If the short shot warps, then either orientation or shrinkage causes the warp. If there is uniform nominal wall and no standing cores, the warp is most likely orientation. Try fast to slow injection rates. Raising mold temperature 20–30 °F (10–15 °C) may also help. If non-uniform nominal wall, find the designer or client and explain the importance of uniform nominal wall. Note: With non-uniform nominal wall you are now battling a law of physics: different cooling rate for thick vs. thin, and you will continue to have problems even if you do find a cooling method to minimize the amount of warp! If there is no warp on the 99% full part then the warp is due to packing pressure or pressure gradient.
Pressure or pressure gradient	Run a gate seal study. Note if there is a difference in amount of warp between samples with and without gate seal. If you have warp with gate seal, try lower 2nd stage pressure with gate seal. If this works, run with lower pressure and gate seal. If lower 2nd stage pressure does not provide a well-packed part, do a DOE to find what 2nd stage time and pressure provides the best compromise running with gate unseal. Running with gate unseal will allow some plastic to backflow out of the cavity to provide less of pressure gradient from end of fill back to the gate. Different gate location may help.
Non-uniform cooling/non uniform shrinkage	This can be due to either mold steel cooling or non-uniform wall sections. To see if it is from the mold: allow plenty of time for the mold steel to come to a uniform temperature, usually 20–30 minutes. Use this time to check water routing and GPM to assure turbulent flow, Reynold's number > 5,000. On start up, save the first part. Compare first part warp to parts after 30 minutes of steady state molding. If warp is less for the first part than others, the warp is due to poor cooling in the mold. Fix the cooling in the mold by adding cooling lines, increasming water flow, or by replacing existing stMeel with a metal that has higher thermal conductivity. Cores especially need optimum cooling and often require high

Possible causes	Possible remedies
	thermal conductivity steel. If it is due to a thick section on the part, the part will need to be cored out. Using a higher thermal conductivity steel may help. Uniform nominal wall is the correct answer. If nothing can be done you can try an extended cycle time. This uses the mold as a shrink fixture and cuts profit margins.
Non-uniform wall thickness	Thicker sections stay hot longer and will shrink more than thinner sections. Taper the junction. Running different temperatures on the mold halves may help, however care must be taken not to damage the mold. As above, the correct answer is uniform nominal wall.
Warp for semi-crystalline resins, high shrink	Semi-crystalline resins shrink 4–6 times more than amorphous resins.
Molecular orientation	Same as above!
Pressure or pressure gradient	Same as above!
Non-uniform cooling/non uniform shrink	Same as above, plus: Try a filled or nucleated version of the resin if available. Also, if there is a differential in shrinkage in flow direction vs. cross-flow, there is the possibility of using two fillers such as glass and mica to try and compensate.
Non-uniform wall thickness	Same as above!
Crystallinity	If all the above fails, try (providing the mold will not be harmed) an amorphous resin such as General Purpose Polystyrene (GPPS) or High Impact Polystyrene (HIPS), ABS or Polycarbonate/ABS blend. If the amorphous parts do not shrink, you have proven that it is a temperature-cooling rate-crystallinity issue! Try a nucleated or filled version of this semi-crystalline resin.

Weld Lines

See also "Flow Lines," "Knit Lines"

Rarely is this due to cold flow fronts, polymer fountain flows with the flow front regenerating with each millimeter of flow distance. The issues are: few molecular chains cross this boundary, air is often trapped as the flow fronts meet due to the shape of the flow front and shear migration to the flow front of the lights and volatiles (additives) in the resin.

Possible causes	Possible remedies
No molecular chain entanglement	Increase injection rate. Raise the mold temperature. Try a higher melt flow rate resin, check properties and performance. Improve venting. Provide a flow channel at the weldline to aid venting and provide some movement at the weldline. This will help with some chain entanglement. Try resin without filler. Change gate location to put weldline in a non-stressed area. If multi gated part, try blocking some gates; get permission first! Add a flow overflow tab. This allow a vent and some chains to entangle
Trapped air	Improve venting, use porous steel. Provide a flow channel to aid venting. Vent core pins. If flow fronts are folding back onto themselves try different flow rate to change fill pattern. Add flow leaders to direct flow path. Add a flow overflow tab. This provides a vent and some chains to entangle.
Low pressure at weld line	Increase 2nd stage pressure. Increase injection rate. Raise melt temperature as last resort.

Worm Tracks

See also "Jetting"

15.3 Two Stage Molding Set-Up

(Bradley G. Johnson)

To best compete in the injection molding industry, or any other industry for that matter, it is important that you make the most consistent parts, in the shortest time, with the least resources possible. To best accomplish this end, a good understanding of plastic behavior, as well as an understanding of the mold and molding machine is needed. What follows is a proven methodology for setting up an injection molding process. It has been called "scientific molding" [1] or "fully decoupled molding^{SM}" [2]. It consists of the following steps:

1. Set clamp
2. Get parts 95% full @ maximum injection velocity
3. Optimize injection velocity
4. Set hold pressure
5. Set hold time
6. Minimize cycle
7. Optimize other parameters

Here, this methodology will be referred to as two-stage molding. The two stages refer to the injection portion of the set-up, with the first stage controlled by velocity and the second stage controlled by pressure. Two-stage molding is intended for processors using molding machines with closed-loop controls.

Note: The barrel temperatures and mold or mold coolant temperature should be set mid-range of what the material supplier recommends for this initial set-up. A check of the purge melt temperature should be within 5 °F of the setting of the front zone barrel temperature. If it is not, the rear and middle zones should be adjusted until the purge temperature is within five degrees.

Set Clamp

This is the step that technical people are most likely to get right without knowing anything about plastics. However, it is also an area that is not always optimized with respect to minimizing cycle time. The primary objectives here are:

- Open and close the mold fast, but smoothly
- Open only as much as needed for the parts to drop
- Eject the parts fast without damaging them
- Minimize ejector strokes (normally shouldn't need more than one)
- Set mold protection to optimize protection and cycle time
- Set clamp force as needed based on the projected area of the part and the material being molded (typically around 3 t/in.^{2} of projected area).

The machine manuals typically provide good instructions on how to determine machine settings to accomplish these objectives. Most have either controls that will ramp speeds when accelerating or decelerating or that allow profiled speed settings during mold and ejector forward and backward movements. Sudden starts or stops of the clamp can cause the molding machine to jump around or cause damage if the plates slam

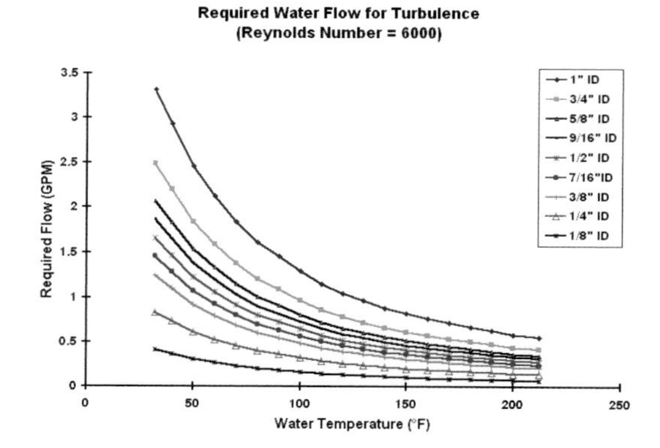

Figure 15.26 Required flow vs. water temperature for turbulent condition

together. Sudden starts when ejecting can damage the plastic parts. An underutilized feature of many molding machines is "ejection-on-the-fly". This feature allows for ejection to start before the mold is completely open and will shorten cycle time in most instances.

When running a mold for the first time, it is a good idea to dry-cycle it first. This is done by disabling the injection unit and cycling without injecting plastic. Water, or other cooling medium, should be circulated through the mold and, if there is a hot runner manifold, the heaters should be turned on. The flow rate of the water through each cooling circuit should be checked to guarantee that it has sufficient turbulence (see Fig. 15.26). Dry-cycling should initially be done at slow speeds, with low pressure and clamp force to minimize the chance of damaging anything that is not installed or fit correctly. It is especially important to start with a low mold protection pressure. As the mold cycles, listen for unusual sounds that may indicate something is binding or galling. Visually inspect for this when the mold is open. Gradually increase speeds and pressures until you minimize the dry-cycle time as much as possible without compromising mold protection. Record this dry-cycle time so that it can be compared to what you can do when injecting plastic into the mold.

Get Parts 95% Full @ Maximum Injection Velocity

Initial Settings:

Transfer position:	0.5 to 1.0 in.
Shot size:	To fill about 75% of the part
Injection velocity:	Mid-range
Injection pressure limit:	Mid-range
Hold (2nd stage) pressure:	0
Hold time:	At least a couple seconds longer than anticipated gate seal time (typically 10–20 s)
Cooling timer:	Set so that total cycle time is about 25% higher than quoted

> Note: Some machines will have a minimum hold pressure even though they allow a setting of zero. If this hold pressure is enough to cause the screw to continue its forward motion during the hold time, then set hold time to zero and set screw rotation delay to the original hold time. You must remember to reverse these two times when you get to Step 4 (setting hold pressure).

The goal here is to get the plastic flow front (in the first cavity to fill) just short (generally 95 to 99% full) of hitting the last place in the mold to fill, when injecting the plastic as fast as possible. It is usually not advisable to set the injection velocity at the maximum setting for the first shot into the mold, especially with molds that are sensitive to being overpacked. A mold is overpacked when too much plastic is forced into the mold with too much pressure. This can sometimes break fragile mold inserts or make the part difficult to eject from the mold. Gradually adjust the shot size, injection velocity, and injection pressure limit until the part is just short. When at the maximum velocity, make sure that the injection pressure limit is at least 2,000 psi (at nozzle) above the pressure it takes to get the part 95% full. This should guarantee that the pressure does not limit the velocity, that is, that the velocity and pressure are not coupled.

Another sign that a mold is being overpacked is a thin film of plastic showing at the parting line of the mold or between inserts. This thin film is called flash. A mold, especially when new, should not flash unless it is overpacked. That is why we fill the mold only approx. 95% full, instead of all the way full. Consider Humpty Dumpty:

Humpty Dumpty sat on a wall.
Humpty Dumpty had a great fall;
All the King's horses, and all the King's men
Cannot put Humpty together again [3]

If you think about it, the fall is not what really broke Humpty. It was that sudden stop when he hit the ground. Analogously, the pressure inside a mold typically does not build up very fast until the plastic hits that last place to fill in the mold. If you hit that last place too fast, the result is a sudden pressure spike in the mold and, quite often, flash. The practice of transferring to hold pressure when the part is slightly short (about 95% full) allows the plastic melt to slow down and, in effect, acts as a safety net. The slower velocity results in a slower pressure build up in the mold, which prevents flash. If Humpty Dumpty would have landed on a safety net, or at least a bush, he might still be together today.

Once you get the part(s) 95% full, there are a few things to look for before proceeding to the next step. It is possible that the mold may need some fixes before it will be ready to run in production. In most cases, it is better to get the mold design and construction correct up front, before qualifying parts and releasing the mold to production. Some things to look for include:

- Flash – As was already discussed, short shots should typically not have flash. If flash occurs at the parting line, it is likely that either the clamp pressure is not set high enough or that the molding machine does not have enough clamp force for the mold being run. If the flash occurs because the slides are being pushed back, the root causes are likely either insufficient clamp force or a poor fit of the wedge block and the slide. If flash is just getting between inserts or around ejector pins, there is too much clearance or the mold walls are being deflected. Any of these problems should be addressed before proceeding. If the problems are minor, you may be able to finish the process set-up and fix the flash problem afterwards.
- Burning – If the burn is on the surface and is caused by trapped air, the venting in the mold needs to be addressed. These types of burns can eat away the steel and cause major problems. If the burn is just underneath the surface and is caused by high shear rates due to the fast velocity, this

should be noted as a possible limit to the maximum injection velocity. Burns caused by high shear rates are most common with PVC.

- Jetting, Gate Blush, or Flow Lines – These defects can be caused by either too fast or too slow a velocity in certain areas of the mold. They can generally be processed out, but, it is usually better to modify the mold or gate design. At this point, these defects should just be noted.
- Cavity Balance – If running a multi-cavity mold, ideally, the parts will all fill at exactly the same time. Since this is not practical, the goal should be to get the weights of the short shots all within 5% of each other.

Optimize Injection Velocity

The part is now 95% full at the maximum velocity. If none of the problems noted in the previous section are present, is this the optimum velocity to run in production? It could be. There are other quality problems that may be more difficult to detect. This step will detail how to determine the optimum velocity and explain why this velocity is usually best.

The viscosity of plastic melts decreases with increasing velocity. Viscosity is defined as resistance to flow and is, therefore, a very important property when filling the mold. By running a series of injection velocities, a plot of viscosity versus velocity can be obtained. We will actually use a flow parameter that is directly proportional to injection velocity, called shear rate, instead of injection velocity in our graph.

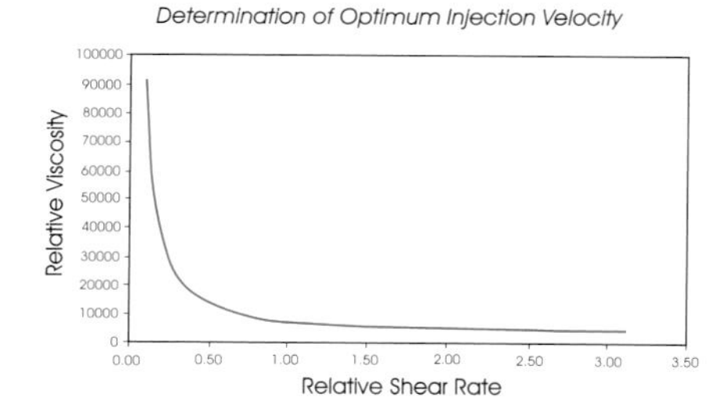

Figure 15.27 Typical relative viscosity vs. relative shear rate

It should be noted that Fig. 15.27 is a plot of relative viscosity versus relative shear rate. Each axis would have to be multiplied by a constant in order to get actual viscosity and actual shear rate. This method is used because the value of the constants will change depending on the geometry of the mold and molding machine. The derivation of the constants for the simple cases of flow through a cylindrical and a rectangular channel are shown in Section 15.5. However, for the purpose of obtaining the optimum injection velocity, the shape of the curve is all we need. As you can see, as the shear goes from low to high (velocity also goes from low to high), the viscosity initially drops off quickly and then tends to flatten out.

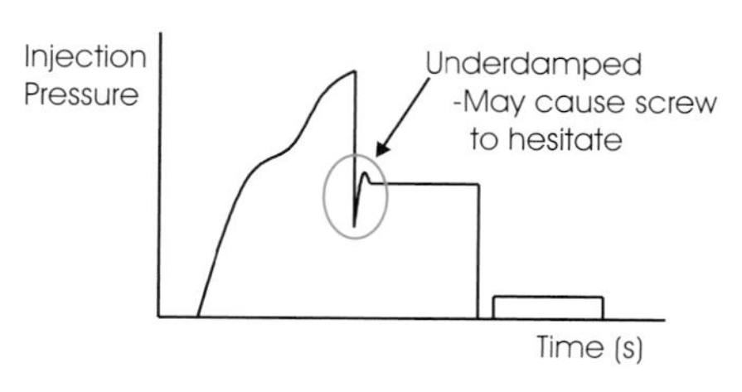

Figure 15.28 Typical injection pressure during injection molding processing

The point at which the viscosity curve starts to flatten out usually represents the best injection velocity. In general, materials process easier at lower viscosities. One of the main reasons is that the pressure drop is minimized across the mold because it is directly proportional to viscosity. It has also been reported that there is less variation in the process when selecting a velocity on the flat portion of the viscosity curve [4]. Lower viscosities also correspond to faster fill times and, therefore, generally faster cycle times. However, there are reasons that the "optimum" point may not be the one with the fastest velocity. Some machines may not be able to respond fast enough when transferring from first stage to second stage (velocity control to pressure control) when the screw is traveling at very high speeds [4]. The result can be an underdamped (see Fig. 15.28) or inconsistent transfer to hold pressure. At fast velocities, shear heat could be

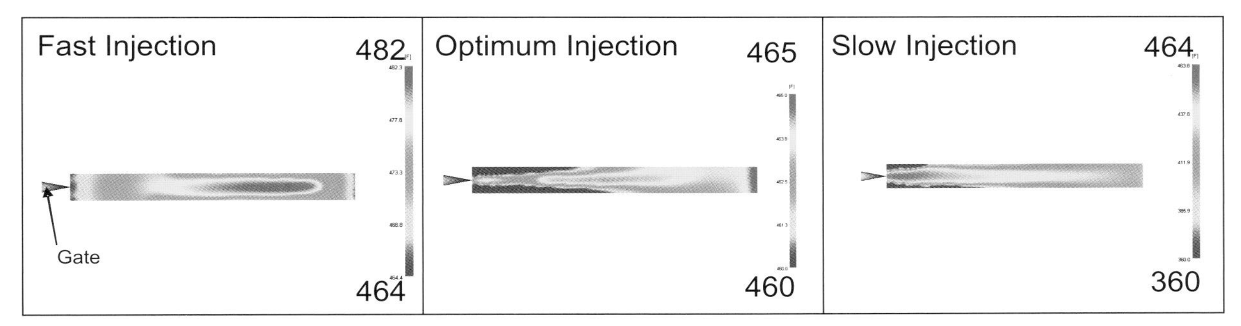

Figure 15.29 Comparison of predicted melt temperature distribution as effected by injection rate

generated as the plastic flows through the part, shown in the flow analysis bulk temperature plots in Fig. 15.29. At the fast velocity, the temperature rise in the rectangular part shown was about 18 °F. This could cause problems because of excessive shrinkage or burning at the hot spot. Generally, it is desired to have the same temperature throughout the part.

To obtain the points needed to generate the relative viscosity graph, a series of injection velocities is run and the transfer pressure and fill time recorded for each velocity. As shown in Section 15.5, the equations are:

Relative viscosity $\eta_r = P_{hyd}t$ (15.3)

Relative shear rate $\gamma_r = 1 / t$ (15.4)

where

t = fill time, and

P_{hyd} = Hydraulic pressure at transfer (injection pressure at the nozzle can be substituted here and is used in the following example)[1].

The plot shown in Fig. 15.30 is an example of the injection pressure for a "fill only" part. Hold pressure was not applied to this shot, but it will not affect the two values (fill time and transfer pressure) which are needed. Some molds are difficult to run without hold pressure, so it is advisable to add enough hold pressure to at least fill the parts in such cases. The fill time for the shot shown is 1.25 s and the transfer pressure is 6690 psi. These values are typically readouts on the machine controller.

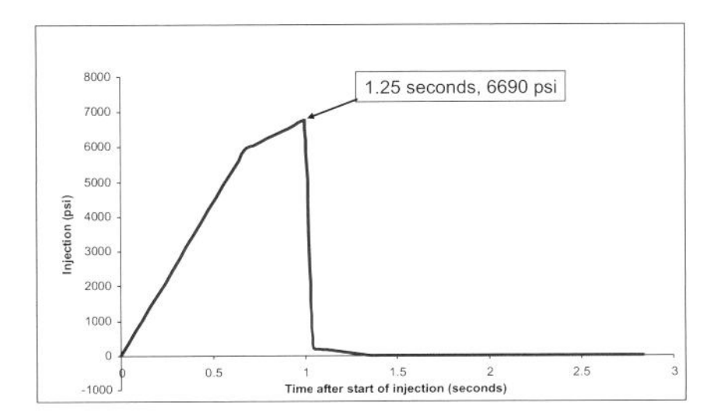

Figure 15.30 Pressure graph resulting from only filling a part with no hold pressures

Figure 15.31 shows the data and graph used to determine the "optimum" injection velocity. Starting with the part about 95% full with the maximum velocity, a sequence of slower injection velocities are run until one fill time is greater than ten seconds. Fill time and transfer pressure should be recorded for each velocity. Although the parts will get slightly shorter if no hold pressure is applied as you slow down, it is best if the shot size and transfer position settings are kept constant during this procedure. If the parts get drastically shorter when the velocity is decreased, chances are you are running the mold in a machine that has too much injection capacity. The optimum injection velocity is determined, based on where the blue horizontal line intersects the relative viscosity curve ($1.18\ s^{-1}$ which

Determination of Optimum Injection Velocity
Note: Yellow highlighted areas are user-defined.

Hydraulic Pressure		Fill Time		Injection Velocity		Rel Shear Rate	Rel Viscosity
1370	psi	0.32	sec.	5	in/sec.	3.13	438
1310	psi	0.35	sec.	4.5	in/sec.	2.86	459
1236	psi	0.37	sec.	4	in/sec.	2.70	457
1177	psi	0.40	sec.	3.5	in/sec.	2.50	471
1022	psi	0.47	sec.	3	in/sec.	2.13	480
919	psi	0.55	sec.	2.5	in/sec.	1.82	505
825	psi	0.67	sec.	2	in/sec.	1.49	553
751	psi	0.85	sec.	1.5	in/sec.	1.18	638
669	psi	1.25	sec.	1	in/sec.	0.80	836
671	psi	2.45	sec.	0.50	in/sec.	0.41	1644
710	psi	4.10	sec.	0.3	in/sec.	0.24	2911
810	psi	8.00	sec.	0.20	in/sec.	0.13	6480
870	psi	10.50	sec.	0.15	in/sec.	0.10	9135

OPTIMUM VELOCITY= 1.5 in/sec.

Determination of Optimum Injection Velocity

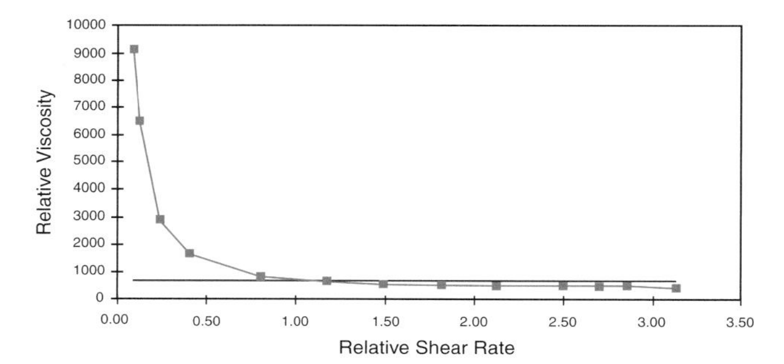

Figure 15.31 Data and graph utilized to determine an optimum injection speed

corresponds to a velocity of 1.5 in./s in the example shown). The formula to determine where to draw the horizontal line (the optimum viscosity) is as follows:

$$\text{Optimum viscosity} = (\text{Maximum viscosity} - \text{minimum viscosity}) \times 0.025 + \text{minimum viscosity} \qquad (15.5)$$

This equation was developed by Phillips Chemical Company [4]. It is based simply on what seemed to work when deciding where the curve started to flatten out after collecting data from numerous mold trials. You can just look at the graph and subjectively pick the point where the viscosity stops decreasing sharply, but the equation provides a more consistent, quantifiable method. However, the graph should still be looked at to make sure that it does level off, since a number can be calculated whether it does or not. If the curve does not level off, it is because the molding machine cannot inject fast enough.

After the "optimum" velocity is set on the machine, the transfer position should be adjusted to regain the original 95 to 99% full part(s). The parts should be as full as possible without hitting the last place to fill repeatedly. If there are defects present on the short shots such as jetting, gate blush, or flow lines that could not be fixed by modifying the mold or part design, a profiled injection velocity may have to be considered to eliminate the defects. This is sometimes necessary, but it is not desirable because these defects tend to reoccur as the mold is run in production. Profiled velocities are especially difficult to repeat if the same mold is run in several different molding machines.

Set the Hold Pressure

The goal here is to identify a hold pressure window and to set the pressure in the middle of the window. As a general rule, this transfer pressure usually amounts to between 50 and 70% of the transfer pressure. The bottom and top of the window are determined as follows:

Minimum hold pressure: Part(s) should be completely filled, should not have "unacceptable" sink marks, or any other signs of under-packing.

Maximum hold pressure: Part(s) should not have "unacceptable" flash, push marks, or any other signs of overpacking.

By setting the hold pressure in the middle of this window, the chance of experiencing these visual defects during production is minimized. It is especially important not to run with flash on your parts, because continued running of parts will cause the mold to wear and make the problem worse. It is acceptable to operate not exactly in the middle of this window if other criteria, such as dimensional requirements, dictate this. However, tooling adjustments should be considered before making deviations that would require the hold pressure to be set near either the top or the bottom of the window.

If it is discovered that there is a very small hold pressure window or, worse, no window, some reasons could be:

- Insufficient clamp tonnage on the machine,
- Mold not polished correctly,
- Insufficient draft angle to release the part,
- Thick sections in the part,
- Poor gate location (flowing from thin-to-thick),
- Poor fit of mold inserts,
- Cavity imbalance

These problems need to be addressed before releasing the mold to production.

For simplicity, the process is set up with one constant hold pressure. There are studies that show a decreasing hold pressure profile is beneficial [5]. This is primarily beneficial to pack the part out more evenly across the part and is generally only needed in special cases.

Set Hold Time

The hold pressure is typically applied until the gate freezes. If the gate and runners have been sized correctly, the gate will not freeze until enough plastic has been pushed into the cavity to compensate for the shrinkage of the plastic. Once the gate is frozen, the continued application of hold pressure only wastes energy and, possibly, increases cycle time. The parts will have less variation if the gate freeze is accomplished on every shot [6]. Therefore, a time slightly longer than the gate freeze time (typically ½ to 1 s) is usually set on the hold timer. The most commonly used method of determining the gate freeze time is by graphing of hold time versus part weight as shown in Fig. 15.32.

When collecting the data for this graph, it is recommended to start with a hold time thought to be longer than the gate freeze and work down to 1 s longer, collecting parts for each hold time setting. As time is taken off the hold timer, the same time should be added to the cooling timer. This will keep the overall cycle time and residence time in the barrel constant, both of which can affect part weight. It is also important to weigh only the parts, not the runner. For multi-cavity molds, it is preferable to measure each cavity individually as gate seal time variation between cavities can indicate tooling problems.

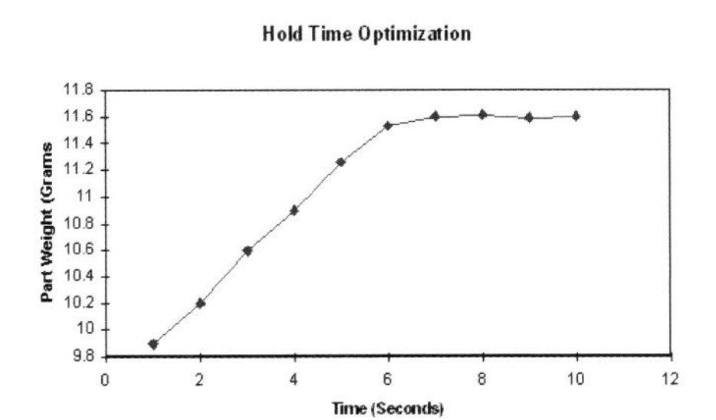

Figure 15.32 Growing part weights with increasing fill time until the gate has frozen (in this case, 6 s)

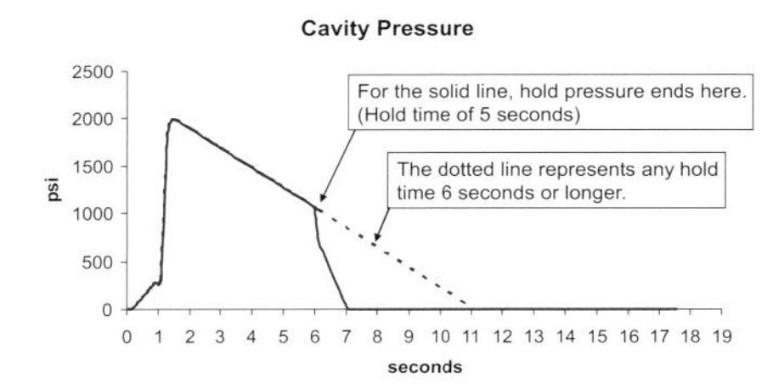

Figure 15.33 Cavity pressure plot illustrating the effect of removing hold pressure prior to gate freeze (in this case, 5 s)

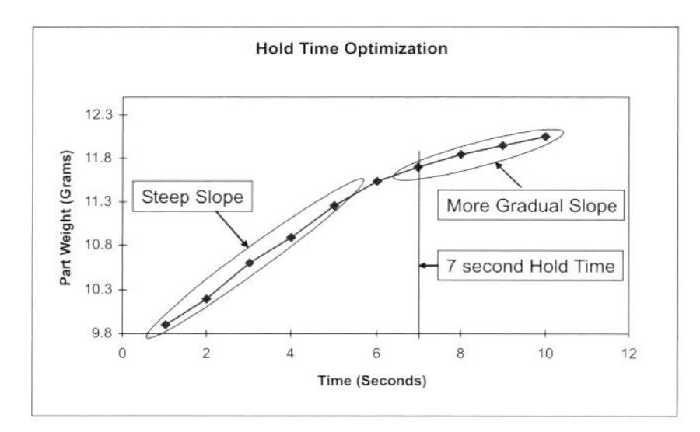

Figure 15.34 Closer examination of the hold time graph to determine proper hold time based on part weights

If a cavity pressure transducer is located in the part cavity near the gate, the gate freeze can be determined more quickly. Figure 15.33 shows a cavity pressure curve for a hold time of 5 s. It can be seen that, when hold time is set at 5 s, the pressure drops quickly because the gate is not frozen and material can flow out of the part into the runner. If the hold time was set at 6 s or longer, the pressure would continue to decrease smoothly as shown in the dotted line.

For hot runner molds, the gate commonly does not freeze. Therefore, pack time may have to be extended to minimize back flow through the open gate. This back flow would result in the cavity pressure loss conditions shown in Fig. 15.33 for the unfrozen gate. These pressure variations can be minimized by choosing a hold time on the portion of the curve where the weight is not changing as fast. This could be done by more closely inspecting the change in pressure with pack time as shown in Figure 15.34.

Minimize Cycle

As anyone in the manufacturing business knows, time is money. This step has the goal to minimize the cooling time on the press. Two things can limit how low this time can be set:

- The time needed so that parts consistently maintain the desired shape when ejected from the mold or,
- The time it takes for the screw to recover.

The second reason is usually only a problem with thin-walled, fast cycling parts. If there are high quantities to be run, such cases usually justify new or revised molding machines. A more powerful screw motor would allow the screw to recover faster; also, the implementation of a shut-off nozzle could allow screw recovery to continue while the mold is opening and closing. For example, if a new screw motor can reduce a 4 s cycle to 3 s, production rate is increased by 25% (four million parts can be made in the time it used to take to make three million parts). The money saved by making more parts in less time will pay for new equipment.

The time needed so that parts consistently maintain the desired shape when ejected from the mold is what usually limits the cycle. The cooling timer is reduced until signs of the part(s) being too warm are seen and then it is set slightly higher. The time required for the part to be cool enough to eject will vary depending on the geometry of the part, material, mold design and construction, and mold temperature. A rule of thumb is that the part surface should be at least 30° below the heat deflection temperature (HDT) at 66 psi when ejected from the mold [7]. A table of HDT values is shown in Table 15.1 [8]. However, it is always better to have specific data for the specific grade being molded.

Table 15.1 Heat deflection Temperature for Selected Resins

Resin	Deflection Temp @ 66 psi, °F
ABS	203
HIPS	185
LDPE	113
HDPE	186
PA6	356
PA6 GF	406
PC	280
PMMA	215
PP	204
PP copolymer	184
PPS GF	500
PS gp	180
PVC	156
SAN	225

Optimizing Other Parameters

Other things that should be checked and, possibly, adjusted include:

- Cushion
  The cushion is the amount of plastic left in front of the screw when the hold time ends. If the screw ever goes all the way forward (zero position) during hold, the cavity pressure does something very similar to what was described for the 5 s hold time in Fig. 15.34. In other words, if the screw position goes to zero, the effective hold pressure is zero at that point, because at this point the applied pressure on the screw is pushing on metal, rather than on plastic. The cushion should be run as small as possible because the material left in the cushion is being kept hot longer and is, therefore, more susceptible to degradation. The cushion will vary shot-to-shot primarily due to variations in how the non-return valve shuts off every shot. Generally, 0.1 to 0.25 in. of cushion is acceptable.
- Injection pressure limit and fill time limit
  These two are grouped together because they work together for only one purpose: to protect the mold from being overpacked, especially in

multi-cavity molds. Under normal processing conditions, neither of these two parameters should ever have an effect on the process. The injection pressure limit should be set about 2000 lb/in.^{2} (200 lb/in.^{2} hydraulic pressure) above the pressure needed to reach the transfer position. The pressure needed to reach the transfer position will vary shot-to-shot, primarily due to non-return valve shut-off variations and plastic viscosity variations. If monitored closely, this pressure can sometimes be set lower on molds that are very sensitive to overpacking. Care must be taken that the injection pressure limit never affects the fill time readout under normal conditions.

The fill time limit can be set very close to the actual fill time. On a good closed-loop control machine, the fill time readout does typically not vary more than 0.02 to 0.03 s, so the fill time limit can be set as close as 0.04 s above the actual fill time on molds very sensitive to being overpacked. If the mold is not that sensitive, 20% above the actual fill time is sufficient.

Figure 15.35 shows the cavity pressure, injection pressure, and injection velocity curves for three cases in a four-cavity mold: 1) standard operation with the two limits set correctly; 2) a shot where one gate gets blocked with

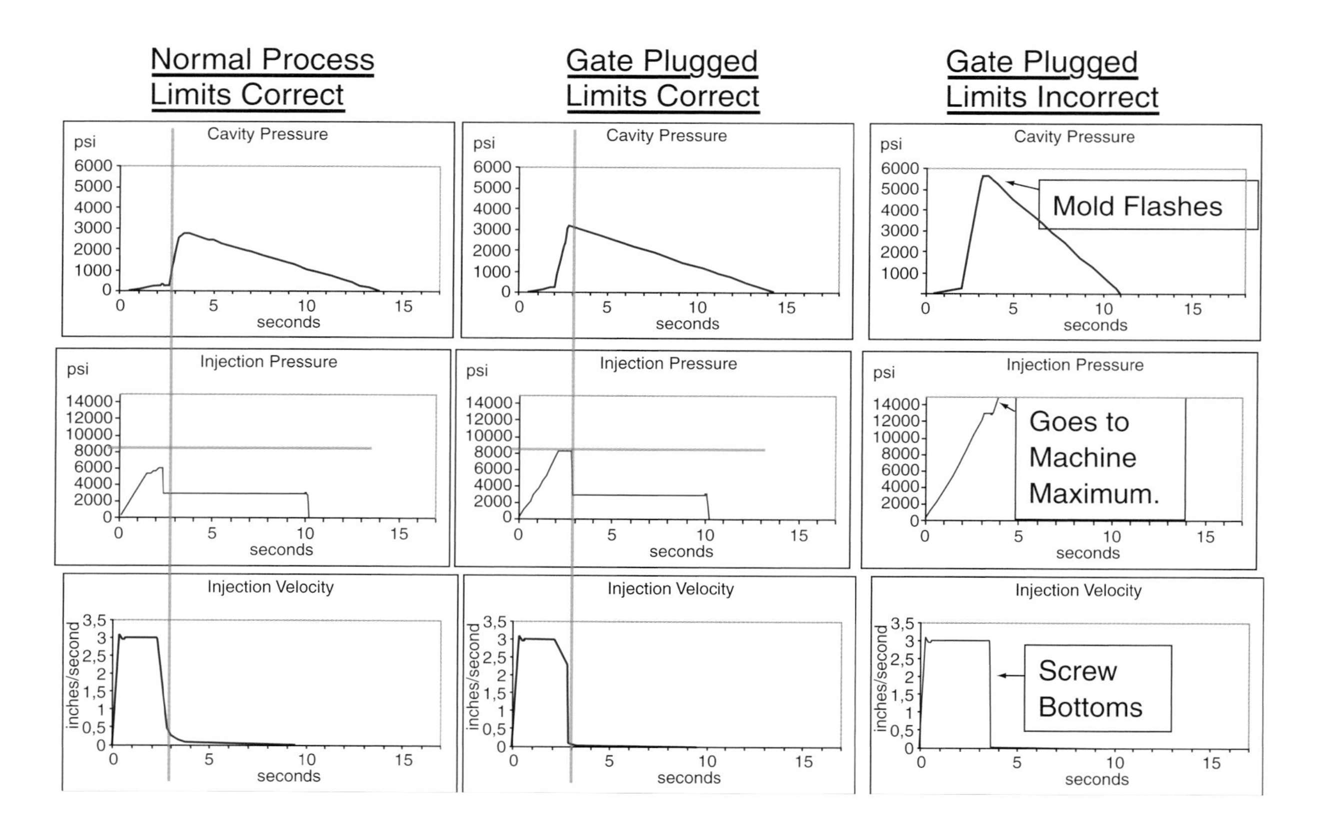

Figure 15.35 Cavity pressure, injection pressure, and injection velocity curves for three cases in a 4-cavity mold. Proper alarm times and pressures will protect the mold and molding machine

the two limits set correctly; 3) a shot where one gate gets blocked with both limits set incorrectly. The red lines represent the two limits. It should be noted that the cavity pressure was being measured in one of the three cavities that did not have the blocked gate. It is the spike of pressure inside the part cavity that can potentially do damage to the mold.

Final Thought

After this step, you should have a fairly robust process. This author also recommends that designed experiments be performed prior to doing capability runs for production. The designed experiments will give an indication of how the various process parameters (temperatures, speeds, pressures, times, etc.) will affect the important parameters of the part (dimensions, strength, visuals, etc.). Decisions can then be made on whether changes should be made to part design, mold design, or process set-up. Especially for parts with tight dimensional tolerances, several iterations of mold modifications are generally required before a mold is released to production. The processor should resist the temptation to please short-sighted managers and release a mold to production before it has been developed properly, with parts being made on the edge of a process window.

15.4 Intensification Ratio (R_i)

In all-hydraulic injection units, hydraulic power is converted and multiplied into plastic pressure. The law of physics involved is

$$F = P \times A$$

That is, force (F) is equal to pressure (P) times area (A). The large hydraulic piston pushes on the screw which in turn pushes on the non-return valve which acts like a piston to push plastic through the nozzle, sprue bushing, runner gate and finally into the mold. The hydraulic piston has a large surface area (for example 100 cm^2), while the non-return valve during injection forward acts like a smaller area piston (for example 10 cm^2). This causes the hydraulic pressure to be converted to melt or plastic pressure in the barrel of the injection unit. In this specific case, hydraulic pressure is "intensified" or multiplied by a factor of 10. That is 800 psi of hydraulic pressure provides ~ 8,000 psi of melt pressure inside the nozzle of our machine. Often, this is called the machine's intensification ratio and explains how hydraulic pressure can provide tens of thousands of psi plastic pressure inside the nozzle. Today you can buy machines with intensification ratios ranging from 6:1 to 43:1. *NOT all machines are 10:1.* It is the plastic pressure that pushes plastic into the sprue, runner, gate and mold cavity, *NOT* the hydraulic pressure. If your plant has different machines, most likely they have different intensification ratios. That is, 800 psi pack and hold pressure on one machine may be packing out the part with 8,000 psi plastic pressure, but on another machine 800 psi hydraulic

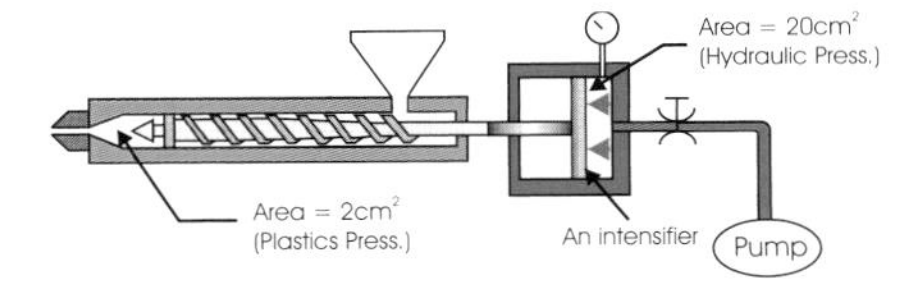

Figure 15.36 Intensification ratio developed by a 10:1 difference in cross sectional area of the machine's hydraulic cylinder and the melt injection screw

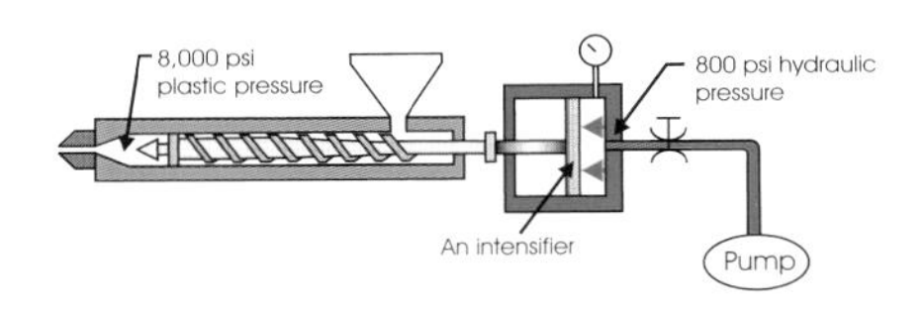

Figure 15.37 Pressure intensification of 10:1 from hydraulic to melt pressure

pressure may be 10,200 psi plastic packing pressure in the nozzle. You will not make the same part even when you are using the same hydraulic pressure! Figures 15.36 and 15.37 provide some illustration.

15.5 Characterizing Flow Behavior in an Injection Mold

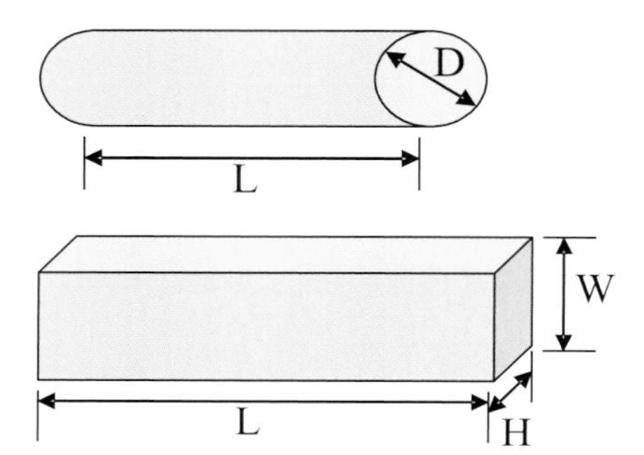

Figure 15.38 Typical injection molding melt behavior

D = Diameter of cylindrical flow channel
W = Width of retangular flow channel.
H = Height of retangular flow channel (smallest dimension).
L = Length of flow channel.
t = Time to flow through a given channel length (L).
ΔP = Pressure drop. = $P_{nozzle} - P_{flow\ front} = P_{hyd}Ri$
P_{nozzle} = Melt pressure in the nozzle required to fill the mold.
$P_{flow\ front}$ = Pressure at leading edge of flow at instant before fill = 0.
P_{hyd} = Hydraulic pressure required to fill mold.
Ri = Intensification ratio.

For rectangular channel:
$\eta = \Delta PH^2 t/12L^2$
$\dot{\gamma} = 6L/Ht$

For cylindrical channel:
$\eta = \Delta PD^2 t/32L^2$
$\dot{\gamma} = 8L/Dt$

Using the variables defined in Fig.15.38, it can be shown that:

$$\eta = k_1(P_{hyd}t), \text{ and} \tag{15.6}$$

$$\dot{\gamma} = k_2\ (1 / t) \tag{15.7}$$

where k_1 and k_2 are geometric constants for a given system.

Therefore, for a given injection mold and molding unit, we can define the following:

$$\text{relative viscosity, } \eta_r = P_{hyd}t, \text{ and} \tag{15.8}$$

$$\text{relative shear rate, } \dot{\gamma}_r = 1 / t \tag{15.9}$$

15.6 List of Amorphous and Semi-Crystalline Resins

Abbreviation	Full common name	Type of resin
ABS	Acrylonitrile-butadiene-styrene	Amorphous
Acetal (POM)	Acetal or polyoxymethylene	Semi-crystalline
Acrylic (PMMA)	Polymethyl methacrylate	Amorphous
ASA	Acrylonitrile-strryene-acrylate	Amorphous
CAP	Cellulose acetate propionate	Amorphous
Crystal PS	Crystal polystyrene (GPPS)	Amorphous
CPVC	Chlorinated polyvinyl chloride	Amorphous
EVA	Ethylene vinyl acetate	Semi-crystalline
EVOH	Ethylene vinyl alcohol	Semi-crystalline
GPPS	General purpose polystyrene	Amorphous
HDPE	High density polyethylene	Semi-crystalline
HIPS	High impact polystyrene	Amorphous
IMPS	Impact polystyrene	Amorphous
LCP	Liquid crystal polymer	Semi-crystalline
LDPE	Low density polyethylene	Semi-crystalline
Nylon (PA)	Polyamide	Semi-crystalline

Abbreviation	Full common name	Type of resin
PA	Polyamide, nylon	Semi-crystalline
PAI	Polyamide-imide	Amorphous
PB	Polybutylene	Semi-crystalline
PBT	Polybutylene terephthalate	Semi-crystalline
PC	Polycarbonate	Amorphous
PC/ABS	PC/ABS blends	Amorphous
PE	Polyethylene	Semi-crystalline
PEEK	Polyetheretherketone	Semi-crystalline
PEI	Polyetherimide	Amorphous
PEK	Polyetherketone	Semi-crystalline
PES	Polyethersulfone	Amorphous
PET	Polyethylene terephthalate	Semi-crystalline, clear
PETG	Polyethylene terephthalate glycol	Amorphous
PMMA	Polymethyl methacrylate, acrylic	Amorphous
POM	Acetal, polyacetal	Semi-crystalline
PP	Polypropylene	Semi-crystalline
PPE	Polyphenylene ether	Amorphous
PPO/Styrene	Polyphenylene- styrene	Amorphous
PPS	Polyphenylene sulfide	Semi-crystalline
PS	Polystyrene	Amorphous
PS syndiotactic	Syndiotatic polystyrene questra	Semicrystalline
PSU or PPSU	Polysulfone polyphenylenesulfone	Amorphous
PTFE	Polytetrafluoroethylene	Semi-crystalline
PU	Polyurethane	Amorphous
PVC	Polyvinylchloride	Amorphous
SAN	Styrene acrylonitrile	Amorphous
TPO	Thermoplastic olefins	Semi crystalline
TPU	Thermoplastic polyurethane	Amorphous

References

[1] Groleau, R. J., Injection Molding & Process Control (1995), RJG Associates, Traverse City, MI

[2] Sloan, J., Injection Molding Magazine, October (1997), 118

[3] Wright, Blanche Fisher, The Real Mother Goose (1944), Checkerboard Press, New York

[4] Mertes, S., Carlson, C., Bozzelli, J., Groleau, M., SPE ANTEC tech. Papers (2001) 620

[5] Bandreddi, M., Nunn, R., Malloy, R., SPE ANTEC tech. Papers (1994) 348

[6] Carender, J., Managing Variation for Injection Molding (2003), Advanced Process Engineering, Corvallis, OR

[7] Mora, A., Molding: Machine Start Up and Shut Down Procedures (2002), SPE Plastics Technician's Toolbox, Book 6, 20

[8] Carender, J., *Injection Molding Reference Guide*, Fourth Ed. (1997), Advanced Process Engineering, Corvallis, OR

Index